GUIDE TO LOAD ANALYSIS FOR DURABILITY IN VEHICLE ENGINEERING

GUIDE TO LOAD ANALYSIS FOR DURABILITY IN VEHICLE ENGINEERING

Editors

P. Johannesson
SP Technical Research Institute of Sweden, Sweden

M. Speckert
Fraunhofer Institute for Industrial Mathematics (ITWM), Germany

WILEY

Library of Congress Cataloging-in-Publication Data

Guide to load analysis for durability in vehicle engineering / editors, Par Johannesson, Michael Speckert ; contributors, Klaus Dressler, Sara Loren, Jacques de Mare, Nikolaus Ruf, Igor Rychlik, Anja Streit and Thomas Svensson – First edition.
    1 online resource. – (Automotive series ; 1)
    Includes bibliographical references and index.
  Description based on print version record and CIP data provided by publisher; resource not viewed.
    ISBN 978-1-118-70049-5 (Adobe PDF) – ISBN 978-1-118-70050-1 (ePub) – ISBN 978-1-118-64831-5 (hardback) 1. Trucks–Dynamics. 2. Finite element method. 3. Trucks–Design and construction.
I. Johannesson, Par, editor of compilation. II. Speckert, Michael, editor of compilation.
  TL230
  629.2'31 – dc23

                                              2013025948

A catalogue record for this book is available from the British Library.

ISBN: 978-1-118-64831-5

Typeset in 10/12pt Times by Laserwords Private Limited, Chennai, India

1   2014

Contents

Part III LOAD ANALYSIS IN VIEW OF THE VEHICLE DESIGN PROCESS

About the Editors

Pär Johannesson (*SP Technical Research Institute of Sweden, Sweden*) received his PhD in Mathematical Statistics in 1999 from Lund Institute of Technology, with a thesis on statistical load analysis for fatigue. During 2000 and 2001 he worked as a PostDoc at Mathematical Statistics, Chalmers, on a joint project with PSA Peugeot Citroën, where he stayed one year in the Division of Automotive Research and Innovations in Paris. From 2002 to 2010 he was an applied researcher at the Fraunhofer-Chalmers Research Centre for Industrial Mathematics in Göteborg, and in 2010 he was a guest researcher at Chalmers. He is currently working as a research engineer at SP Technical Research Institute of Sweden, mainly on industrial and research projects on statistical methods for load analysis, reliability and fatigue.

Michael Speckert (*Fraunhofer Institute for Industrial Mathematics (ITWM), Germany*) received his PhD in Mathematics from the University of Kaiserslautern in 1990. From 1991 to 1993 he worked at TECMATH in the human modelling department on optimization algorithms. From 1993 to 2004 he worked at TECMATH and LMS in the departments for load data analysis and fatigue life estimation in the area of method as well as software development. Since 2004 he has been working at the department for Dynamics and Durability at Fraunhofer ITWM as an applied researcher. His main areas of interest are statistical and fatigue-oriented load data analysis and multibody simulation techniques.

Contributors

Klaus Dressler (*Fraunhofer ITWM, Kaiserslautern, Germany*) received his PhD in Mathematical Physics from the University of Kaiserslautern in 1988. From 1990 to 2003 he led the development of load data analysis and simulation software for the vehicle industry at TECMATH and LMS International. In that period he initiated and organized the cooperation workgroups 'load data analysis' and 'customer correlation' of the German automobile companies AUDI, BMW, Daimler, Porsche and Volkswagen. Since 2003 he has been the manager of the department for Dynamics and Durability at Fraunhofer ITWM with 35 researchers, working on load data analysis and simulation topics. He is also coordinating the Fraunhofer innovation cluster on 'commercial vehicle technology' where leading companies like Daimler, John Deere, Volvo and Liebherr cooperate with Fraunhofer on usage variability and virtual product development.

Jacques de Maré (*Department of Mathematical Sciences at Chalmers University of Technology and University of Gothenburg, Göteborg, Sweden*) received his PhD in mathematical statistics in 1975 from Lund University. He worked at Umeå University from 1976 to 1979 before securing a position at Chalmers University of Technology. He became a professor there in 1995. He was a visiting researcher at the University of North Carolina in 1982, at the University of California, Santa Barbara, in 1989, and at Kyushu University in Fukuoka, in Japan, in 2004. He is a member of the International Statistical Institute and was one of the founders of UTMIS (the Swedish Fatigue Network) and a member of the first board. He is currently working with statistical methods for material fatigue in co-operation with SP Technical Research Institute of Sweden. At Chalmers he has also worked in different ways to bring the mathematical and engineering disciplines closer together.

Sara Lorén (*School of Engineering at University of Borås, Borås, Sweden*) received her PhD in mathematical statistics in 2004 from Chalmers University of Technology: with a thesis entitled 'Fatigue limit, inclusion and finite lives: a statistical point of view'. From 2005 to 2010 she was an applied researcher at Fraunhofer-Chalmers Research Centre for Industrial Mathematics, working with statistical methods for material fatigue. She is currently at the School of Engineering at University of Borås.

Nikolaus Ruf (*Fraunhofer ITWM, Kaiserslautern, Germany*) studied mathematics at the University of Kaiserslautern. He obtained a degree in mathematics in 2002 with a specialty in optimization and statistics, and a doctoral degree (Dr. rer. nat.) in 2008 for his work on statistical models for rainfall time series. He has worked as a researcher at ITWM since

2008 and focuses on the analysis of measurement data from technical systems, in particular regarding the durability, reliability, and efficiency of vehicles and their subsystems.

Igor Rychlik (*Department of Mathematical Sciences at Chalmers University of Technology and University of Gothenburg, Göteborg, Sweden*) is Professor in Mathematical Statistics at Chalmers University of Technology. He earned his PhD in 1986, with a thesis entitled 'Statistical wave analysis with application to fatigue'. His main research interest is in fatigue analysis, wave climate modelling and in general engineering applications of the theory of stochastic processes, especially in the safety analysis of structures interacting with the environment, for example, through wind pressure, ocean waves, or temperature variations. He has published more than 50 papers in international journals, is the co-author of the text book *Probability and Risk Analysis. An Introduction for Engineers*, and has been visiting professor (long-term visits) at the Department of Statistics, Colorado State University; the Center for Stochastic Processes, the University of North Carolina at Chapel Hill; the Center for Applied Mathematics, Cornell University, Ithaca; and the Department of Mathematics, University of Queensland, Brisbane, Australia.

Anja Streit (*Fraunhofer ITWM, Kaiserslautern, Germany*) received her PhD in Mathematics from the University of Kaiserslautern in 2006, with a thesis entitled 'Coupling of different length scales in molecular dynamics simulations'. Since 2007 she has been working in the department for Dynamics and Durability at Fraunhofer ITWM as an applied researcher. Her main areas of work are statistical and fatigue-oriented load data analysis.

Thomas Svensson (*SP Technical Research Institute of Sweden, Borås, Sweden*) received his PhD in mathematical statistics in 1996 from Chalmers, with a thesis entitled 'Fatigue life prediction in service: a statistical approach'. He was a research engineer at SP of Sweden, 1990–2001, Fraunhofer-Chalmers Research Centre for Industrial Mathematics, 2001–2007, and returned to work at SP in 2007. He has been Adjunct Professor in Mathematical Statistics at Chalmers University of Technology since 2010, and a member of the Editorial Board for the journal, *Fatigue and Fracture of Engineering Materials and Structures*. Since 2008, he has been the chairman of UTMIS (the Swedish Fatigue Network).

Series Preface

The automotive industry is one of the largest manufacturing sectors in the global community. Not only does it generate significant economic benefits to the world's economy, but the automobile is highly linked to a wide variety of international concerns such as energy consumption, emissions, trade and safety.

The primary objective of the *Automotive Series* is to publish practical and topical books for researchers and practitioners in industry, and postgraduate/advanced undergraduates in automotive engineering. The series addresses new and emerging technologies in automotive engineering supporting the development of more fuel efficient, safer and more environmentally friendly vehicles. It covers a wide range of topics, including design, manufacture and operation, and the intention is to provide a source of relevant information that will be of interest and benefit to people working in the field of automotive engineering.

In 2006, six leading European truck manufacturers (DAF, Daimler, Iveco, MAN, Scania, and Volvo) commissioned a research project to produce a guide to load analysis oriented towards fatigue design of trucks. The project was run by Fraunhofer-Chalmers Research Centre for Industrial Mathematics (FCC) in collaboration with Fraunhofer ITWM, the SP Technical Research Institute of Sweden, Mathematical Sciences at Chalmers University of Technology, and the industrial partners.

The project included an investigation of the current practice and future needs within load analysis, together with a survey on the state-of-the-art in load analysis for automotive applications. This book, *Guide to Load Analysis for Durability in Vehicle Engineering*, is the result of this research.

The guide presents a number of different methods of load analysis, explaining their principles, usage, applications, advantages and drawbacks. A section on integrating load analysis into vehicle design aims at presenting what methods are useful at each stage of the design process.

The *Guide to Load Analysis for Durability in Vehicle Engineering* covers a topic usually presented in separate works on fatigue, safety and reliability; signal processing, probability and statistics. It is up-to-date, has been written by recognized experts in the field and is a welcome addition to the Automotive series.

Thomas Kurfess
August 2013

Preface

This work is the result of a collaboration between researchers and practitioners with an interest in load analysis and durability but with different backgrounds, for example, mathematical statistics, applied mathematics, mechanics, and fatigue, together with industrial experience of both load analysis problems and specific fatigue type problems. The project started in 2006 when the six European truck manufacturers: DAF, Daimler, Iveco, MAN, Scania, and Volvo, commissioned a research project to produce a *Guide to Load Analysis* oriented towards fatigue design of trucks. The project was run by Fraunhofer-Chalmers Research Centre for Industrial Mathematics (FCC) in collaboration with Fraunhofer ITWM, SP Technical Research Institute of Sweden, Mathematical Sciences at Chalmers University of Technology, and the industrial partners. All the research groups involved have long experience and profound knowledge of load analysis for durability, where the Swedish group (FCC, SP and Chalmers) has the key competencies in statistics and random processes, and the German group (Fraunhofer ITWM) are experts in mathematical modelling of mechanical systems. The complete *Guide* was available in 2009, as planned, after a joint effort of ten staff years.

Transport vehicles are exposed to dramatically different operating conditions in different parts of the world and in different transport missions. The ultimate goal for the manufacturer is to make a design that exactly meets the needs of the customers, neither too strong nor too weak. The requirements need to be converted into, for example, a certain small risk of failure, a proper safety factor, or an economical expected life. In order to make a robust design it is as important to have a good working knowledge of the properties of the customer loads, as it is to have good working knowledge of the mechanical behaviour of the material and structure in question.

In the process of designing a robust and reliable product that meets the demands of the customers, it is important not only to predict the life of a component, but also to investigate and take into account the sources of variability and their influence on life prediction. There are mainly two quantities influencing the life, namely, the load the component is exposed to, and the structural strength of the component. Statistical methods present useful tools to describe and quantify the variability in load and strength. The variability in the structural strength depends on both the material scatter and the geometrical variations. The customer load distribution may be influenced by, for example, the application of the truck, the driver behaviour, and the market.

The development of information technology and its integration into vehicles have presented new possibilities for in-service measurements. Further, the design process has also moved to the computer. Both these tasks, together with demands for lightweight design and fuel efficiency, require a refined view on loads and lead to a renewed interest in load analysis.

During 2006 an initial one-year project was carried out, with the aim of preparing the ground for a *Guide to Load Analysis*. The project included an investigation of the current practice and future needs within load analysis, together with a survey of the state of the art in load analysis for automotive application.

The main project that developed the *Guide* in 2007–2009 also included several seminars at the companies, with the aim of spreading the knowledge within the companies. The themes of the seminars were *Basics on load analysis* in 2007, *Methods for load analysis* in 2008, and *Load analysis in view of the vehicle design process* in 2009.

The *Guide* presents a variety of methods for load analysis but also their proper use in view of the vehicle design process. In Part I, *Overview*, two chapters present the scope of the the book as well as giving an introduction to the subject. Part II, *Methods for Load Analysis*, describes useful methods and indicates how and when they should be used. Part III, *Load Analysis in View the Vehicle Design Process*, offers strategies for the evaluation of customer loads, in particular the characterization of the customer populations, which leads to the derivation of design loads, and finally to the verification of systems and components. Procedures for generation and acceleration of loads as well as planning and evaluation of verification tests are also included. All through the book, the methods are accompanied by many illustrative examples.

To our knowledge there is no other comprehensive text available covering the same content, but most of the results and methods presented in this *Guide* are distributed in books and journals in various fields. Partial information on load analysis for durability is mainly found in journals on mechanics, fatigue and vehicle design as well as in text books on fatigue of engineering materials, but also in conference and research papers in other areas, such as signal processing, mathematics and statistics.

Our intended readership is those interested in designing for durability. The audience is probably advanced design engineers and reliability specialists. Especially, people interested in durability, fatigue, reliability and similar initiatives within the automotive industry, are the target group. The *Guide* should provide a better understanding of the currently used methods as well as inspire the incorporation of new techniques in the design and test processes.

Pär Johannesson
Göteborg, March, 2013
Michael Speckert
Kaiserslautern, March, 2013

Acknowledgements

This book springs from the four-year project (2006–2009) *Guide to load analysis for automotive applications* commissioned by six European truck manufacturers: DAF, Daimler, IVECO, MAN, Scania, and Volvo. The project was run by Fraunhofer-Chalmers Research Centre for Industrial Mathematics (FCC) in Gothenburg, Sweden, together with Fraunhofer ITWM in Kaiserslautern, Germany, SP Technical Research Institute of Sweden in Borås, Sweden, and Mathematical Sciences at Chalmers University of Technology in Gothenburg, Sweden.

We are most grateful for the financial support from the industrial partners, as well as the valuable feedback on the *Guide* during the project. Among the many people involved, we are especially grateful to Peter Nijman at DAF, Christof Weber at Daimler, Massimo Mazzarino at IVECO, Manfred Streicher at MAN, Anders Forsén at Scania, and Bengt Johannesson at Volvo.

The Swedish Foundation for Strategic Research has supported the Swedish research teams through the Gothenburg Mathematical Modelling Centre (GMMC), which is gratefully acknowledged.

Part One

Overview

1

Introduction

The assessment of durability is vital in many branches of engineering, such as the automotive industry, aerospace applications, railway transportation, the design of windmills, and off-shore construction. A fundamental element of the discussion is the very meaning of *durability*. A rather general definition is the following:

> Durability is the capacity of an item to survive its intended use for a suitable long period of time.

In our context, durability may be defined as the ability of a vehicle, a system or a component to maintain its *intended function* for its *intended service life* with *intended levels of maintenance* in *intended conditions of use*.

The analysis of durability loads is discussed with truck engineering in mind, however, most of the contents are applicable also to other branches of industry, especially for applications in the automotive context. Properties of loads that cause fatigue damage are emphasized rather than the properties of extreme crash loads or acoustic loads. The fatigue damage mechanisms are assumed to be similar to those encountered in metal fatigue, but a few comments concerning rubber and composite material are given in Section 2.1.5.

In vehicle engineering the purpose of load analysis is:

- to evaluate and quantify the customer service loads;
- to derive design loads for vehicles, sub-systems, and components;
- to define verification loads and test procedures for verification of components, sub-systems, and vehicles.

The *Guide* is divided into three parts, where the introductory part sets the scope. Part II, *Methods for Load Analysis*, presents different methods with the aim of providing an understanding of the underlying principles as well as their usage. It is important to know where and when each method is applicable and what merits and disadvantages it has. Part III, *Load Analysis in View of the Vehicle Design Process*, is organized according to the bullet list above, and describes what methods are useful in the different steps of the vehicle engineering process.

Guide to Load Analysis for Durability in Vehicle Engineering, First Edition. Edited by P. Johannesson and M. Speckert.
© 2014 Fraunhofer-Chalmers Research Centre for Industrial Mathematics.

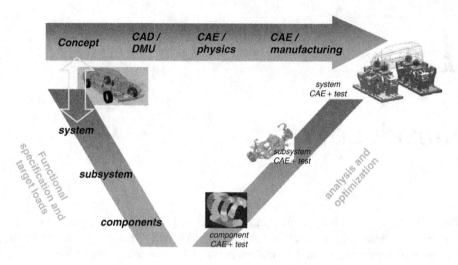

Figure 1.1 The vehicle engineering process

1.1 Durability in Vehicle Engineering

In vehicle engineering the aim is to design a vehicle with certain physical properties. Such properties can be specified in the form of 'design targets' for so-called 'physical attributes' such as durability, NVH (Noise Vibration Harshness), handling, and crash safety. Design variants are analysed, optimized, and verified by means of physical tests and numerical simulations for the various attributes. An often used view of the vehicle engineering process is illustrated in Figure 1.1, and can be summarized as follows:

1. Concept for the new vehicle (class of vehicles, market segment, target cost, size, weight, wheel base, etc.).
2. Overall targets and benchmarks are defined for the physical properties of the vehicle (performance, durability, safety (crash), acoustics, vibration comfort, etc.).
3. Target cascading: Design targets for the sub-systems and components are derived (chassis suspension, engine, transmission, frame, body, etc.); those targets are again related to different physical attributes (durability, NVH, handling, crash, etc.).
4. Design of components, sub-systems and the full vehicle.
5. Design verification and optimization by means of physical tests and numerical simulations on the various levels for the various attributes.
6. Verification on vehicle level.

Especially for trucks, durability is one of the most important physical attributes for the customer, and therefore durability needs to be highlighted in the development process. The vehicle engineering process in Figure 1.1 needs to be implemented with respect to load analysis for durability. The process illustrated in Figure 1.2 is frequently used in the automotive industry.

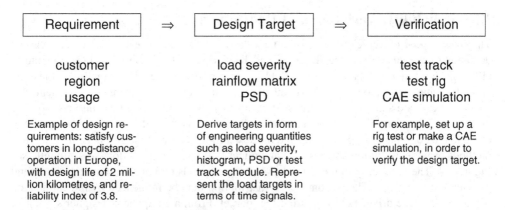

Figure 1.2 An implementation of the vehicle engineering process with respect to load analysis

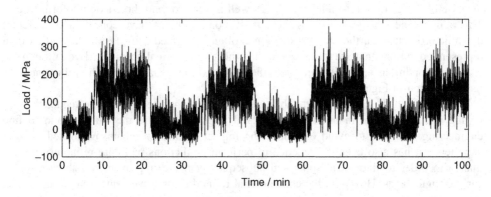

Figure 1.3 A measured service load of a truck transporting gravel

Metal fatigue and other durability phenomena are degradation processes in the sense that an effect builds up over time. A certain force applied to a structure once or a few times may cause no measurable effect, but if it is applied a million times, the structure may fail. Loads in durability engineering need to be studied with regard to the fatigue phenomenon as well as with regard to the vehicle dynamics and the variation in customer usage.

Loads may be displacements (linear or rotational), velocities, accelerations, forces, or moments. They may represent road profiles, wheel forces, relative displacements of components, frame accelerations, or local strains. When we talk about *load signals*, we mean one- or multi-dimensional functions of time as they appear in the vehicle, for example, during customer usage, on test tracks, in test benches, or in virtual environments. Figure 1.3 shows an example of a measured service load, where a stress signal has been recorded for about 100 minutes on a truck transporting gravel. There we can observe different mean levels as well as different standard deviations of different parts of the load. The changes in the mean level originate from a loaded and an unloaded truck while the changes in the standard deviation derive from different road qualities.

1.2 Reliability, Variation and Robustness

The overall goal of vehicle design is to make a robust and reliable product that meets the demands of the customers; see Bergman and Klefsjö [22], Bergman *et al.* [23], O'Connor [172], Davis [64] and Johannesson *et al.* [126] on the topic of reliability and robustness. In order to achieve this goal it is important not only to predict the life of a component, but also to investigate and take into account the sources of variability and their influence on life prediction. There are mainly two quantities influencing the life of the component, namely, the load the component is exposed to, and the structural strength of the component. Statistical methods provide useful tools to describe and quantify the variability in load and strength, see Figure 1.4. The variability in the structural strength depends on both the material scatter and geometrical variations. The customer load distribution may be influenced by the application of the vehicle, the driver behaviour, and the market. From a component designer's point of view, the varying vehicle configurations on which the component, for example, a bracket, is to be used are yet another variation source. For example, for trucks, the same design may well be used on semi-trailer tractors as well as on two- and three-axle platform trucks. This adds to the load variation, as these truck configurations have considerably different dynamic properties. Further, the verification is often performed using test track loads, which represent conditions that are more severe than those of a normal customer. Even though the test track conditions are well controlled, they also exhibit variation, which is illustrated by its distribution in Figure 1.4.

The conventional strategy for reliability improvement has been to utilize feedback from testing and field usage in order to understand important failure mechanisms and to find engineering solutions to avoid or reduce the impact of these mechanisms. Based on past experience it has also been the practice to perform predictions of future reliability performance in order to find weak spots and subsequently make improvements already in the early design stages. However, the conventional reliability improvement strategy has strong limitations, as it requires feedback from usage or from testing. Thus, it is fully applicable

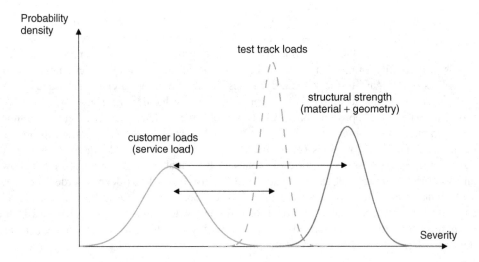

Figure 1.4   Distributions of customer loads, test track loads and structural strength

only in the later stages of product development when already much of the design is frozen and changes incur high costs. Therefore, we propose putting effort into load analysis also in the early design stage, and not primarily in the verification process. In this context, understanding load variation is an important aspect of engineering knowledge.

In industry, the method of *Failure Mode and Effect Analysis* (FMEA) is often used for reliability assessments. Studies of FMEA have indicated that the failure modes are in most cases triggered by unwanted variation. Therefore, the so-called *Variation Mode and Effect Analysis* (VMEA) has been developed, which takes the quantitative measures of failure causes into account; see Johannesson *et al.* [127], Chakhunashvili *et al.* [54] and Johannesson *et al.* [125]. The VMEA method is presented at three levels of complexity: basic, enhanced and probabilistic. The basic VMEA can be used when we only have vague knowledge about the variation. The sensitivity and variation size assessments are made by engineering judgements and are usually made on a 1–10 scale. When better judgements of the sources of variation are available, the enhanced VMEA can be used. The probabilistic VMEA can be used in the later design stages where more detailed information is available. It can, for example, be more detailed material data, finite element models for calculating local stresses, and physical experiments in terms of load and strength. The different sources of uncertainty can be measured in terms of statistical standard deviation. The load-strength model described in Section 7.6 is an implementation of the probabilistic VMEA for the application of fatigue and durability problems. Both FMEA and VMEA are methods well suited for use in the framework of *Design for Six Sigma* (DfSS). The above topics are further discussed in Bergman *et al.* [23] and Johannesson *et al.* [126].

1.3   Load Description for Trucks

Here we give a description of the typical features of loads for the truck application, and discuss the so-called load influentials. The particular durability loads which affect trucks are governed by their applications. The application decides where the truck will be used and how it may be used. The main factors governing the loads are

- *The vehicle utilization*, that is the particular use of the truck, given the utilization profile described by, for example, the transport mission and yearly usage.
- *The operational environment*, that is, the road conditions and other environmental conditions that the truck will experience.
- *The vehicle dynamics*, for example, the transfer of external road input to local loads will be affected by the particular tyres and the suspension of the truck.
- *The driver's behaviour*, that is, the driver's influence on the load such as speed changes, braking, and the ability to adapt to curves.
- *Legislation*, for example, the speed limits, and allowed weight and size of trucks, in different regions and countries.

Loads that will act on a truck can be described by using the above load influentials, that is, by making a description of the vehicle utilization, the operational environment, the vehicle dynamics, and so on. One such approach is given in Edlund and Fryk [87]. The different load influentials are preferably described as simply as possible, for example, by classifying the types of roads, or by describing each road by some few parameters. Such approaches have been developed especially for the vertical road input, see for example, Bogsjö [30], Bogsjö *et al.* [33], Öijer and Edlund [175, 176] and the references therein, but also for lateral loads, see for example, Karlsson [132].

It is desirable to separate the load description into a vehicle-independent load environment and a description of the vehicle-dependent load influentials. The *vehicle usage* and the *vehicle dynamics* can then be connected to the *vehicle independent load environment* description, in order to compute the *load distribution for the customer population of interest for a specific vehicle*, see the schematic view in Figure 1.5. Here, the vehicle usage is the vehicle utilization together with the driver's behaviour, both of which are dependent on the specific vehicle. The load environment is independent of the specific vehicle and includes the operational environment as well as legislation.

The vehicle utilization may be described and classified, by for example

- *Transport cycle* (Long distance – Distribution – Construction).
- *Transport mission* (Timber – Waste – Trailer – Distribution – and so on).
- *Yearly usage*.
- *Pay load or gross combination weight*.

The operational environment may be described by a number of influential variables, such as

- *Road surface quality* (Smooth – Rough – Cross-country).
- *Hilliness* (Flat – Hilly – Very Hilly).
- *Curve density* (Low – Moderate – High).
- *Altitude* (Sea level – High altitudes).
- *Climate* (Temperature, humidity, dust, etc.).

The driver's behaviour also causes variations in the load. The origin of the variation is the driver's influence on the way of driving the vehicle, such as speed changes, braking, and acceleration. A specific driver may be characterized by his or her load severity, while a population of drivers may be described by the distribution of their load severities.

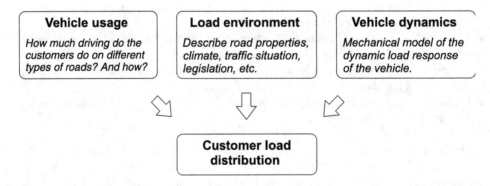

Figure 1.5 The customer load distribution can be described in terms of the vehicle-independent load environment together with the vehicle usage and the vehicle dynamics

Further, the loads can be classified according to their origin, namely *external excitations*, for example, coming from the road, and *internal excitations*, for example, coming from the engine and transmission.

1.4 Why Is Load Analysis Important?

Lack of durability is not only a problem for customers, also the producers suffer. Failures reduce company profitability through call-backs, warranty costs and bad will. In other words, good durability leads to good quality, company profitability and customer satisfaction; see Bergman and Klefsjö [22]. In order to make a good durability assessment there are many influences that need to be considered and most of those are not fully known beforehand. This is illustrated by Figure 1.6 showing a schematic view of engineering fatigue design.

The numerical procedures for calculating stresses and strains of mechanical systems are nowadays excellent and quite accurate, however, the calculations are surrounded by uncertainties. On the input side, loads are approximated by simplifications of the service environment; material strength is represented by empirical characteristics; geometry is given by specifications where defects like scratches, inclusions, pores and other discontinuities are neglected because of lack of information. On the output side, the stresses and strains are further processed using empirical fatigue models, such as the Wöhler curve, the Palmgren-Miner rule, and Paris' law. These rough models introduce model errors and their parameters are empirically determined, often from quite limited information, for example, data in the literature on similar materials, a number of fatigue tests, or previous experience. Thus, in order to evaluate the output of the fatigue assessment, it is necessary to reflect on the

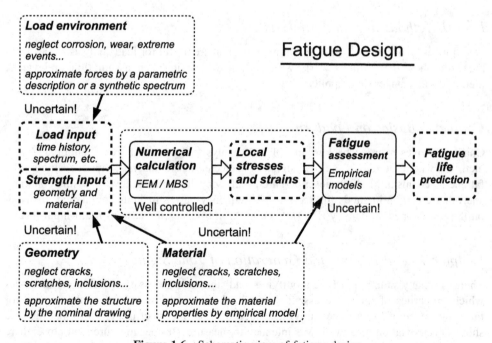

Figure 1.6 Schematic view of fatigue design

uncertainties in load as well as the uncertainties in strength defined by material and geometry input. However, it should be noted that also the numerical procedures may have significant model errors, especially for non-linear modelling of, for example, welded joints in FEM (Finite Element Models) and tyres in MBS (Multi-Body Simulation). Moreover, load analysis is not only important when analysing the load input, but also for the numerical simulation process, the evaluation of measurements, and the physical verification tests.

The *Guide* is mainly devoted to the load input problem; how should the service environment be evaluated and represented in the design process? However, in order to correctly understand and treat the load information some basic knowledge about the other pieces is necessary. Further, methods are developed which handle the overall uncertainty problem by using the load-strength model, which is presented in Chapter 7.

1.5   The Structure of the Book

The material is organized into three parts.

Part I Overview

Part I contains, apart from the introduction, Chapter 2 presenting some basic concepts of fatigue assessment and how to apply those to different kinds of loads. It is indicated how the type of system or component affects the choice of suitable load analysis methods to be applied. Finally, it is emphasized that fatigue prediction is affected by a number of sources of variation and uncertainty, which need to be treated and quantified in a reasonable way.

Part II Methods for Load Analysis

Part II gives an account of the different methods that are useful for load analysis. Apart from presenting how the methods work, we also aim to describe their assumptions, relevance, merits, disadvantages, and applicability.

Chapter 3 Basics on Load Analysis

Chapter 3 gives a broad background of load analysis. Section 3.1 treats amplitude-based methods, where the rate of the load signal is neglected in the analysis, thus focusing on the fatigue mechanism. Methods described are rainflow cycle counting, level crossing counting, and other counting methods. In Section 3.2 frequency-based methods are studied, focusing on the power spectral density (PSD). Section 3.3 introduces the case of multi-input loads.

Chapter 4 Load Editing and Generation of Time Signals

There are many situations where modifying load signals is necessary. Section 4.1 discusses which properties of loads are essential for durability, and how to define the criteria for the equivalence of loads. Frequently, measured data are incorrect in the sense that the data show some deviation from what was intended to measure. Besides measurement noise, there are essentially three types of disturbances, namely offsets, drifts and spikes. Methods for inspection and correction of load signals are treated in Section 4.2. Editing of load signals in the time domain is studied in Section 4.3, where amplitude-based methods such as hysteresis filtering are considered, as well as frequency-based methods such as low or high pass filtering. Load editing in the rainflow domain is the topic of Section 4.4, especially rescaling, superposition, and the extrapolation of rainflow matrices are discussed. In some cases the time signal is not available, but only, for example, the rainflow matrix. Section 4.5 presents methods for generating load signals from condensed load descriptions.

Chapter 5 Response of Mechanical Systems

When analysing loads it is necessary to consider the mechanical structure that the loads act on. The role in durability applications of multi-body simulations, 'from system loads to component loads', and finite element models, 'from component loads to local stress-strain histories', are reviewed in Section 5.2 and Section 5.3, respectively. The issue of invariant system loads is addressed in Section 5.4, that is, the question of getting realistic excitations before measurements on prototypes have been made.

Chapter 6 Models for Random Loads

Load signals in customer usage vary in a more or less unpredictable manner. The load variability can be modelled by using random processes, which are treated in Chapter 6. Statistical modelling of load signals and their durability impact, in terms of damage, are discussed in connection with range-pair counts and level crossing spectra. Two main classes of random loads are treated: Gaussian loads, which model the frequency content, and Markov loads, which model the turning points of a load. The main topic is to compute the expected damage of a random load. Furthermore, the uncertainty in a measured damage number is treated.

Chapter 7 Load Variation and Reliability

The reliability of a component depends on both the load it is subjected to and its structural strength. The sources of variability in load and strength are discussed, and different reliability approaches are reviewed. Our recommendation is to use a second-moment reliability method. Thus, a load-strength model, adopted to the fatigue application, is developed in Section 7.6. The safety factor can then be formulated in terms of a reliability index. In Section 7.6.9 a compromise between statistical modelling and engineering experience is proposed by combining a statistically determined safety factor with a deterministic safety factor based on engineering judgement.

Part III Load Analysis in View of the Vehicle Design Process

The idea of Part III is to present load analysis in view of the vehicle design process, and describe which methods are appropriate in the different stages of design. Recall the vehicle design process presented in Figure 1.2 on page 5, which also represents the structure of Part III.

A brief description of the tasks to be solved may start at the end of the process, namely the verification of the final design. A question that arises is: 'How many specimens should be tested with which loads, such that a given reliability target can be verified?' First, the reliability target needs to be formulated in terms of engineering quantities. It may be given as a safety factor based on engineering experience, for example, by using in-house standards at the company. However, we promote the use of safety factors derived by using the load-strength interference, see Figure 1.4, thus including statistical modelling in order to take care of the uncertainties in load and strength.

It is important to follow the reliability requirements throughout the design process. The design and verification loads should thus be determined with respect to the customer population that the vehicle is aimed for. Customer loads may, for example, be obtained from measurement campaigns on public roads, either with professional test drivers along a planned route, or by selecting suitable that of customers. It is often practical to define a design load that is more severe than a typical customer, and the concept of a severe target customer, say, the 95%-customer, is widely used. The design load is often represented as driving schedules on the proving ground.

Finally, the task is to derive verification loads for testing, and relate the corresponding test results to the reliability target. As has been illustrated above, a statistical point of view should be taken in the design process, which is especially the case when performing and evaluating the verification tests. However, it is also important to use previous experience and engineering judgement, for example, in matters of how to accelerate testing without changing the failure modes.

Chapter 8 Evaluation of Customer Loads

The main task of Chapter 8 is to assess the customer load distribution. Apart from defining the load of interest (e.g. the load on the steering arm), it is important to define which population it represents, e.g. all potential customers, a specific application (e.g. timber trucks), or a specific market (e.g. the European market). In this context, principles of survey sampling are discussed. Further, the uncertainty in the calculated load severity is evaluated. In Chapter 8 we discuss three strategies for estimating the customer load distribution:

- *Random sampling*: Choose customers randomly, however, not necessarily with equal probabilities, and measure their loads.
- *Customer usage and load environment*: Estimate the proportion driven on different road types, and combine this with measurements from the different road types.
- *Vehicle-independent load description*: Define models for customer usage, road types, driver influence, and legislation, which can then be combined with a model for the vehicle dynamics.

Chapter 9 Derivation of Design Load Specifications

The topic of Chapter 9 is to derive loads for design and verification purposes. The basic specification is the severity of the load, which needs to be related to the design approach taken. Load time signals can be derived using simple synthetic loads, random load models, modification of measured signals, standardized load sequences, test track measurements, or can be defined through an optimized mixture of test track events.

Chapter 10 Verification of Systems and Components

Chapter 10 is devoted to the verification process; principles of verification, generation and acceleration of loads, and planning and evaluation verification of tests. Three verification approaches are discussed:

- *Highly Accelerated Life Testing*, HALT, based on the idea that failures give more information than non-failures and give rise to improvements regardless of severities that exceed what is expected.
- *Load-Strength analysis based on characterizing tests*. Strength and load properties are investigated by characterizing experiments. Uncertainties are evaluated within a statistical framework to verify the design against reliability targets by means of established safety factors.
- *Probability-based formal procedures*, with test plans based on formal consistent rules that, by experience, give safe designs. Typically, a low quantile in the strength distribution is verified by testing.

2

Loads for Durability

We discuss the basic engineering methods used for fatigue and load analysis, as well as some special features that are important when designing for durability. The classic Wöhler and Palmgren-Miner models for fatigue prediction are presented for loads with increasing complexity. A way to consider fatigue is to view it as caused by load cycles, and different ways to count and plot load cycles are discussed. Depending on the use and safety demands of the systems and the components, different design strategies are reviewed. Further, different kinds of mechanical systems require different load analysis methods, and these principles are reviewed. Finally, the role of load uncertainties, caused by scatter and lack of knowledge, in fatigue prediction, is emphasized for various stages of design.

2.1 Fatigue and Load Analysis

A short introduction to fatigue and load analysis is given which introduces some basic concepts for high cycle fatigue (HCF), i.e. the fatigue regime of some ten thousand or more cycles to failure, that are needed for the next sections. These topics will be revisited and explained in more detail in Chapter 3 and Appendix A.

2.1.1 Constant Amplitude Load

The simplest kind of load condition is the constant amplitude load, see Figure 2.1a. A common model for the high-cycle fatigue damage is the SN-curve, also called the Wöhler curve

$$N = \begin{cases} \alpha S^{-\beta}, & S > S_f \\ \infty, & S \le S_f \end{cases} \tag{2.1}$$

where N is the number of cycles to fatigue failure, and S is the stress amplitude of the applied load. The material parameters are: α, describing the fatigue strength of the material; β, the damage exponent; and S_f, the fatigue limit.

Guide to Load Analysis for Durability in Vehicle Engineering, First Edition. Edited by P. Johannesson and M. Speckert.
© 2014 Fraunhofer-Chalmers Research Centre for Industrial Mathematics.

2.1.2   Block Load

The next generalization is to consider block loads, i.e. blocks of constant amplitude loads following after each other, see Figure 2.1b. The Palmgren-Miner [183, 161] damage accumulation hypothesis then states that each cycle with amplitude S_i uses a fraction $1/N_i$ of the total life. Thus the total fatigue damage is given by

$$D = \sum_i \frac{n_i}{N_i}$$ (2.2)

where n_i is the number of cycles with amplitude S_i. Fatigue failure occurs when the damage D exceeds one.

2.1.3   Variable Amplitude Loading and Rainflow Cycles

The loads that a vehicle experiences in service are seldom constant amplitude loads or block loads. In Figures 2.1c and 2.1d, two so-called variable amplitude loads are shown. The first

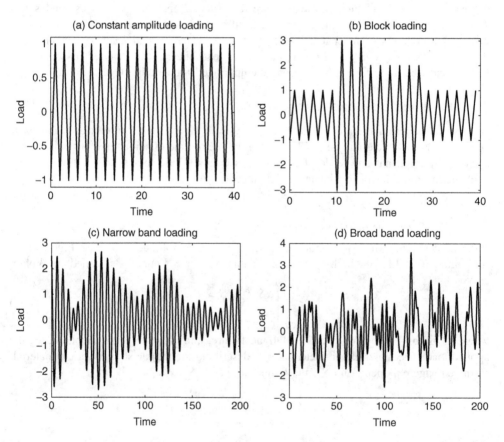

Figure 2.1 Different types of loads. (a) Constant amplitude load, (b) Block load, (c) Variable amplitude load, narrow band, (d) Variable amplitude load, broad band

one is a narrow band load, the second a broad band load. For an example of a real load, see Figure 1.3 on page 5, which shows a measured service load of a truck for 100 minutes.

One way to deal with varying amplitude loads is to form load cycles and then use damage accumulation methods on the counted cycles, cf. Equation (2.2). The load cycles are formed by pairing the local maxima with the local minima, using some kind of cycle counting algorithm. There are many definitions of cycle counting procedures in the literature, see Collins [58].

The rainflow counting method is generally accepted as being the best cycle counting procedure to date, and has become the industrial de facto standard. It was first presented by Endo in 1967, see Endo *et al.* [89, 90, 157]. There are now several versions of the rainflow counting algorithm, which are reviewed in Section 3.1.3, where the 4-point algorithm is explained in detail. Here the definition by Rychlik [198] is illustrated in Figure 2.2, which is especially useful for understanding the statistical and mathematical properties of rainflow cycles.

With regard to the damage accumulation, there are many theories in the literature, see Fatemi and Yang [92] for a review. The most popular one is the simple linear Palmgren-Miner, damage accumulation rule; Palmgren [183] and Miner [161], which in combination with rainflow cycle counting can be seen as the industrial state of the art for engineering applications. The validity of the rainflow cycle method has been studied by, for example, Dowling [76], and Jono [129]. The conclusion of Dowling's confirmation experiment was:

> ... the counting of all closed hysteresis loops as cycles by means of the rain flow counting method allows accurate life predictions. The use of any method of cycle counting other than range pair or rain flow methods can result in inconsistencies and gross differences between predicted and actual fatigue lives.

The range-pair method was independently developed in 1969 by de Jonge [68, 69], and extracts the same cycles as the rainflow method. Further, in Jono [129] it is experimentally

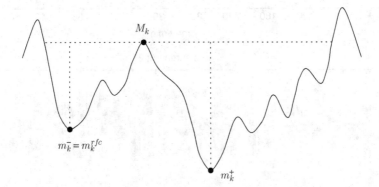

Figure 2.2 The definition of the rainflow cycle, as given by Rychlik [198]. From each local maximum M_k one should try to reach above the same level, in the backward (left) and forward (right) directions, with as small a downward excursion as possible. Thus, the maximum of the two minima m_k^- and m_k^+, representing the smallest deviation from the maximum M_k, is defined as the corresponding rainflow minimum $m_k^{\mathrm{rfc}} = \max(m_k^-, m_k^+)$. The k:th rainflow cycle is defined as $(m_k^{\mathrm{rfc}}, M_k)$

shown that the Palmgren-Miner rule works well if the damaging events are the rainflow cycles of the plastic strain.

2.1.4 Rainflow Matrix, Level Crossings and Load Spectrum

The main part of load analysis for durability is connected to the fatigue life regime. We will here demonstrate some basic procedures for load analysis and introduce the rainflow matrix, load spectrum, level crossings and rainflow filter. These topics will be revisited in Chapter 3, where they are extended and explained in more detail. The analysis is here exemplified using two measured signals, from two different trains of the same type, running from Oslo to Kristiansand in Norway, see Figure 2.3.

The first step in the analysis is to extract the peaks and valleys of the signal, which are here called *Turning Points* (TP). It is also customary to remove small cycles from the measured signal that may originate from measurement noise or cause negligible damage. This is of particular importance in fatigue testing, where it is often necessary to accelerate the testing and hence reduce the time of testing, but still have an appropriate load signal giving the correct damage. In our case the sample frequency of the signals is 200 Hz,

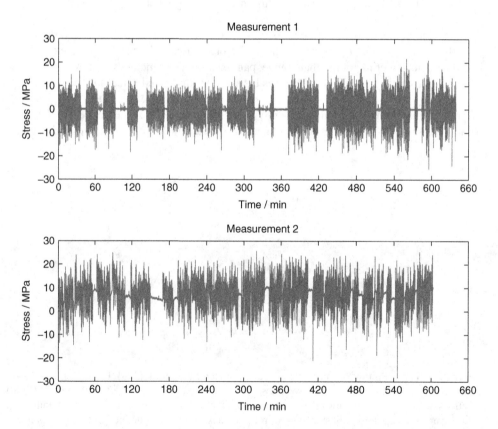

Figure 2.3 Stress signals from two trains in Norway, measured near a weld just above the boogie

resulting in about 7.5 million sample points, which is reduced to about 500 000 cycles. The proper way to remove small cycles is to use the so-called rainflow filter, which removes the turning points in the signal that constitute rainflow cycles with ranges smaller than a given threshold. By applying a rainflow filter with a threshold range of 4 MPa, the number of cycles is reduced to about 25 000, which means an acceleration by a factor of 20, but in this case keeping 99.8% of the original damage (based on the Palmgren-Miner rule and a damage exponent of $\beta = 4$).

From the rainflow filtered signals we extract the rainflow cycles, and obtain the rainflow matrices presented in Figures 2.4 and 2.5. The two figures represent the same set of rainflow cycles, but are presented in different ways. In Figure 2.4 the min-max format is used, which means that the x-axis is the minimum of the cycle, and the y-axis is the maximum of the cycle, and the colour represents the frequency of occurrence. The min-max format is the most convenient format for further statistical or mathematical analysis of the rainflow matrix, for example, extrapolation of the rainflow matrix or generation of time signals, see Sections 4.4 and 4.5. The most common way in fatigue applications is to present the rainflow matrix in amplitude-mean (or range-mean) format, where the x-axis represents the mean value of the cycle, and the y-axis the amplitude (or range) of the cycle, see Figure 2.5. In the amplitude-mean format it is easier to relate to the fatigue properties of the load, where the amplitude distribution is the most important one. However, it should be kept in mind that the two formats represent the same cycle information, only presented in different manners.

We can observe some differences in the rainflow matrices from the two measurements. Measurement 2 has a higher mean value, and also a wider rainflow matrix, due to the fluctuating mean value of the signal, compared to measurement 1. The fluctuating mean value of measurement 2 was found to be caused by a drift in the measuring device. The stresses were calculated from strain gauges that are sensitive to temperature variations. The temperature fluctuations at the measurement point give rise to strains but in this case not to stresses in the structure. Consequently, the fluctuating mean is an artificial stress component. This effect had already been removed from measurement 1 by removing a moving average

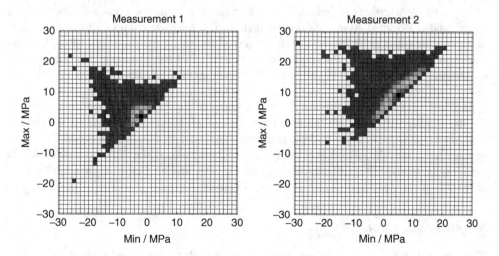

Figure 2.4   Rainflow matrices from the two Norway measurements, presented in min-max format

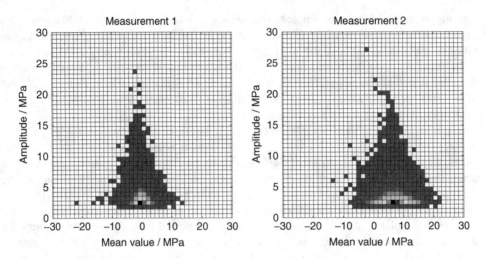

Figure 2.5   Rainflow matrices from the two Norway measurements, presented in amplitude-mean format

from the time signal, but it had not been removed from measurement 2. Methods for data inspection and correction are presented in Section 4.2.

Since the rainflow matrix may be hard to interpret, due to its two-dimensional nature, simpler characteristics like level crossings and rainflow amplitudes are useful. The level crossing spectrum can be calculated directly from the time signal, but also from the rainflow matrix. The two measurements have quite different level upcrossing spectra, see Figure 2.6a, which reflects the previous observation that measurement 2 has a higher mean value than measurement 1. Furthermore, we can observe that the shape of the tails of the level crossings seems to be different.

Another useful one-dimensional characteristic of a signal is the load spectrum, which here and by most authors is defined as the distribution of rainflow cycle amplitudes in the signal, and is obtained from the rainflow matrix. The load spectrum is also called range-pair count, see Section 3.1.4. It is often presented as the cumulative count of cycles with amplitudes larger than a given amplitude, as a function of the amplitude. For the Norway measurements there seem to be differences for small amplitudes, while for large amplitudes the two measurements have about the same characteristics, see Figure 2.6b.

The load spectrum can also be illustrated using a histogram, see Figure 2.7a, but in the linear scale we do not see the important large amplitude cycles. However, we can put weight according to the damage of the corresponding cycle amplitudes, see Figure 2.7b. This damage histogram shows how the damage is distributed among the different amplitudes. In our case the damage is distributed among many cycles, and the largest cycle contributes to about 2% of the total damage, where we used damage exponent $\beta = 4$.

2.1.5   Other Kinds of Fatigue

There are still many areas in fatigue and load analysis that are not well understood, mainly due to the difficulty of modelling the physical phenomenon in question. Such problems,

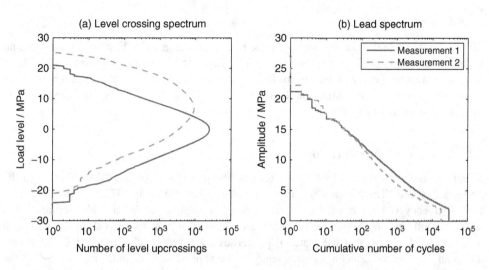

Figure 2.6   Level upcrossing spectrum and load spectrum from Norway measurements

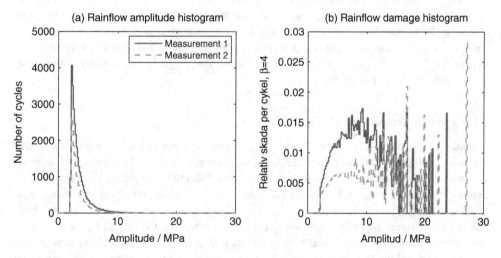

Figure 2.7   The rainflow cycle amplitude histogram and the corresponding damage histogram for Norway measurements

for example, are the durability assessment of non-metallic materials like rubber, ceramics, composite materials, glue joints, and plastics. Other problematic topics are high temperature fatigue and environmental effects like dust, humidity and corrosion. These topics will not be specifically covered in the *Guide*, but many of the methods presented are still useful. We will here briefly review load analysis in these situations, mainly pointing out the difficulties and the special care that needs to be taken. Note that the effects above can easily override other fatigue phenomena, which leads to the important topic of balanced model complexity, and where special efforts in fatigue modelling should be made.

2.1.5.1 Low Cycle Fatigue

The presentation in the *Guide* focuses on the high cycle fatigue (HCF) phenomenon, and no special attention is paid to low cycle fatigue (LCF). However, many of the methods and results are also applicable in the case of LCF, for example, rainflow cycle counting and other counting methods, load editing methods, response of mechanical systems, random load models, and reliability methods.

2.1.5.2 Non-metallic Materials

Materials like rubber are sensitive not only to the amplitudes of the load, but also to the deflection speed. There exists a strain rate effect, but the main effect of the increased load frequency is heating of the material. When modelling loads on rubber components, and especially when making accelerated tests, it is important to keep the temperature at a reasonable level, so that the material properties are representative of service conditions. This can be done by reducing the frequency of the applied load, or in testing by cooling the component. Rubber is also sensitive to environmental effects, and exhibits degrading due to ageing. For variable amplitude loads, rainflow counting together with Palmgren-Miner still seems to be a good engineering method. Surveys on rubber fatigue are presented in Mars and Fatemi [153], treating fatigue approaches, and in Mars and Fatemi [154], discussing factors affecting the fatigue life. For brittle materials, like ceramics, plastics and composites, the fracture strength is the most important property, while cumulative fatigue is rarely applicable in practice. For plastics and glue joints the phenomenon of ageing of the material may result in a significant loss in strength.

2.1.5.3 High Temperature Fatigue

Sometimes it is necessary to the consider the thermomechanical load, though only a small number of automotive components experience thermomechanical load cycles that result in significant low-cycle fatigue. Mostly it is components related to the engine, like cylinder heads, exhaust manifolds, and crankcases. In this case, the low-cycle fatigue problem is connected to the start-stop cycles, rather than to the combustion cycle. Consequently, it is the slowly varying temperature that is most important to model correctly. Compared to room temperature fatigue, there are several complicating factors that need to be considered when modelling the loads and making life assessments, like ageing of the material, creep, cyclic viscoplastic behaviour, stress-strain behaviour, and a proper fatigue criterion. Thermomechanical fatigue assessments for the automotive industry have been considered in for example, Charkaluk *et al.* [55], Thomas *et al.* [233], applied to cast-iron exhaust manifolds, and with an example on an aluminium cylinder head. The most fundamental model for low cycle fatigue is the Coffin-Manson relationship between plastic strain and fatigue life (see Tavernelli and Coffin [231]), which corresponds to the Wöhler curve for the high cycle fatigue.

2.1.5.4 Environmental Effects

Many engineering structures experience some form of alternating stress and are in addition exposed to harmful environments during their service life. Loads due to environmental effects, like corrosion and dust, are difficult to model, but may be important to consider depending on the component and on the application environment. Corrosion fatigue is the mechanical degradation of a material under the joint action of corrosion and cyclic load, see, for example, Roberge [197]. The most important effect of a corrosive environment is that the fatigue strength decreases and especially that the fatigue limit disappears. Hence, for load analysis it is also important to consider cycles below the conventional fatigue limit. An important relation in corrosion is the Arrhenius equation, modelling the time-dependent degradation process. Other environmental effects like dust and humidity are discussed in Karlberg *et al.* [131].

2.2   Loads in View of Fatigue Design

The appropriate method for load analysis in a certain situation is tightly coupled to the type of design approach that is applicable. A useful categorization of vehicle design approaches when considering load analysis is

- *Fatigue life* – Cumulative damage,
  - design for a finite life, typically structural components.
- *Fatigue limit* – Maximum load,
  - design for an infinite life, typically engine components.
- *Sudden failures* – Maximum load,
  - design for rare events, typically structural components.
- *Safety critical components* – 'Zero' failure,
  - design for high reliability, typically steering components.

2.2.1   Fatigue Life: Cumulative Damage

The design concept of fatigue life and cumulative damage is typically appropriate for structural components, such as the frame, the cabin, and the axles of a truck. The fatigue regime considered is finite fatigue life, using, for example, the Wöhler curve for life prediction together with the Palmgren-Miner damage accumulation hypothesis. Questions that arise are how cycles should be counted for variable amplitude loads, what load information is relevant, and how the load should be filtered when making accelerated fatigue tests.

It is generally agreed that in most cases fatigue can be treated as a rate independent process. Thus, the most important properties of the local loads for fatigue analysis are the values and configuration of the local extremes. Therefore, the load can be seen as a sequence of cycles formed by combining local maxima with local minima. However, the transformation from external loads to local loads may be highly rate dependent, which is discussed in the following section on system response.

2.2.2   Fatigue Limit: Maximum Load

In this category we typically think of engine and transmission components. These components experience millions of cycles in quite a short period of time compared to the design life, and cannot be allowed to contain any growing cracks. Thus they need to be designed using the fatigue limit philosophy, and consequently the maximum load is the most important load characteristic.

The main issue in load analysis here is how the maximum load should be determined, but also the interpretation of the maximum load. For engine and transmission components the loads are mainly internal loads coming from the engine itself, and can thus be controlled to a large extent. The interpretation of the maximum load could be a certain high load quantile when applying the maximum engine torque.

2.2.3   Sudden Failures: Maximum Load

The maximum load may also be used as a design criterion for overloads that originate from misuse, rare events, or abnormal use. In this case the vehicle should not collapse suddenly, but may experience global plastic deformations, and may be severely damaged. When considering overloads the interpretation of the maximum load could be the load that appears on average once or a few times during the design life. This maximum load may be estimated from load measurements using the statistical extreme values theory.

2.2.4   Safety Critical Components

A special category of components are the so-called safety critical components, for example, the steering knuckle, which in principle are not allowed to fail. In practice the 'zero failure rate' needs to be interpreted as a very small risk of failure. Special care needs to be taken when considering the safety critical components, and the load analysis methods need to be adopted accordingly, for example, by using the load-strength approach described in Section 7.6.

2.2.5   Design Concepts in Aerospace Applications

The requirements and resources in the aerospace field differ from other industrial sectors. In general the aerospace industry can afford to use more advanced methods compared to other fields. To a large extent this is due to relatively simple crack geometries and the possibility of inspections in service, but also due to a difference in financial and safety constraints. The design concepts have been developed over the years, and the evolution is often described as (see Wanhill [242] and Stephens *et al.* [223]):

- *Infinite life* – Design for an infinite life using the fatigue limit approach, however, historically using a static design approach.
- *Safe-life* – Design for a finite life using cumulative fatigue damage.
- *Fail-safe* – Design to be inspectable in service, in addition to safe-life design, in order to easily detect damage before safety is compromised. Generally, the requirement is that if one part fails, the system should not fail, and this includes concepts like redundancy, multiple load paths, and crack stoppers built into the structure.
- *Damage tolerance* – Design for a finite life using crack propagation methods, where a largest allowed initial crack is specified. This is a refinement of the fail-safe design.

2.3 Loads in View of System Response

Besides describing methods available for load analysis it is important to sketch the context of application of the methods. It is necessary to decide in specific applications which kind of method is best suited for a certain task. Thus, the complete engineering design process is considered and some requirements on the methods are pointed out at various stages of the process.

As can be seen from the schematic vehicle engineering process in Figure 1.2, the loads acting on specific components of a vehicle depend, on the one hand, on the external target load setting, which is assumed to depend on the envisaged customers but not on the specific vehicle, and, on the other hand, on the path the load is taken from outer excitations, which for trucks and cars depend on the road profile, through the tyres, the suspension and bearings to a local spot of the vehicle. The latter dependency is the topic to be discussed in the following. Taking this view we can talk about of loads on different levels, namely

- *system level* (road profiles, wheel forces, etc.);
- *sub-system level* (section forces or accelerations at the mount points between frame and suspension);
- *component level* (section forces at the steering knuckle); and
- *local level* (local stresses or strains at a certain spot of a component).

This situation is illustrated in Figure 2.8. In connection with these level categories it is useful to introduce the following load type categories, namely the representation of loads in the following domains:

- *time domain* (time signals);
- *frequency domain* (PSD functions, etc.); and
- *amplitude domain* (histograms such as rainflow, range-pair and level crossing).

The time domain still contains the full information, whereas the frequency domain or the amplitude domain describes special aspects suitable in different applications.

As was mentioned previously, it is commonly agreed that at the local level, many fatigue-related phenomena can be described based on a rate-independent load description, that is, the series of local maxima and minima of the stresses is important but the time scale no longer is (a more detailed description is given in Chapter 3 and Appendix A). In this case the amplitude domain description often is sufficient.

Service loads → (input)	mechanical system → /component	effect
road profile	vehicle	forces on bearings
wheel forces	wheel	local stress
engine vibration	axle	fatigue
...	cabin	comfort & acoustics
...	...	...

Figure 2.8   Loads on different levels

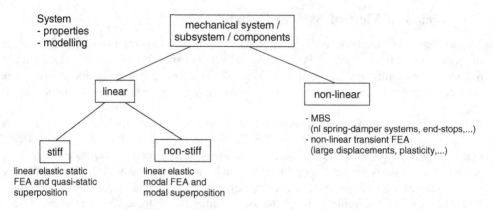

Figure 2.9   Different groups of mechanical systems

At the component level we already have a different situation depending on the specific properties of the component itself. For a compact stiff part like a knuckle, the situation is different from the case of a non-stiff vibrating structure like a cabin. The mechanical properties of systems and components may be categorized, according to Figure 2.9.

Depending on the category of the component, different load properties become more or less important. This is roughly sketched in Figure 2.10. With the exception of the case to the far left (linear stiff components), the full time domain is needed for an adequate description of the load. In that case, the frequency and amplitude domain aspects alone are not sufficient. Nevertheless they are and should be used in a combined mode for an insight into the loads and an analysis of them with respect to their various impact on the component. On the sub-system or even the system level the same remarks apply. The most suitable modelling and the analysing method for loads thus depend heavily on the mechanical system the loads are acting on.

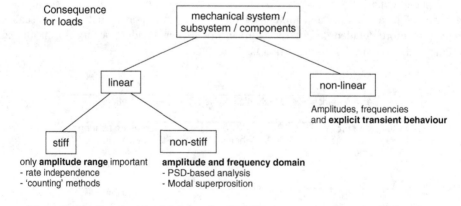

Figure 2.10   Required load information for different groups of mechanical systems

2.4   Loads in View of Variability

Every vehicle is unique in the sense that no other vehicle has exactly the same mechanical and material properties. There is a variation due to the manufacturing process, both at a macroscopic level, like the geometry of components, and at the microscopic level, like the grain structure at a critical point of the structure. This variation can be represented by a strength distribution, which is illustrated in Figure 1.4 on page 6.

Every vehicle is also unique in the exact way it is to be used. The load that a truck is exposed to depends on the type of truck (e.g. construction or distribution), which market it is used in, who it is sold to (a small or a large company), who is driving the truck (the owner or an employee), and so on. All these load influentials that were discussed in Section 1.3, contribute to the variability of the service loads, which can be represented by a customer load distribution, see Figure 1.4.

2.4.1   Different Types of Variability

In order to make a reliability statement of a component, both the variability of its strength and the variability of the customer loads are necessary to evaluate. The *Guide* treats load analysis; therefore we will here focus on the variability connected to the loads.

The variation is different depending on which entity we study. First of all, the load time signal itself is a physical quantity that varies over time. Further, there is a variation between different repeated measurements. The load signal is often condensed into, for example, a rainflow matrix, which will not be identical for two repeated measurements. Recall the two repeated measurements on the train running in Norway; see Figure 2.3. It is useful to characterize the load by a measure of its severity, which should be connected to fatigue life. In such measured severities there is also variability between repeated measurements.

When considering variability of loads, the discussion should be on two levels, on the level of an individual customer, and on the level of a population of customers. For an individual customer, two measurements of the same stretch of road will not give exactly the same load signal; there will be a random variation between different load measurements, even though the conditions are the same. A population of customers can, for example, be all potential customers, a specific market, or a certain application. A population can be described by a statistical distribution that could be characterized by its mean and standard deviation.

An important question is what population of customers is the target for our design. It could be randomly chosen customers from all segments of vehicles and markets, a randomly chosen customer of a specific type of vehicle, a randomly chosen customer in Brazil, a specific group of drivers, or a certain severe driver, for example, the 95%-driver that is represented on the test track or in the test rig.

When constructing models for the variability, it is useful to classify the types of variation, see for example, Melchers [159], and Ditlevsen and Madsen [74]. In a discussion of variation it is useful to distinguish between the following three kinds of uncertainties

- *Scatter*, or physical uncertainty, which is associated with the inherent random nature of the phenomenon, for example, the variation in load severity between different operators.
- *Statistical uncertainty*, which is associated with the uncertainty due to statistical estimators of physical model parameters based on available data, for example, the uncertainty in the estimated 95%-customer, due to a small sample of customers.

- *Model uncertainty*, which is associated with the use of one (or more) simplified relationship to represent the 'real' relationship or phenomenon of interest, for example, a finite element model used for calculating stresses is only a model for the 'real' stress state.

The first one is also called *aleatory uncertainty* and refers to the underlying, intrinsic uncertainties that cannot be avoided, for example, the scatter in fatigue life and the variation within a population of customers. The last two types are so-called *epistemic uncertainties* that can be reduced by means of additional data, more information, better modelling or refined parameter estimation.

2.4.2   Loads in Different Environments

Loads are acquired and used in different environments, ranging from real customer loads, test track loads, to the loads applied on a test rig. These different environments are combined with different kinds and magnitudes of variations in the loads.

2.4.2.1   Customer Loads

It is desirable to design the vehicle for the real usage by a customer population. In order to do that we need to evaluate the load influentials described in Section 1.3, both their typical values and their variations. Consequently, it is necessary to obtain this information from the customers in some way. Here, there are several options, such as a questionnaire of customer usage, measurements of service loads on some specific customers, and on-board-logging devices on a large number of customers. All these options require careful planning in order to obtain useful data. In this context the statistical planning of experiments are well suited. These issues are discussed in Chapter 8.

2.4.2.2   Test Tracks and Test Drivers

It is customary to translate the load requirements into test track cycles. This topic is treated in Chapter 9. Even though the test track is a very controlled environment and the measurements are obtained by using trained test drivers, the test track loads contain a significant amount of variability. There is scatter both between different test drivers and within a driver (i.e. between repeated measurements from the same driver). An example is shown in Figure 2.11, where three drivers have been driving 5 laps each on the same test track with the same vehicle and with the same driving instructions, see Section 8.3.1 for more details. It is also possible to acquire loads on specific public roads using test drivers, where more or less the same issues as for test track measurements arise.

2.4.2.3   Test Rigs

Also in controlled fatigue life experiments there are unwanted load variations. For example, the control system of the test facility is not perfect, and neither is the attachment of the test specimen, which may introduce unwanted stress components. Note that a 5% error in the applied load may well decrease the life by 30% or more. Further, if we want to relate

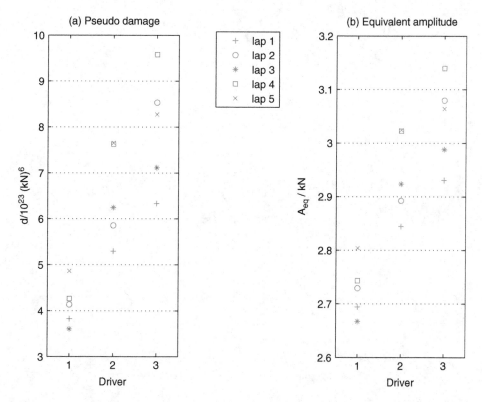

Figure 2.11 (a) Pseudo damage, d, and (b) equivalent amplitude, A_{eq}, for 3 drivers and 5 laps each on a test track, using damage exponent $\beta = 6$

test results to design specifications, it may also contain the uncertainties in the translation from a customer load distribution to fatigue test specifications. This translation typically involves determination of load specifications, as well as the mechanical constraints for the component or system that is tested. Verification tests are discussed in Chapter 10.

2.5 Summary

Basic concepts of reliability and loads are put in place and the ground is prepared for a more thorough discussion of methods for treating loads in different situations. As is pointed out, it is the case that on some levels the dynamic properties of the vehicle are crucial and in some instances they are not. Also the amount of information available affects the approach, and the uncertainty in the load information needs to be considered when evaluating the results.

Part Two
Methods for Load Analysis

3

Basics of Load Analysis

This chapter introduces the basic methods of load analysis for durability. By loads we mean any physical quantity that reflects the excitation or the behaviour of a system or component over time. The most typical loads are forces, torques, stresses, strains, displacements, velocities, and accelerations. However, other loads may be pressure or flow in hydraulic devices, rotational speed, temperature, or even state variables on the CAN bus or in some electronic control unit.

Typically the load is a measured time signal. An example of such signals is given in Figure 3.1, showing the vertical wheel forces measured on the front left wheel of a truck. These signals will be used as examples demonstrating the counting methods. Another example of a load signal is given in Figure 1.3, showing the vertical load measured on a truck transporting gravel. However, loads may also be numerically calculated, for example, the output of an FEM or a MBS calculation, which is treated in Chapter 5. The load can also be defined synthetically, for example, a block load or a standard load sequence like CARLOS, see [113] for a review of standard load sequences. Further, it can be the outcome of a random process with certain properties, see Chapter 6.

In a mathematical setting, the load is described by a function $x(t)$, where t ranges over the time interval of interest. Typically, the values of x are known only at certain discrete points in time (samples) $x_k = x(t_k)$, for $k = 1, 2, \ldots, N$. Due to the measurement processes the samples are mostly equally spaced with a sampling time $\Delta t = t_{k+1} - t_k$ and a sampling rate $f_s = 1/\Delta t$.

In the following, such a set of data is called a time signal, a time series, or a load history. Depending on the application there are different aspects which have to be taken into account during analysis. Therefore, we distinguish between rate independent methods (also called amplitude-based methods) and rate-dependent methods (also called frequency-based methods). The rate-independent methods only depend on the sequence of the local extreme values in the signal without regarding the shape of the time signal between the local extremes, thus focusing on the local fatigue load analysis. The rate-dependent methods are useful for system load analysis.

The aim of the methods for load analysis is to understand and describe the load properties that are important for durability analysis. Therefore, each method extracts some important

Guide to Load Analysis for Durability in Vehicle Engineering, First Edition. Edited by P. Johannesson and M. Speckert.
© 2014 Fraunhofer-Chalmers Research Centre for Industrial Mathematics.

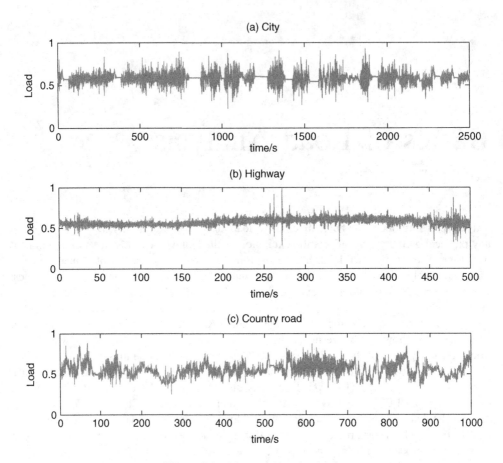

Figure 3.1   Measured service loads

durability feature of the load. The different methods reduce the load information in different ways and to different degrees, suitable for different situations.

However, before going into details of load analysis one has to ensure that the data to be analysed are correct. As data often come from some measurement equipment, this cannot be taken for granted at all. Frequently, measured data are incorrect in the sense that the data show some deviation from what was intended to be measured, mostly due to problems during the measurement process or properties of the equipment. Besides measurement noise, there are essentially three types of disturbances, namely offsets, drifts, and spikes. If present, these types of errors are such that they have to be corrected in order to obtain valuable results during a later analysis. Details of data inspection and correction are found in Chapter 4. Further, data may also be obtained from a mathematical analysis such as a multi-body simulation, which might suffer from numerical artefacts that need to be corrected. The bottom line is that unless offsets, drifts, spikes, noise, and other artefacts are appropriately corrected, the analysis result may be misleading, or in the worst case the results may simply be wrong.

The amplitude-based methods are treated in Section 3.1, being the fundamental methods for durability. In Section 3.2, the frequency-based methods, needed for system analysis, are dealt with. These two sections treat load analysis for a one-dimensional load signal. In practice there are often several load signals acting simultaneously on the system. However, at a certain critical point of the structure the load signals can in many cases be condensed into a one-dimensional load signal. Multi-input loads are the topic of Section 3.3.

3.1 Amplitude-based Methods

This section deals with methods operating in the so-called amplitude domain of the signals under consideration. Consequently, the rate or frequency of the load signal will be of no relevance. Instead, only the sequence of local extreme values (also called turning points) is considered by these algorithms. To start with, some global characteristics of the signal may be computed, for example, the global minimum/maximum, the global range of the signal, the number of mean crossings, and the irregularity factor. These simple methods, resulting in a single number, can be used in order to characterize or compare some properties of the loads. The results are particularly useful for the pre-analysis of data in the phase of verification and correction of data after or during a measurement campaign (see Section 4.2).

To get more detailed information about the durability impact of the load, there are several methods counting events in the load signal. The counting methods can be divided into three categories:

- **Peak value counting.** This method summarizes the load as the histograms of the local minima and local maxima.
- **Level crossing counting.** This method counts the number of times the load signal crosses each load level.
- **Cycle counting.** This category of methods counts cycles in the load signal by using some cycle counting principle. A cycle consists of a minimum and a maximum, and there are different methods for pairing the local minima with the local maxima. The simplest principle is to form a cycle of two consecutive turning points, which is the **Markov counting** procedure. A drawback of the Markov counting is that the large amplitude cycles may be missed. Therefore, the **rainflow cycle counting** principle extracts hysteresis loops in the load signal, which is motivated by fatigue and material models. Thus, for rainflow cycles, a large amplitude cycle may be formed so that there are several smaller oscillations between the cycle minimum and maximum. The rainflow cycle method is the most widely used method for evaluating fatigue damage, and is recommended for durability assessments. The **range-pair count** is the histogram of the rainflow ranges.

Load analysis is often performed on outer loads, even though the justification is based on the local loads that govern the fatigue phenomenon. In the presentation, we will start in Section 3.1.1 by discussing and giving reasons for the assumptions for doing load analysis on outer loads. The preparation of the signals, like extracting turning points, rainflow filtering and the discretization, is presented in Section 3.1.2, together with cycle definitions and some more terminology needed. In the presentation of the counting methods we will start with the rainflow cycle counting, Section 3.1.3, being the most important one, and the range-pair count, Section 3.1.4. In Section 3.1.5, Markov counting is described together with range

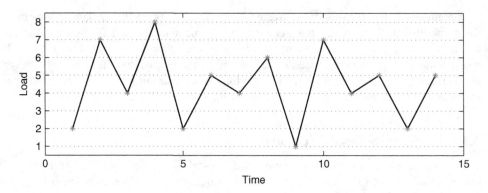

Figure 3.2   A simple load used for demonstrating the different counting methods

counting (Section 3.1.6). Level crossing counting is presented in Section 3.1.7, interval crossing counting in Section 3.1.8, and peak value counting in Section 3.1.10. Each method is exemplified using a short constructed signal, see Figure 3.2, with the aim of illustrating the principle of the methods. Realistic service loads, see Figure 3.1, are also used to illustrate how the results can be presented and interpreted. In Section 3.1.11 further examples are presented.

Another important topic is to define a simple measure of the severity of the load. The concepts of pseudo damage and equivalent loads are discussed in Section 3.1.12. Further, Section 3.1.13 is devoted to the special case of rotating components, where a dedicated rainflow counting variant and the time-at-level counting are presented.

3.1.1   From Outer Loads to Local Loads

Fatigue is a local phenomenon depending on the applied local stress or strain histories $\sigma(t), \epsilon(t)$. However, in load analysis we typically deal with measured or calculated outer loads $L(t)$, e.g. representing forces. If the local stress is (at least approximately) related to the outer load by a relation like $\sigma(t) = c \cdot L(t)$ or a nonlinear but monotone relation like $\sigma(t) = \varphi(L(t))$, then the stress history can be mapped onto the outer load history. Therefore, the counting procedures can be applied to the outer load and the results can then be back-transformed to the domain of the local load. Note that this is valid for cycle counting methods as well as for level crossing and peak value counting.

Using linear static elasticity to describe the relation between external loads and local stress, we get a linear relationship between the loads $L_i(t)$ acting on the component and the stresses and strains $\sigma(t), \epsilon(t)$ at a local spot x of the component in the form

$$^e\sigma(x, t) = c^{(1)}(x) \cdot L_1(t) + c^{(2)}(x) \cdot L_2(t) + \cdots + c^{(n)}(x) \cdot L_n(t). \tag{3.1}$$

Here, $^e\sigma$ denotes the elastic stress (sometimes called pseudo stress) calculated by using linear elastic analysis. In case of stiff components (eigenfrequencies much higher than load frequencies), the coefficients $c^{(i)}(x)$ are the results of several static FE computations (unit load cases). See Chapter 5 for more details.

Since linear elastic analysis typically overestimates stresses in notches when local yielding (plastic deformation due to the high local stresses) occurs, correction schemes are applied in order to get a better estimate of real stresses (without having to do a much more demanding elastic-plastic, non-linear analysis). The oldest and still best known scheme is Neuber's formula (see Neuber [170])

$$\sigma \cdot \epsilon = \frac{{}^e\sigma^2}{E},$$ (3.2)

where E denotes Young's modulus. If we combine the linear FE-analysis with Neuber's rule in a uniaxial situation (one outer load and a locally uniaxial stress state) and use the material law $\epsilon = g(\sigma)$ (see Section 3.1.3), we get a relation of the form $\sigma(t) = \varphi(L(t))$ as introduced above.

3.1.2 Pre-processing of Load Signals

The methods described below count some kind of events that are considered relevant for fatigue. Before counting these events, it is advisable to remove irrelevant information from the load signal. As has been mentioned, the fatigue damage process can be modelled as being rate-independent. Further, the transformation from local to outer loads is assumed to be such that there is no frequency response of the structure. Therefore, the value and order of the local extremes in the load contain all the information that is necessary for the counting methods. Further, it is also customary to remove small oscillations from the load, since their contribution to the fatigue damage can be considered negligible. Moreover, if the values of the turning points are changed by only a small amount, the fatigue impact can be expected to be more or less unchanged. Thus, the turning points may be discretized to fixed load levels, which can be practical for the analysis and presentation of the results. In this way the results of the counting methods can be stored in the form of a vector or a matrix, where each element counts the number of events of the corresponding type.

3.1.2.1 Turning Points Filter

The local extremes of the load signal contain all the information necessary for the counting methods. Therefore, the load signal can be reduced to a sequence of turning points (TP)

$$\text{TP}(L(t)) = \{L(t_1^0), L(t_2^0), \ldots, L(t_{N_0}^0)\}$$ (3.3)

where $t_1^0, \ldots, t_{N_0}^0$ are the time points of the local extremes.

Example 3.1 (Turning Points Filter)
The first step is to extract the turning points of the load. Figure 3.3a shows a short load signal, sampled at a frequency of 30 Hz, where the turning points of the load are marked with stars. The number of data points is reduced from 122 to 32, a reduction by a factor of 4. The method is also called min-max filter. □

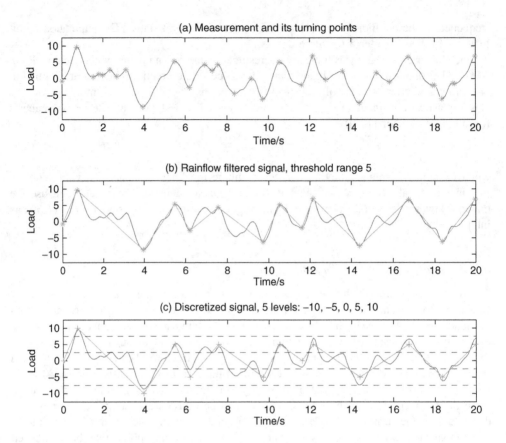

Figure 3.3 Example of (a) turning points filter; (b) rainflow filter; and (c) discretization

3.1.2.2 Rainflow Filter

The omission of small cycles is performed using the rainflow filter (also called the hysteresis filter), where all oscillations below a given threshold (or gate) are removed. This removes all turning points that correspond to rainflow cycles with the ranges below the given threshold, h, resulting in

$$\mathrm{TP}(L(t), h) = \{L(t_1^*), L(t_2^*), \ldots, L(t_{N_*}^*)\} \tag{3.4}$$

where $t_1^*, \ldots, t_{N_*}^*$ are the time points of the local extremes.

Example 3.2 (Rainflow Filter)
It is recommended that small oscillations from the signal are removed by using the rainflow filter. Figure 3.3b shows an example of the rainflow filter. The rainflow filtered turning points are presented together with the original load signal, where the remaining turning points are marked with stars. The number of turning points is reduced from 32 to 14, a reduction of about a factor 2.   □

The rainflow filtering can significantly reduce the number of turning points, which can be of importance both for testing purposes and numerical simulation. The choice of threshold range should be related to the fatigue properties of the load. One methodology is comparing the remaining damage in the signal for different choices of the threshold range, which is discussed in detail in Section 4.2. A rule of thumb is to use a threshold range equal to 5% of the global range of the load signal.

3.1.2.3 Discretization

It is customary, though not necessary, to discretize the signal levels of the load, and extract its turning points. The load range is then divided into a number of classes or bins, and each value of the signal is identified with the bin it belongs to. The classes are often equally spaced on the range of the measured signal, or the expected range of the signal that is to be measured. The load signal is thus reduced to a sequence of classes

$$\text{DTP}(L(t)) = \{\text{class}(L(t_1^d)), \text{class}(L(t_2^d)), \ldots, \text{class}(L(t_{N_d}^d))\} \tag{3.5}$$

$$= \{z_1, z_2, \ldots, z_{N_d}\} \tag{3.6}$$

which is a sequence of turning points taking values $1, 2, \ldots, n$, and $t_1^d, \ldots, t_{N_d}^d$ are the time points of the local extremes. Note that the number of discretized turning points may be smaller than in the original signal, since two consecutive turning points may be in the same class and are thus no longer any turning points. The sequence of turning points in the load domain is

$$\text{TP}_d(L(t)) = \{u_{z_1}, u_{z_2}, \ldots, u_{z_{N_d}}\} \tag{3.7}$$

where $u_1, u_2, \ldots, u_n$ are the discrete load levels.

Example 3.3 (Discretization)
Often the signal levels are discretized. Figure 3.3c shows an example of the discretization, where five load levels, $u = (-10, -5, 0, 5, 10)$, have been used. The dashed lines indicate the borders between the five classes of load levels   □

The discretization can act on the load signal itself or on the turning points of the signal; the result is the same. Note that two consecutive turning points may be classified into the same bin. Therefore it is necessary to run the turning points filter also after the discretization in order to ensure that the result really is a sequence of turning points. Before discretization, it is therefore recommended that the signal is rainflow filtered with a threshold equal to (at least) the discretization step.

In the following the resulting sequence of turning points will be denoted by

$$\text{TP}(L(t), h) = \{x_1, x_2, \ldots, x_N\} \tag{3.8}$$

where h is the threshold range for the rainflow filter. The resulting discretized sequence of turning points will be denoted

$$\text{DTP}(L(t)) = \{z_1, z_2, \ldots, z_N\} \tag{3.9}$$

where z_k takes values $1, 2, \ldots, n$.

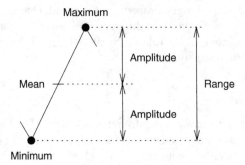

Figure 3.4   The definition of the amplitude, the range and the mean of a cycle

3.1.2.4   Cycles

A cycle is a pair consisting of a minimum and a maximum, where the range (or amplitude) is the most important characteristic for fatigue evaluation. In fatigue applications a cycle is often represented as a range-mean pair. The definition of the amplitude, the range, and the mean of a cycle is

$$\text{amplitude} = (\text{Max} - \text{Min})/2$$

$$\text{range} = \text{Max} - \text{Min}$$

$$\text{mean} = (\text{Max} + \text{Min})/2$$

see also Figure 3.4.

When we discussed cycles above, it has been full cycles, meaning that the load goes from a minimum to a maximum and then back again to the minimum. Sometimes it is convenient to use the terminology half-cycle (or reversal), which is defined as the range minimum to maximum or maximum to minimum. Hence, a (full) cycle consists of two half-cycles.

3.1.3   Rainflow Cycle Counting

The idea behind the rainflow cycles is to count hysteresis loops in the load. However, prior to the description of the counting algorithm and its many variants, we will spend some effort on the description of hysteresis models, which are important concepts trying to model the local stress and strain phenomena leading to fatigue in metal components. The hysteresis models serve as the main motivation for the rainflow count method.

3.1.3.1   Hysteresis Models and Rate Independence

This section introduces some basic and simple aspects of hysteresis models. The goal is not to give a complete introduction to the subject, which would be outside of the scope of this *Guide*, but rather to give an extended motivation for the rainflow method, which is fundamental in load analysis for fatigue. As was seen above, we often find a relation of the form $\sigma(t) = \varphi(L(t))$, where φ is a monotonic transformation from outer load to stress, which is valid at least approximately. This justifies 'back-transforming' local stress analysis methods to load analysis. Hence, it is worthwhile spending at least some effort on the local stress-strain mechanisms governing metal fatigue.

Stresses and strains are the governing local variables with respect to fatigue. Depending on the specific application, there are different practically successful models in use (constitutive laws), which describe the evolution of strain tensors over time, given the stress tensor or vice versa. They all have in common that the relation cannot be expressed in a simple functional form like $\sigma(t) = f(\epsilon(t))$. It exhibits hysteresis cycles and memory effects, which are essential for the description of fatigue. There are a couple of different mathematical models to describe the complex phenomena (e.g. ratchetting or many others), but their description is beyond the scope of this *Guide*.

Since we are primarily interested in giving the reason for using the amplitude-based counting methods, we concentrate on the simplest case of a locally uniaxial stress state and refer to Chaboche and Lemaitre [53] (continuum mechanics) and Brokate and Sprekels [42] (hysteresis operators) for a more detailed description. Although it is known that none of these models is fully satisfactory in all important applications, the common basis, which is the locally uniaxial Masing memory model, serves as a good starting point for the explanation of the counting algorithms.

The Masing memory model is based on the cyclic stress-strain curve $\epsilon = g(\sigma)$, which can be represented by the Ramberg-Osgood relation

$$g(\sigma) = \frac{\sigma}{E} + \left(\frac{\sigma}{K'}\right)^{\frac{1}{n'}}. \tag{3.10}$$

The model describes the evolution of hysteresis cycles in the stress-strain plane given either the stress or the strain time history. It is an empirical model for a uniaxial stress-strain state, and has been proven to give reasonable results for metal components such as steel or aluminium under loads which are well below the static limit.

The following set of rules describe the evolution of the stress time history given the strain time history.

- Masing rule: Starting with $\epsilon = 0$, calculate $\sigma = g^{-1}(\epsilon)$ until the first local maximum or minimum of ϵ is reached. The stress-strain path thus follows the cyclic stress-strain curve. At a turning point of the strain, the succeeding increments in stress are calculated according to Masing's rule, Masing [156]

$$\frac{|\Delta\epsilon|}{2} = g\left(\frac{|\Delta\sigma|}{2}\right), \quad \Delta\sigma = sign(\Delta\epsilon)|\Delta\sigma| \tag{3.11}$$

  until the next turning point, another hysteresis branch, or the cyclic stress-strain curve is reached.

For the continuation of the stress-strain path the following three memory rules apply:

- Memory rule 1: if the hysteresis branch meets the cyclic stress-strain curve, follow the cyclic stress-strain curve.
- Memory rule 2: if the hysteresis branch meets another hysteresis branch, follow that branch.

● Memory rule 3: if the hysteresis branch meets the cyclic stress-strain curve in the opposite quadrant, follow the mirrored cyclic stress-strain curve.

This set of rules is sufficient to calculate the complete stress-strain path given either by the strain time history or the stress time history. An example is given in Figure 3.5. As will be explained below, the cycles defined by this model can easily be identified in the corresponding stress or strain time history without explicitly constructing the stress-strain path.

According to the definition of the Masing memory model the hysteresis cycles only depend on the sequence of the local extreme values of the stress or strain signal. If we now assume that the hysteresis cycles in the stress-strain path are the most relevant events for fatigue, load analysis methods need only consider the local extremes.

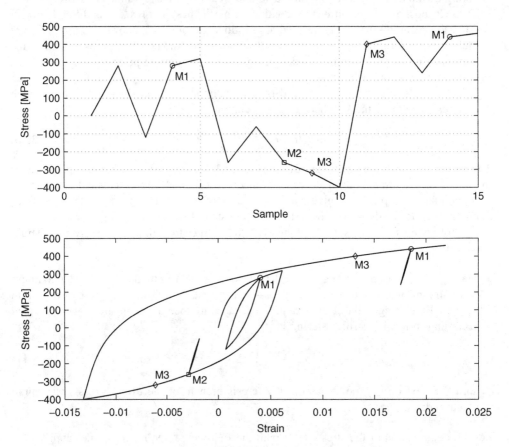

Figure 3.5 A stress signal and its hysteresis cycles. There are three closed cycles. The points of application of the different memory rules are marked explicitly. Note that the hysteresis branch starting at the left bottom corner does not meet the cyclic stress-strain curve before the point marked M3 at about 400 MPa. The points marked M1 or M2 are the points, where the 3 cycles are closing.

3.1.3.2 From Outer Load to Local Load

What has been said so far about hysteresis models primarily applies to local stress or strain histories $\sigma(t)$, $\epsilon(t)$. However, in load analysis we typically deal with measured or calculated outer loads, for example, forces. If the local stress $\sigma(t)$ is related to the outer load $L(t)$ by a relation like $\sigma(t) = c \cdot L(t)$ or a nonlinear but monotone relation like $\sigma(t) = \varphi(L(t))$, then the stress-strain path can be mapped to a load-strain path, where it is a local strain in both cases. The corresponding load-strain cycles open and close at the same time as the underlying stress-strain cycles. Consequently, the local stress-strain cycles have outer load-strain cycles as counterparts, which can be identified by outer load data analysis. Therefore, a load data analysis method should be constructed in such a way, that it pays attention to the load-strain cycles.

3.1.3.3 Rainflow Cycle Algorithms

There are a number of definitions for this method, starting with Endo *et al.* [91] or Murakami [168], where the name of the method is explained by an analogy with rain flowing over a pagoda roof. Other early publications are de Jonge [68, 69] on the equivalent range-pair-range counting procedure, and Dowling and Socie [78]. In Clormann and Seeger [56], the relation to the Masing memory model and the difference between the first and second run (see below) are pointed out. Another definition is given in Rychlik [198], which is very useful for statistical analysis, and is described in Chapter 1, see Figure 2.2 on page 17. The 4-point counting algorithm is presented in de Jonge [68, 69], Brokate *et al.* [43], and Dressler *et al.* [80], and will be described in detail below. It is algorithmically very simple and pays special attention to the open cycles (residual). There are also standards for rainflow counting: AFNOR [3], which is identical to the 4-point rule, and ASTM [7], which works on 3 points and differs with respect to the treatment of the residual. There are no ISO or DIN definitions of rainflow cycle counting.

What all the definitions have in common is that they interpret the signal as a stress (or strain) history and count the number of hysteresis cycles which result from an application of the Masing memory rules. The main result is a set of cycles that can be stored in a rainflow matrix *RFM*, where *RFM*(i, j) is the number of hysteresis cycles starting in bin i and ending in bin j. The definitions vary in the treatment of the so-called residual. This is the part of the signal that belongs to the stress-strain path after removal of all closed cycles.

3.1.3.4 Some Comments on Rainflow Cycles

A full rainflow cycle (or closed hysteresis cycle) emerges from the cyclic stress-strain curve or another hysteresis branch. The cycle starts at a certain point, reaches another level, and then returns to the starting point. In Section 3.1.3 this has been explained in detail in the course of the Masing memory rules and the corresponding example (see Figure 3.5). There are two local extremes in the time signal that define a cycle (the starting and the turning point). Thus, if there are N_h full cycles in the stress-strain path, the signal contains $2N_h + N_r$ reversals, where N_r denotes the number of remaining reversals in the stress-strain path after eliminating all full cycles. In the 4-point algorithm for counting rainflow cycles as described below, the full cycles and the remaining residual are found.

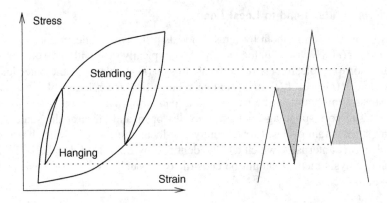

Figure 3.6 Hysteresis loops in the stress-strain plane

It is possible to divide the set of rainflow cycles into two groups depending on whether the rainflow minimum occurs before or after the maximum. The two different kinds of cycles occur on an up-going or a down-going hysteresis arm, and are called hanging and standing rainflow cycles, respectively, see Figure 3.6.

3.1.3.5 The 4-point Algorithm

Because of the importance of the rainflow cycle counting method, the 4-point algorithm is now explained in detail. We start with a sequence of discretized turning points z_k, for $k = 1, \ldots, N$, taking values $1, \ldots, n$, that is the min-max filtering and discretization has already been done. The rainflow matrix RFM and the residual RES are empty at the beginning (i.e. the elements of the rainflow matrix are zero, and the length of the residual is $r = 0$). The algorithm works with a stack of 4 points, which is initialized with the first 4 points of the signal $s = [s_1 = z_1, s_2 = z_2, s_3 = z_3, s_4 = z_4]$. Next we apply the following counting rule

(c) if $\min(s_1, s_4) \leq \min(s_2, s_3)$ and $\max(s_2, s_3) \leq \max(s_1, s_4)$,
  then the pair (s_2, s_3) is a cycle,

that is we detect a cycle, provided the interval spanned by the inner points of the stack is contained in the interval spanned by the corners of the stack. If this is the case, we store the cycle in the matrix $RFM(s_2, s_3) = RFM(s_2, s_3) + 1$ and remove both points from the stack. Now the stack has to be refilled. The way this is done reflects the memory rules from the hysteresis model. If possible, the stack is filled with points from the past, that is from the residual. In detail, this means (k denotes the next point of the signal z):

(r1) if $r = 0$, **then** $[s_1 = s_1, s_2 = s_4, s_3 = z_k, s_4 = z_{k+1}]$, and $k = k + 2$
(r2) if $r = 1$, **then** $[s_1 = RES_r, s_2 = s_1, s_3 = s_4, s_4 = z_k]$, $k = k + 1$, and $r = 0$
(r3) if $r \geq 2$, **then** $[s_1 = RES_{r-1}, s_2 = RES_r, s_3 = s_1, s_4 = s_4]$, and $r = r - 2$

Then the counting rule is applied again. If the counting condition (c) is not fulfilled, then

(r4) $r = r + 1$, $RES_r = s_1$, $[s_1 = s_2, s_2 = s_3, s_3 = s_4, s_4 = z_k]$, and $k = k + 1$.

This is repeated until the last point of the time signal is reached and (c) does no longer apply. The result of this procedure is the rainflow matrix *RFM*, containing all closed cycles, and the residual *RES*, containing the remaining sequence of turning points.

Example 3.4 (Rainflow Counting: Simple Load)
Applying the 4-point algorithm to the signal in Figure 3.2 gives the following closed cycles

$$\{(7,4),\ (5,4),\ (2,6),\ (4,5)\}$$

together with the residual

$$(2,\ 8,\ 1,\ 7,\ 2,\ 5).$$

We can observe that there are two hanging cycles (the maximum occurring before the minimum), and two standing cycles (the minimum occurring before the maximum). Further, the cycle (5, 4) hangs inside the cycle (2, 6). The rainflow matrix becomes

		to							
		1	2	3	4	5	6	7	8
	1	0	0	0	0	0	0	0	0
	2	0	0	0	0	0	1	0	0
	3	0	0	0	0	0	0	0	0
from	4	0	0	0	0	1	0	0	0
	5	0	0	0	1	0	0	0	0
	6	0	0	0	0	0	0	0	0
	7	0	0	0	1	0	0	0	0
	8	0	0	0	0	0	0	0	0

where the first column indicates the "from" load level and the first row indicates the "to" load level, and the other figures the number of counted cycles with the corresponding sizes. The two hanging cycles are below the diagonal, and the two standing cycles are above the diagonal. Note that the largest load cycle is in the residual, which is treated below. □

3.1.3.6 The Residual

The residual consists of a part with increasing amplitudes followed by a part with decreasing amplitudes. It is often quite short, but it can be as long as the original signal, in case no closed cycles are found. However, for a discretized load signal the length of the residual is limited by $2n - 1$ where n is the number of classes.

Especially for short signals, used repeatedly in laboratory test or computer simulation, it is of the utmost importance to pay attention to the residual. The methods also differ with respect to the following aspect. If the signal (interpreted as a local stress history) is applied

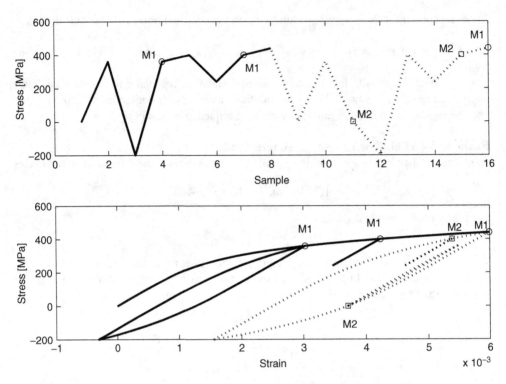

Figure 3.7 A stress signal and its corresponding stress-strain path. The signal is $[0, 360, -200, 400, 240, 440]$. The solid lines represent the first run, the dotted line the second run through the signal. As can be seen from the stress-strain path, only the small cycle $(400, 240)$ exists in both runs. The larger cycle $(360, -200)$ exists only in the first run and the cycles $(0, 360)$ and $(440, -200)$ exist only during the second run

once (first run) to a certain component, some hysteresis cycles will result from that. If the signal is applied a second time, some other cycles will be created, as can be seen from Figure 3.7. If the signal is applied further, the same cycles as during the second run will be created. Since we typically assume that a load signal can be applied multiple (say, N_b) times before a component fails, it might be important to distinguish between the cycles of the first run (which exist only once) and the ones from the successive runs through the history which will exist $N_b - 1$ times.

In that sense the 4-point-counting algorithm is the most elegant one because it takes these considerations into account. It stores those cycles which belong to the first as well as to the second run in the rainflow matrix *RFM* and in addition it stores the remaining short residual, from which the cycles of the first run and the cycles from the successive runs can easily be derived.

If we denote the cycles belonging to the first run only by RFM_1, the cycles belonging to the second run only by RFM_2 and the ones closing in the first as well as the second run by RFM_0, then, according to the Palmgren-Miner rule (see Palmgren [183] and Miner [161] for more details), we can calculate the separate damage contributions $d_i = d(RFM_i)$, for

$i = 0, 1, 2$ of the three groups of cycles and get the following expression for the number N_b of repeats until failure:

$$1 = d_1 + (N_b - 1) \cdot d_2 + N_b \cdot d_0 \tag{3.12}$$

$$\Rightarrow d = \frac{1}{N_b} = \frac{d_0 + d_2}{1 - d_1 + d_2}. \tag{3.13}$$

If the number N_b of repeats until failure is large, it is clear that d_1 is of minor importance, however, d_2 should not be ignored since it always contains the largest cycle.

For the 4-point algorithm, RFM contains all cycles, which are closing in the first as well as in the second run through the signal, that is $RFM = RFM_0$. A simple extra rule applied to the residual gives the cycles closing in the first run only. Starting with the 4-point stack $s = [RES_1, RES_2, RES_3, RES_4]$, the following counting condition (c1) is checked:

(c1) if $u_{s_2} \cdot u_{s_3} < 0$ **and** $|u_{s_4}| \leq |u_{s_2}| \leq |u_{s_3}|$ **then** (s_2, s_3) **is a cycle,**

where u_{s_2} and u_{s_3} are the physical values of the load corresponding to the bin values s_2 and s_3, respectively. If (c1) is fulfilled, then s_2, s_3 are deleted and the stack is refilled in the same way as described above. This procedure gives all cycles closing in the first run only, that is, the matrix RFM_1. Getting the cycles RFM_2, which exist in the second run only, is achieved by constructing an intermediate signal y by doubling the residual in the form $y = (RES, RES)$ and applying the 4-point rule to that signal.

Some authors (e.g. ASTM [7]) propose circumventing the treatment of the residual by starting rainflow counting with the value of highest absolute load. This ensures that the residual in the stress-strain path is just a part of the cyclic stress-strain curve and all cycles lie inside the largest enveloping cycle, which starts and ends at the highest absolute load. This procedure changes the order of the load and is not fully correct in the sense of material memory, but the difference from the first and second run counting as introduced above is small in most cases.

3.1.3.7 Summary of Rainflow Counting

To summarize, the result of the 4-point rule is the tuple $[RFM, RES]$, where RFM contains the cycles RFM_0 closing in all runs. The cycles RFM_1 and RFM_2, which close at the first and second run, respectively, can easily be extracted from RES. The clear separation of the three types of cycles will be used during some of the advanced operations like superposition and extrapolation in the rainflow domain (see the following chapters, especially Section 4.4).

Example 3.5 (Rainflow Counting Residual: Simple Load)

The largest load cycle is in the residual, and thus needs to be treated. The cycles in the residual can be counted by doubling the residual and making a rainflow count, resulting in the cycles

$$\{(2, 5), \ (7, 2), \ (1, 8)\}.$$

These cycles represent the so-called second run count, which is the extra cycles that would be formed each time the load signal is repeated.   □

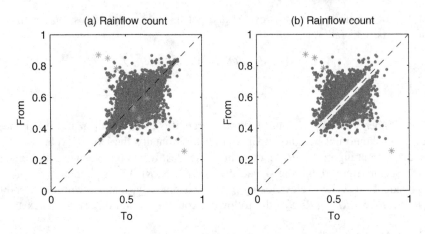

Figure 3.8   Rainflow count of a vertical wheel force (a) all rainflow cycles; (b) with rainflow filter

Example 3.6 (Rainflow Counting: Vertical Wheel Force)
The vertical wheel force for the country road, Figure 3.1c, will be used to demonstrate the
rainflow counting, and show how the results can be visualized. First, the turning points of the
signal are extracted, and the rainflow cycles are counted. The result is shown in Figure 3.8a,
where each dot represents a full cycle. The stars represent the cycles in the residual, counted
by doubling the residual, according to the recommended second run counting. The hanging
cycles are in the upper left part of the graph (the maximum occurring before the minimum),
and standing cycles are in the lower right part (the minimum occurring before the maximum).
However, prior to the analysis it is recommended that a rainflow filter is applied, here with
threshold range 1/31 (as we later will discretize to 32 levels). Now all "small" amplitude
cycles are removed, see Figure 3.8b.

Next, we will calculate the rainflow matrix. The rainflow filtered signal is discretized to
32 levels, equally spaced between 0 and 1, and the rainflow cycles counted and classified
into the rainflow matrix, which is a 32 by 32 matrix. The result can be presented in a matrix
plot, Figure 3.9a, where the row corresponds to the starting point (from) of a cycle and the
column is the turning point (to) of a cycle. The shading indicate the count itself, and the
axis represent the indices of the matrix. The values 1 to 32 can be mapped to the physical
values, see Figure 3.9b. However, note that the y-axis goes from 1 to 0. In Figure 3.9c and
Figure 3.9d, the y-axis is flipped, so that the graphs are plotted in the normal xy-coordinate
system.

The hanging and standing cycles are often not distinguished in the damage calculation,
and this information can thus be removed by combining the cycles, see Figure 3.10(a).
Further, it is also possible to display the rainflow matrix by using a contour-plot, see
Figure 3.10(b).

In fatigue applications it is common to present the rainflow cycles in a range-mean plot
(either separately for the standing and hanging cycles or combined), where the rows in
the matrix indicate the range of a cycle and the column indicates the mean value, see
Figure 3.11a. Further, it is also possible to display the rainflow matrix by using a 3D-
histogram, see Figure 3.11b, where the height represents the number of cycles.   □

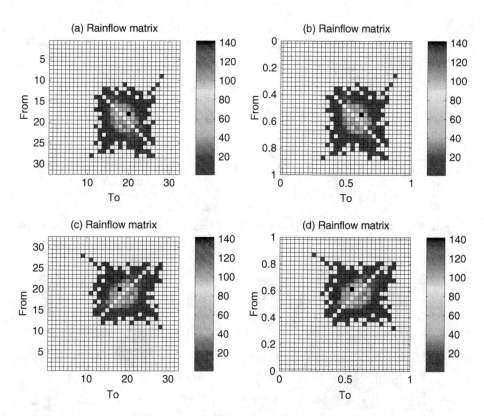

Figure 3.9 Rainflow matrix of a vertical wheel force (a) indices 'matrix'-plot; (b) 'matrix'-plot; (c) indices 'xy'-plot; (d) 'xy'-plot

3.1.4 Range-pair Counting

Besides the rainflow counting method, this is the only one with a direct relation to hysteresis cycles. The range-pair counting extracts the cycle ranges of the rainflow cycle counting

$$\{rp_1, \ rp_2, \ rp_3, \ \ldots\} \qquad \text{with} \qquad rp_i = |\mathrm{rfc}_i^{from} - \mathrm{rfc}_i^{to}|. \tag{3.14}$$

For a discretized load, it summarizes all cycles of a certain range or amplitude and forgets about the mean value. It can thus be obtained from the rainflow count by simply summing over all cycles with the same amplitude

$$f^{rp}(k) = \sum_{|i-j|=k} RFM(i, j) + \text{contribution from } RES. \tag{3.15}$$

The result is a vector, where the index indicates the range (or amplitude) and the value indicates the number of cycles. It can be plotted as a histogram. However, it is often presented in cumulative form, where at level k all cycles with an amplitude greater than k are shown

$$n^{rp}(k) = \sum_{i>k} f^{rp}(i). \tag{3.16}$$

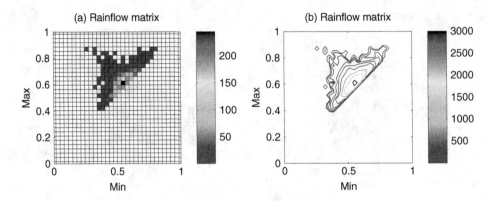

Figure 3.10   Rainflow matrix of a vertical wheel force (a) 'min-max'-plot; (b) contour plot

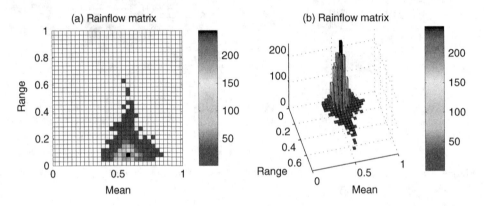

Figure 3.11   Rainflow matrix of a vertical wheel force (a) range-mean plot; (b) range-mean plot as 3D-histogram

Note that the load spectrum introduced in Section 2.1.4 is the same as the range-pair count.

Example 3.7 (Range-pair Counting: Simple Load)
The range-pair counting gives the ranges

$$\{3, \ 1, \ 4, \ 1, \ 3, \ 5, \ 7\}$$

and the range-pair histogram becomes

k	1	2	3	4	5	6	7
$f^{rp}(k)$	2	0	2	1	1	0	1

where the first row indicates the load range, and the second the number of counted cycles with the corresponding range. The cumulative range-pair count becomes

k	0	1	2	3	4	5	6	7
$n^{rp}(k)$	7	5	5	3	2	1	1	0

□

Sometimes it is more convenient to have a continuous version of the range-pair count defined as follows. For a sequence of amplitudes $S_1, S_2, \ldots, S_{N_c}$ the number of rainflow ranges greater than a level s, $s \geq 0$, is called the cumulative range-pair count

$$N^{rp}(s) = \text{"number of amplitudes } S_i > s\text{"}. \qquad (3.17)$$

For example, if three amplitudes $100, 50, 120$ have been found, then

$$N^{rp}(s) = \begin{cases} 3 & 0 \leq s < 50 \\ 2 & 50 \leq s < 100 \\ 1 & 100 \leq s < 120 \\ 0 & s \geq 120 \end{cases}$$

Similar to the rainflow count, there are a couple of definitions which differ in their treatment of the residual. There are the following possibilities:

1. contribution from $RES = 0$
2. contribution from $RES = \sum_{|i-j|=k} RF M_2(i, j)$
3. contribution from $RES = \sum_{|i-j|=k} RF M_1(i, j) + RF M_2(i, j)$
4. contribution from $RES = $ according to German DIN
5. find min-max cycles in the residual and find ranges of the cycles

The German DIN version [72] tries to find pairs of corresponding ranges in the residual. If, for example, the residual looks like $(0, -100, 100, 0)$, then it finds two pairs of cycles, namely $(0, -100, 0)$ and $(0, 100, 0)$, but it neglects the large range $(-100, 100)$ which leads to a cycle during the second run. This approach is generally non-conservative because the largest second run cycle is typically missing. In contrast to this, the ASTM definition, ASTM [7], essentially gives the "second run" count.

Example 3.8 (Range-pair Counting: Vertical Wheel Force)
Figure 3.12 shows a comparison of the vertical wheel forces for three measurements in Figure 3.1, where a truck has been driven in a city, on a highway, and on a country road. To the left the level crossing plot is shown, to the right the range-pair plot. For both types of plots the "second run count" is used, which means that the contribution from the residual has been taken from $RF M_2$. □

3.1.5 Markov Counting

The most straightforward method for counting cycles in the load is to pair consecutive local minima and maxima. This is called Markov counting or min-max cycles. The Markov counting thus defines the following half-cycles

$$\{(x_1, x_2), (x_2, x_3), \ldots, (x_{N-1}, x_N)\}. \qquad (3.18)$$

In the case of a discretized load, the resulting cycle count can be stored in a matrix. For each bin i it counts the number of transitions to another bin j. The result is the so-called Markov matrix, $\boldsymbol{F}^{\text{Markov}}$, containing the number of transitions.

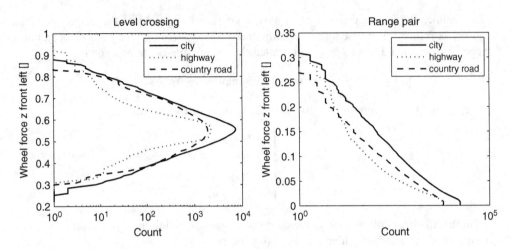

Figure 3.12 Comparison of three different road types for the vertical wheel force (normalized). The overall appearance of the histograms is rather similar. In this case, the maximum amplitudes on the country road are somewhat smaller than on the highway and in the city

Example 3.9 (Markov Counting: Simple Load)
The Markov counting of the short sequence of turning points in Figure 3.2 gives the half-cycles

$$\{(2, 7), \ (7, 4), \ (4, 8), \ (8, 2), \ (2, 5), \ (5, 4),$$
$$(4, 6), \ (6, 1), \ (1, 7), \ (7, 4), \ (4, 5), \ (5, 2), \ (2, 5)\}$$

and the Markov matrix becomes

		to							
		1	2	3	4	5	6	7	8
	1	0	0	0	0	0	0	1	0
	2	0	0	0	0	2	0	1	0
	3	0	0	0	0	0	0	0	0
from	4	0	0	0	0	1	1	0	1
	5	0	1	0	1	0	0	0	0
	6	1	0	0	0	0	0	0	0
	7	0	0	0	2	0	0	0	0
	8	0	1	0	0	0	0	0	0

where the first row and column indicate the load level, and the other figures the number of counted half-cycles with that corresponding size.   □

Example 3.10 (Markov Counting: Vertical Wheel Force)
An example of Markov counting of the vertical wheel force for the country road, Figure 3.1c, is shown in Figure 3.13a, where each dot represents one cycle. The Markov matrix is presented in Figure 3.13b, where the signal has been discretized to 32 levels equally spaced between 0 and 1, and prior to the discretization the signal has been rainflow filtered with a threshold equal to one discretization step.   □

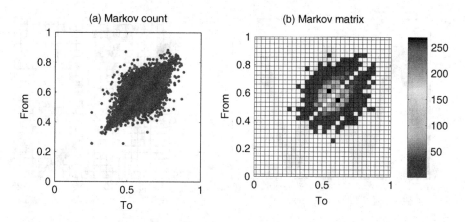

Figure 3.13 Markov count and Markov matrix

Example 3.11 (Markov Counting: Sinus with Superimposed Noise)

To use the Markov count for fatigue damage calculation is often a bad approximation, since the larger hysteresis loops, found over several consecutive cycles, are lost. This is illustrated by the following example where we have two different signals with similar Markov matrices but important differences in the rainflow domain. From a "Markov" point of view, the signals look equivalent, see Figure 3.14, but from a fatigue point of view the signals differ considerably because the first one leads to two large cycles and one large half-cycle, whereas the second one contains only one large half-cycle. In fact, it can be shown that the damage from a Markov count always underestimates the rainflow damage. However, for narrow band loads the result of the Markov counting and rainflow counting becomes similar. □

If the Markov matrix is normalized, it can be interpreted as an estimation of the probability for transitions from one bin to another. Such a Markov transition matrix defines a Markov chain, which is a useful class of random processes for load analysis. From such a description of a Markov load, the theoretical rainflow matrix can be computed explicitly. It is also easy to simulate a load history from a Markov transition matrix. More about Markov loads can be found in Chapter 6.

3.1.6 Range Counting

The range counting extracts the cycle ranges of the Markov counting

$$\{r_1, \ r_2, \ \ldots, r_{N-1}\} \qquad \text{with} \qquad r_i = |x_{i+1} - x_i|. \tag{3.19}$$

Starting from the Markov matrix this method counts the number of transitions of a certain size. The result is a histogram or vector f^{rc}, where $f^{rc}(k)$ is the number of ranges of size k

$$f^{rc}(k) = \sum_{|i-j|=k} F^{\text{Markov}}(i, j). \tag{3.20}$$

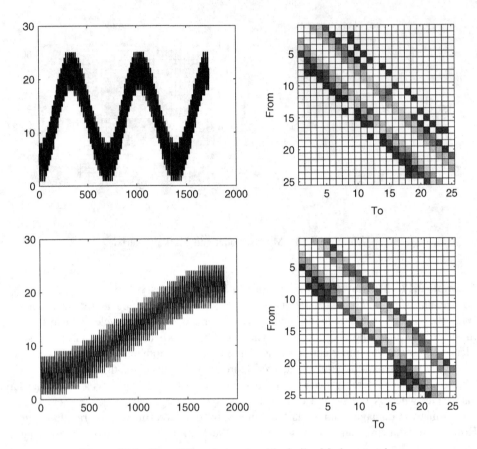

Figure 3.14   Two different signals with similar Markov matrices

In contrast to the Markov counting, the range count does not consider its dependence of the starting and end points nor the orientation (upward/downward), only the range is extracted.

Example 3.12 (Range Counting: Simple Load)
The range counting gives the following ranges

$$\{5, \ 3, \ 4, \ 6, \ 3, \ 1, \ 2, \ 5, \ 6, \ 3, \ 1, \ 3, \ 3\}$$

and the range histogram becomes

k	1	2	3	4	5	6	7
$f^{rc}(k)$	2	1	5	1	2	2	0

where the first row indicates the load range, and the second the number of counted half-cycles with the corresponding range.   □

3.1.7 Level Crossing Counting

This method counts the number of times the load up- and/or down-crosses each load level

$$N^{lc}(u) = \text{"Number of times the load crosses load level } u\text{"},$$

$$N^+(u) = \text{"Number of times the load up-crosses load level } u\text{"}.$$

The level crossing count can be derived directly from the load signal (or sequence of turning points), but it can also easily be derived from both the Markov matrix or the rainflow matrix. For example, from the rainflow matrix the symmetric level crossing count is calculated as

$$N^{lc}(k) = 2 \cdot \sum_{i \le k < j} (RFM(i, j) + RFM(j, i))$$

$$+ \text{ contribution from } RES. \tag{3.21}$$

Example 3.13 (Level Crossing Counting: Simple Load)
The level crossing counting gives the following up-crossing result

k	1	2	3	4	5	6	7
$N^+(k)$	1	4	4	7	4	3	1

where the first row indicates the load level, and the second row the number of upcrossings (or more precisely, the levels between 1 and 2 have been up-crossed 1 time, the levels between 2 and 3 have been up-crossed 4 times, and so on).   □

There are different definitions of the level crossing count. The symmetric version counts both up- or downcrossings of each load level, whereas the German DIN [72] counts the up-crossings for all levels with physical value greater than or equal to zero and the down-crossings below zero. If the signal starts and ends at the same level, then the symmetric result is just twice the DIN result.

Example 3.14 (Level Crossing Counting: Vertical Wheel Force)
The result is usually presented in a plot of level over logarithmic count. Some examples are given in Figure 3.12.   □

The level crossing plot gives an overview of the distribution of the signal across its range and is frequently used for comparison of different signals in combination with range-pair plots. The range-pair count shows the distribution of amplitudes and the level crossing count gives a hint of the distribution of the mean values and information on the maximum and minimum of the signal. Consequently, the level crossing plot gives the opportunity to evaluate the load with regard to the yield strength of the material. Further, it can be shown that the level crossing counting gives an upper bound for the rainflow damage, while the Markov counting gives a lower bound, see Chapter 6 for details.

3.1.8 Interval Crossing Counting

A generalization of the level up-crossing counting is the interval up-crossing counting, $N^+(u, v)$, equal to the number of up-crossings of the interval $[u, v]$. More precisely, it is defined as the number of up-crossings of level u, such that the load up-crosses level v before it down-crosses level u. The importance of the interval crossing count is that it contains the same information as the rainflow count. In other words knowing $N^+(u, v)$, it is possible to compute the rainflow count, and knowing the rainflow count, the interval upcrossing count is found by

$N^+(u, v) =$ "number of rainflow cycles with top above v and bottom below u".

3.1.9 Irregularity Factor

This is a scalar value relating the number N_m of upcrossings through the mean level to the number N_p of local maxima

$$\alpha_2 = \frac{N_m}{N_p}. \qquad (3.22)$$

The irregularity factor ranges from 0 to 1, where 0 means very irregular and 1 means very regular (the name is thus a bit misleading).

3.1.10 Peak Value Counting

This method essentially counts the number of maxima and minima as a function of the bins. Three definitions exist in the German DIN [72]. The third one simply counts all maxima and minima for each bin separately. The result consists of two vectors, one for the maxima and one for the minima. These two vectors are plotted in the same way as the level crossing results.

Example 3.15 (Peak Value Counting: Simple Load)
The peak value counting of the load in Figure 3.2 gives the following result

k	1	2	3	4	5	6	7	8
min	1	3	0	3	0	0	0	0
max	0	0	0	0	3	1	2	1

where the first row indicates the load level, the second the min count, and the third the max count.   □

Since there is no close relation to fatigue, it is not used as a standard tool like the rainflow count, but it is useful for the analysis of maximum loads (fatigue limit and engine components) and it serves as the input for statistical extreme value analysis.

3.1.11 Examples Comparing Counting Methods

Here we will present examples illustrating different aspects of the presented counting methods.

Example 3.16 (Wheel Forces in Different Directions)

To illustrate the rainflow counting, real truck measurements in a city are used. The signals have been normalized and units are not shown accordingly. Figure 3.15 shows the time signals for the wheel forces (front left) and the longitudinal acceleration.

Figures 3.16–3.19 show the corresponding rainflow matrices. From-to plots as well as range-mean plots are presented. In addition, the residual is shown as a short time signal. Figure 3.16 shows the longitudinal wheel force, where most of the cycles are located at 0 mean load ($R = -1$, where the R-ratio is the quotient between the lower and the upper value of a cycle). Cycles with higher amplitude tend to have positive mean load but still $R < 0$. The corresponding acceleration is shown in Figure 3.17, which is a more narrow band signal compared to the force signals. In particular, the high amplitude cycles have mean value 0.

For the lateral wheel force in Figure 3.18, we can conclude that large cycles have a mean load of approximately 0 ($R = -1$). The mean load of the small cycles ranges from -0.5 to 0.5. The matrix is symmetric with respect to the mean load level 0. This corresponds to a symmetric distribution of left and right curves during the measurement.

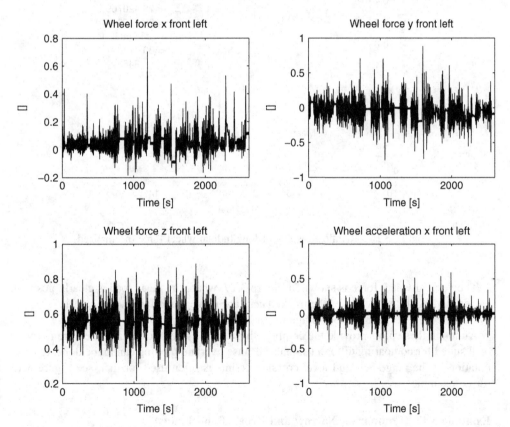

Figure 3.15 Time signals (normalized) of the wheel forces (front left) and the longitudinal acceleration of a truck in a city

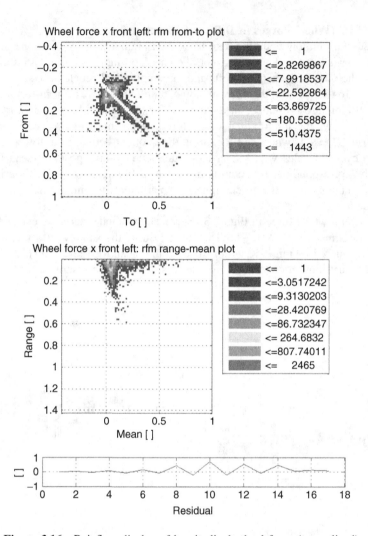

Figure 3.16 Rainflow display of longitudinal wheel force (normalized)

The vertical wheel force is shown in Figure 3.19, where the mean load is clearly positive, corresponding to the weight of the truck. There are no negative forces. The matrix is nearly symmetric with respect to the average mean load.

Although these plots provide the complete hysteresis cycle information, they are not very well suited for comparing different signals because overlaid plotting cannot be done. A combination of the range-pair and level crossing counts is often used for that, see Figure 3.12 for an example.   □

Example 3.17 (Comparing Narrow and Broad Band Loads)
The amplitude of the cycles induced by the upper signal in Figure 3.20 and the value of the local maxima are more or less the same. Thus, counting the maxima leads to a good

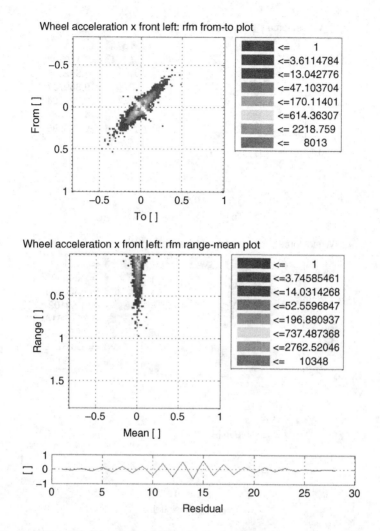

Figure 3.17 Rainflow display of longitudinal wheel acceleration (normalized)

approximation to the rainflow cycles in this case. There is a narrow peak in the frequency spectrum of the signal as can be seen from the PSD function to the right. The fact that the rainflow cycles can be estimated from the local maxima in such a case is related to *Bendat's narrow band approximation*, which is explained in Chapter 6.

In contrast to that, the amplitude of the cycles of the lower signal in Figure 3.20 differs from the local maxima. Using the local maxima as an approximation to the amplitude of the cycles would lead to an overestimation of the damage that increases with increasing irregularity (i.e. decreasing α_2), see Chapter 6 for more details. The rainflow matrices for both signals are shown in Figure 3.21. The mean load of the cycles in the narrow band signal is approximately 0, whereas many cycles in the broad band signal have non-zero mean load. The amplitude of these cycles will be overestimated if the "narrow band approach" (local maximum = amplitude of a cycle) is applied. □

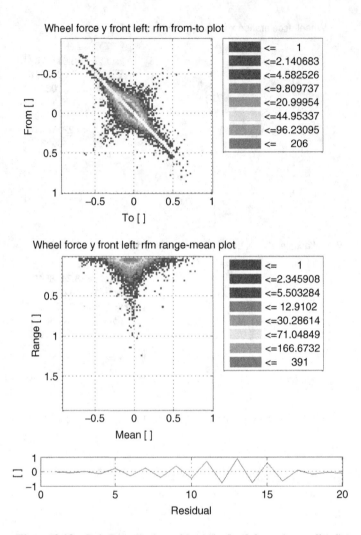

Figure 3.18   Rainflow display of lateral wheel force (normalized)

3.1.12   Pseudo Damage and Equivalent Loads

The preceding sections have presented methods giving a more or less deep insight into the loads with respect to durability. This list will get even longer when frequency-based methods are discussed in Section 3.2. Many specific details or aspects of the loads are considered in these methods, but the amount of data to be analysed for a certain purpose in real life projects is often very large. We are talking here about several hundred files with 50–200 channels in each file. It is hardly possible to plot and analyse rainflow matrices for all channels. Even if we restrict ourselves to the most important ones (let's say 4*6 wheel forces, 8–12 accelerations at the axles or frame, some accelerations at the cabin, some strains at interesting spots), we easily get more than 40 channels to analyse. In such situations, one has to combine rough and simple methods applied to the entire data set with

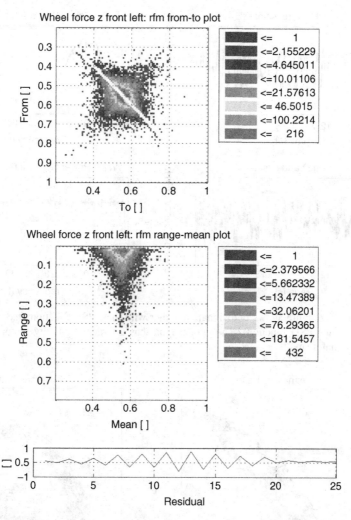

Figure 3.19 Rainflow display of vertical wheel force (normalized). Here the mean load is clearly positive, corresponding to the weight of the truck. There are no negative forces. The matrix is nearly symmetric with respect to the average mean load

more sophisticated methods applied to spots of specific interest, or to those spots that need special attention due to the results of the simple analysis.

Therefore, a simple measure (i.e. a scalar value) describing the severity of the load is needed. It can be used to compare the severity of different load measurements. A single number describing the load severity is also needed to describe the customer load distribution when applying reliability methods like the load-strength method described in Chapter 7. Further, it is also very helpful for comparisons between drivers, missions, markets, vehicle types, and other kinds of load influentials. Statistical methods for making such comparisons, such as ANOVA (see Chapter 8), often rely on scalar measures.

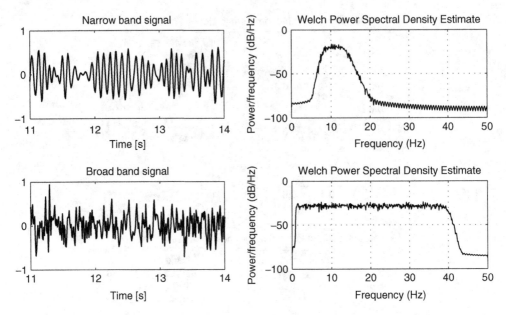

Figure 3.20 The figure shows two signals together with their PSD functions (see Section 3.2 for a definition of the PSD function). The upper signal has a large irregularity factor $\alpha_2 = 0.95$ and the second signal a smaller one $\alpha_2 = 0.75$

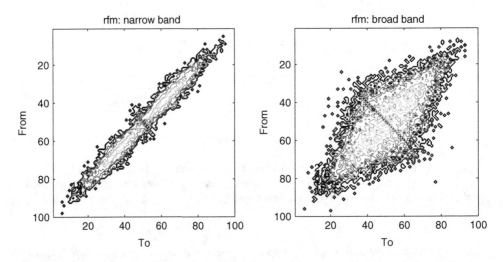

Figure 3.21 Rainflow matrices for the narrow band and broad band signals shown in Figure 3.20

3.1.12.1 Pseudo Damage Number

The construction of the measure of load severity should rely on simple and robust methods that capture the most important aspects of fatigue. The most important and widely used method is what we here call pseudo damage calculation, but is also called duty value. It is based on the concepts of the Wöhler curve, the Palmgren-Miner rule, and the rainflow cycle counting. The load input is the set of rainflow cycles, for example, in the form of a rainflow matrix or a range-pair histogram. The simplest Wöhler curve is given by Basquin's equation

$$N = \alpha \cdot S^{-\beta} \tag{3.23}$$

where the damage contribution of a cycle having amplitude S_i is $1/N_i$. The simplest damage accumulation rule is the Palmgren-Miner rule, which sums the damage contributions of all counted cycles

$$D = \sum_i \frac{1}{N_i} = \frac{1}{\alpha} \cdot \sum_i S_i^{\beta}. \tag{3.24}$$

We want the pseudo damage to be as independent of the material model as possible. Therefore, we define the pseudo damage number as the sum

$$d = \sum_i S_i^{\beta} \tag{3.25}$$

leaving out the proportional constant α, which is irrelevant when comparing the severity of different loads. The only material parameter left is the damage exponent β. The damage is computed as $D = d/\alpha$.

With this method one gets scalar numbers d_{ij} where $i = 1, \ldots, m$ ranges over all measurements (files) and $j = 1, \ldots, n$ over all channels. The size of this array is $m \times n$ which can be handled conveniently in Excel-type programs. Since the lengths l_i of the measurements differ, it is advisable to calculate normalized pseudo damage per kilometre and use the corresponding normalized pseudo damage array $\tilde{d}_{ij}$ for the analysis.

Typically, the parameter β, which defines the Wöhler curve together with α, is given by some company-specific settings. The choice of β should reflect the type of component to be analysed. As a rule of thumb, the following choices can be used, $\beta = 3$, for welded components or in case of crack growth, $\beta = 5$, for typical automotive components with rough surfaces, $\beta = 7$, for components with smooth surfaces. Of course the pseudo damage d or the normalized pseudo damage $\tilde{d}$ depends on the choice of damage exponent β. However, as is indicated by the notion of pseudo damage, the emphasis of the method is not on fatigue life estimation but on the reduction of the data to a scalar number for each file and channel that still reflects the most important fatigue-related aspects. This is fulfilled by the pseudo damage number. It serves very well as a simple tool for a first rough overview of the data. Further, it is also suitable for comparison of the different measurements, for example, different road types, driver influence, different markets, and so on. It is also the basis for describing the customer distribution when making the reliability assessments in Chapter 7.

3.1.12.2 Damage Matrix

Of course, more insight into the damage induced by a load signal can be gained by analysing how the total damage is accumulated from the individual cycles. The first way to look at this is a rainflow matrix type of plot as shown in Figure 3.22. Instead of plotting the number of cycles for each bin, the relative damage contribution is shown (sum over all entries = 100%). In the left matrix the maximum contribution of one bin is approximately 1.5%.

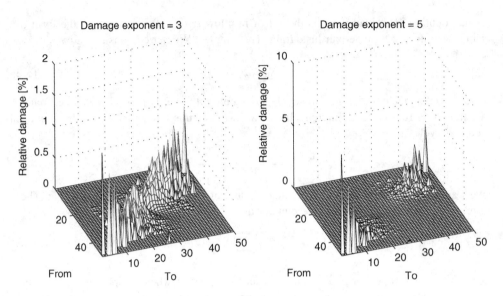

Figure 3.22   Relative damage of each rainflow bin for a vertical wheel force (Figure 3.19)

Many bins close to the diagonal (small amplitudes) do not contribute very much to the total damage in spite of the fact that near the diagonal there are many more cycles. This effect gets stronger if the damage exponent β increases, as can be seen when comparing the matrices on the left and on the right side. Consequently, this example clearly illustrates the influence of the slope β of the Basquin equation on the pseudo damage calculation.

In a similar way, the range-pair plot can be turned into a 'damage per amplitude plot'. In this case, the horizontal axis no longer shows the logarithm of the number of cycles, but the damage contribution of all cycles of a specific amplitude, see, for example, Figure 2.7b. Figures of this type give an impression of the relative damage contribution as a function of the rainflow amplitudes of a signal. In addition, one can calculate the pseudo damage $d(s)$ as a function of time s for a given signal $x(t), t \in [0, T]$, where T is the length of the signal. Here, $d(s), s < T$ is the accumulated pseudo damage of the signal $x(t), t \in [0, s]$. Plotting $d(s)$ shows when the damage accumulation is high, and thus indicates which segments of the load signal contribute the most to the total damage.

3.1.12.3   Equivalent Load Amplitude

Often the pseudo damage numbers are hard to interpret as the engineering unit is not physically understandable. For example: Is $d = 7 \cdot 10^{13}$ kN^5 a reasonable pseudo damage number for a vertical wheel force? This is hard for an engineer to judge. One aim of the equivalent load amplitude, and also the equivalent mileage in the next section, is to define a severity measure that is possible to understand in terms of physical units, and thus easier for the engineers to relate to. It is then possible to judge, for example, if an equivalent load amplitude of 53 kN is reasonable for the vertical wheel load.

The aim here is to construct a simpler load that is equivalent, in terms of damage, to the measured load. Often the choice of equivalent load is a constant amplitude load, which is the simplest possible load, and will be used in the discussions here. However, it is also possible to define more complicated equivalent loads, see, for example, Genet [99]. The equivalent constant amplitude load is characterized by its amplitude, A_{eq}, and by the number of cycles, N_0, it is applied. The damage is evaluated using the Basquin equation, giving

$$D_{eq} = N_0 \frac{1}{\alpha} A_{eq}^{\beta}. \tag{3.26}$$

Further, we want the equivalent load to represent a specified target life, e.g. two million kilometres. Therefore, we need to extrapolate the measured pseudo damage d to the damage corresponding to the target life

$$D_{life} = K \cdot \frac{d}{\alpha} \tag{3.27}$$

where K is the extrapolation factor, that is, the fraction of the target life over the length of the measurement. By solving the damage equivalence $D_{eq} = D_{life}$, we obtain

$$A_{eq} = \left(\frac{K \cdot d}{N_0} \right)^{1/\beta} \tag{3.28}$$

where A_{eq} is the resulting equivalent load amplitude. Note that the parameter α in the Basquin equation is cancelled in the calculations. The interpretation of the equivalent load is consequently that, in terms of damage, applying N_0 cycles of amplitude A_{eq} is equivalent to repeating the measured load until the target life.

The choice of N_0 is arbitrary. However, it should be connected to the fatigue regime that is relevant. If the application is in the high cycle fatigue regime, the number of cycles to failure is often in the order of $10^5 - 10^7$, and a reasonable choice is $N_0 = 10^6$. The low cycle regime is typically $10^2 - 10^4$ cycles, and a proper choice here is $N_0 = 10^3$.

Example 3.18 (Gravel Truck: Equivalent Load)
We will here calculate an equivalent load for the measured service load of a truck transporting gravel, see Figure 1.3. The measured distance is 80 km and the pseudo damage is calculated to $d = 1.01 \cdot 10^{13}$ MPa^5, using the damage exponent $\beta = 5$. Assume that the target life for this segment of trucks is 1 million kilometres, giving an extrapolation factor of $K = 1\,000\,000/80 = 12\,500$. This gives the pseudo damage for the target life $d_{life} = 1.26 \cdot 10^{17}$ MPa^5. Choosing the equivalent number of cycles to $N_0 = 10^6$ gives the equivalent amplitude

$$A_{eq} = \left(\frac{K \cdot d}{N_0} \right)^{1/\beta} = \left(\frac{12.5 \cdot 10^3 \cdot 1.01 \cdot 10^{13} \text{ MPa}^5}{10^6} \right)^{1/5} = 166 \text{ MPa}. \tag{3.29}$$

This can be compared to the largest amplitude of 221 MPa in the measured signal. □

3.1.12.4 Equivalent Mileage

The aim here is similar to that of the equivalent amplitude, however, the measured load is transformed into an equivalent mileage on a predefined reference road. The choice for reference is often the test track.

From the measurement, the pseudo damage, d, is calculated. The corresponding load signal is measured on the reference road, and the pseudo damage normalized by the number of kilometres, $\tilde{d}_{ref}$, is calculated. The equivalent mileage is simply calculated as

$$M_{eq} = \frac{d}{\tilde{d}_{ref}}. \tag{3.30}$$

A drawback of the method is that for the calculation of the equivalent mileage not only the measurement is needed, but also a corresponding reference measurement, which should preferably be made on the reference road with the same vehicle and measurement equipment. An advantage, compared to the pseudo damage number, is that the equivalent mileage is measured in an engineering unit, say, kilometres, that is easy to relate to.

Example 3.19 (Gravel Truck: Equivalent Mileage)
For the same gravel truck as in the example above, there is a reference measurement on the test track, resulting in the normalized pseudo damage $\tilde{d}_{ref} = 9.45 \cdot 10^{12}$ MPa^5/km, using the same damage exponent of $\beta = 5$. This gives the equivalent mileage

$$M_{eq} = \frac{d}{\tilde{d}_{ref}} = 1.06 \text{ km} \tag{3.31}$$

indicating that the test track is 75 times more severe, in terms of distance driven, than the load measurement of 80 km. □

3.1.12.5 An Application

We finish the section on pseudo damage and equivalent loads with an example.

Example 3.20 (Customer Usage)
A typical customer load is often defined by combining measurements from different road types. Here we will use the three measurements in Figure 3.1 and assume that the typical customer usage profile is 45% city driving, 30% highway, and 25% country road, which represents a city distribution truck. The pseudo damage during the design life of, say, $L = 600 \cdot 10^3$ km, can then be calculated as

$$d_{life} = L \cdot (0.45 \cdot \tilde{d}_1 + 0.30 \cdot \tilde{d}_2 + 0.25 \cdot \tilde{d}_3)$$
$$= 600 \cdot 10^3 \cdot (0.45 \cdot 0.00261 + 0.30 \cdot 0.00164 + 0.25 \cdot 0.00112)$$
$$= 600 \cdot 10^3 \cdot 0.00195 = 1.17 \cdot 10^3 \tag{3.32}$$

where $\tilde{d}_1$, $\tilde{d}_2$, and $\tilde{d}_3$ are the damage intensities (per km) for the different road types, and have been calculated from the measurements in Figure 3.1 using damage exponent $\beta = 5$. In terms of equivalent load it becomes

$$A_{eq} = \left(\frac{d_{life}}{N_0}\right)^{1/\beta} = \left(\frac{1.17 \cdot 10^3}{10^6}\right)^{1/5} = 0.259. \tag{3.33}$$

□

The topic of combining customer usage with a description of the load environment will be further discussed in Chapter 8 (Sections 8.5 and 8.6). The example above is continued in Examples 8.15 and 8.16.

3.1.13 Methods for Rotating Components

For rotating components such as shafts or gear wheels in gear boxes, special attention needs to be paid to the periodicity induced by the rotation.

3.1.13.1 Rainflow Counting for Rotating Components

Consider a pair of gear wheels (Figure 3.23) rotating at a constant speed n in units of rpm (number of revolutions per minute) under a constant torque m in units of Nm. Assume we are interested in the fatigue life of the teeth and want to analyse the loads at the base of the teeth as indicated in Figure 3.23.

A rainflow count of the torque signal shows no cycles at all. This is due to the fact, that there are no load cycles in the shaft. But there are load cycles at an individual tooth of the gear wheel. These cycles are induced by the rotation, which gives a force at the tooth once in each revolution. This can be calculated if the torque and the rotational speed signals are combined. To this end, the number of revolutions r of an arbitrary but fixed tooth of the gear wheel

$$r(t) = \frac{1}{60} \cdot \int_0^t |n(s)|ds \qquad (3.34)$$

and the torque m_{gw} at the tooth

$$m_{gw}(t) = \varphi(r(t) - \text{floor}(r(t))) \cdot m(t), \qquad (3.35)$$

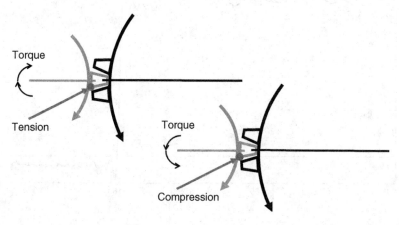

Figure 3.23 A pair of gear wheels. At the spot on the base of one tooth, we have tension or compression when that tooth is in contact depending on the sign of the torque at wheel shaft

are calculated, where floor(r) denotes the largest integer not greater than r and $\varphi : [0, 1] \rightarrow$ [0, 1] is a function describing the contact condition between the tooth and the second gear wheel. Its values are 0 at the boundaries, monotonically increasing to 1 at some point in (0, 1) and monotonically decreasing to 0 again. Different shapes (hat, step function . . .) are possible. This model ensures that there is no contact at the beginning and the end of one revolution and full contact somewhere in between. The normalization $\frac{1}{60}$ in formula (3.34) is used to transform from revolutions per minute, which is the typical unit for rotational speed, to seconds.

In Figure 3.24 there is an example of the gear wheel torque in a gear box of a passenger car, calculated by using the procedure described above. The chosen segment shows a transition from a status where the engine drives the car (positive torque) to the status where the engine slows down the car (negative torque). The rotational speed is nearly constant during this period. This can be seen from the frequency of the gear tooth torque signal, which is also approximately constant.

Now standard rainflow or range-pair counting, as explained in Sections 3.1.3 and 3.1.4, of this signal (rotational rainflow counting) gives the load cycles of a tooth of the gear wheel induced by the torque and the rotation. In Figure 3.25 there is an example of the resulting rainflow matrix. Most of the cycles are located at the 0-column or 0-row. They are induced by the oscillation between no contact (load = 0) and contact of the tooth. There are some cycles in the upper right and lower left triangle of the matrix. These cycles correspond to load variations between positive and negative torque.

3.1.13.2   Time at Level Counting

There also exists a simplified version of this type of counting based on the so-called *time at level count*. The time at level count simply divides the physical range of a signal into a

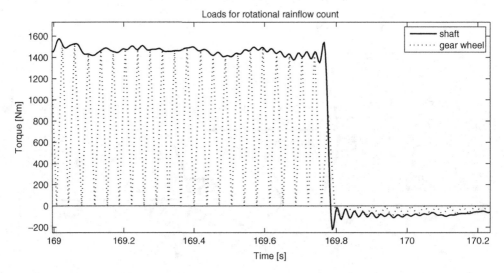

Figure 3.24 Example of a torque signal (solid line) and the corresponding gear wheel torque (dotted line) in a gear box of a passenger car

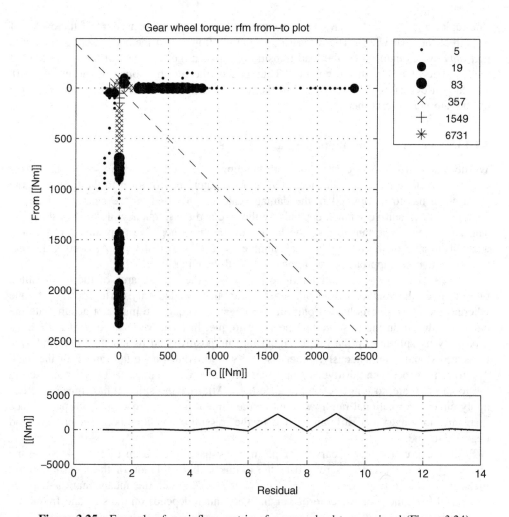

Figure 3.25 Example of a rainflow matrix of a gear wheel torque signal (Figure 3.24)

number of bins and calculates the time spent in each bin. If the sampling rate of the signal is high enough, we can make an approximate calculation of the time at level count by simply counting the number of samples lying in a bin (histogram) and multiplying the result by the length Δt of one time step. Multi-dimensional time at level counting is defined in a similar way and is represented by a multi-dimensional histogram (see Section 3.3.6). If we apply this to a 2-dimensional time at level count where the torque of a shaft is one signal and the rotational speed is the second signal, we get a matrix $t_{i,j}$, where $i = 1 \dots N_m$ ranges over all torque levels m_i and $j = 1 \dots N_n$ ranges over all speed levels n_j. The number of rotations r_i at a certain torque level m_i is then given by

$$r_i = \frac{1}{60} \cdot \sum_{j=1}^{N_n} n_j \cdot t_{i,j}. \tag{3.36}$$

The tuple $(m_i, r_i), i = 1 \ldots N_m$ can be interpreted as the range-pair count of the cycles of a tooth at the gear wheel. The difference to the rainflow-based method described above is that only cycles induced by the load transitions between the free state of the tooth and the loaded state of the tooth are counted. Thus one of the turning points of the cycles is 0. The largest cycles, which correspond to the load transitions between positive and negative torque, are not covered here.

3.1.13.3   Bending at the Tooth base and Pitting

Wöhler curves ($\sigma \sim N -$ curve) for tooth bending used during the analysis of gear boxes may have a slope up to $\beta \sim 9$. Since σ is approximately proportional to the torque m, the same slope has to be applied in the damage calculation based on the gear wheel torque signal and the result very much depends on the size and the frequency of the largest cycles. Thus, it is very important to take the largest cycles from negative to positive torque into account in spite of the relatively small number of cycles of that type. As a consequence, the rainflow-based approach is preferable for tooth bending.

There is also another potential fatigue problem at the contact area of the tooth called pitting. Since the contact area at one side of a tooth is loaded only during positive torque whereas the other side is loaded only during negative torque, we must not accumulate the damage induced during positive and negative torque. Instead, we have to treat both cases separately by splitting up the torque signal into its positive and its negative part and treat the speed signal simultaneously. Otherwise we would overestimate the impact of the load. The transitions between positive and negative torque, which are important when considering the tooth base, are no longer of importance here. After having split up the signals, we can apply either the rotational rainflow count or the time at level approach to both parts, since there are no longer any differences due to the missing transitions between positive and negative torque.

Wöhler curves ($\sigma \sim N -$ curve) for pitting may have a slope up to $\beta \sim 14$. Since the stress σ used in the curve is related to the torque m by a relation of the form $\sigma \sim \sqrt{m}$, the slope of the corresponding $m \sim N -$ curve is $\beta \sim 7$. Again this means that the damage calculated from the gear wheel torque signal very much depends on the size and frequency of the largest cycles.

For more about gear wheels and rotating components, see, for example, Steinhilper and Sauer [222] and Carboni et al. [50].

3.1.14   Recommendations and Work-flow

3.1.14.1   Pre-processing of Load Signals

1. *Data inspection and correction.* The process of acquisition of load data may contain errors. Therefore, the measured load signal needs to be inspected with the aim of detecting measurement errors, such as offsets, drifts, periodic variations, spikes, and noise. If present, these types of errors need to be corrected in order to get valuable results during later analysis. In the worst case the results may otherwise be misleading, or simply wrong. Details about data inspection and correction are found in Section 4.2.
2. *Turning points and rainflow filter.* Only the sequence of local extreme values of the load is of importance for the amplitude-based methods. Therefore, the load signal can be

reduced to a sequence of turning point, without losing any information. Further, small oscillations should be removed by using the rainflow filter, which further reduces the amount of data, but only gives a negligible change in the durability impact of the signal. Turning points and rainflow filter are further explained in Section 3.1.2 and Section 4.3.1.
3. *Discretization.* Often the load levels of the signal are discretized to a number of fixed values. This is explained in Section 3.1.2 and Figure 3.3. This step is necessary, for example, when calculating the rainflow matrix.

3.1.14.2 Durability Evaluation of Load Signals

1. *Rainflow cycle counting.* For durability evaluation, rainflow cycle counting is recommended, which extracts hysteresis cycles in the load, see Section 3.1.3 for details. The rainflow matrix gives a good picture of the durability impact of the load. However, the order of the cycles is lost. Examples of rainflow matrices are found in, for example, Figures 3.16–3.19, where the hanging and standing cycles (see Figure 3.6) are distinguished, and the graph below shows the residual, which is the part of the signal that has not yet formed closed hysteresis loops. The recommended 4-point algorithm for counting rainflow cycles extracts the above information. Often it is not necessary to distinguish between hanging and standing cycles, and the plot can be simplified, see Figures 3.8–3.11 for various ways of plotting the rainflow matrix. In these figures the cycles in the residual have also been counted and included in the rainflow matrix. The rainflow matrix is, however, a two-dimensional structure, which makes it difficult to interpret, and especially to compare different signals. Therefore, one-dimensional characteristics are useful.
2. *Range-pair and level crossing counting.* These two methods are often used together for comparison of signals as they show somewhat complementary load information. The range-pair diagram (Section 3.1.4) shows the distribution of rainflow amplitudes, while the level crossing diagram (Section 3.1.7) gives a hint of the distribution of the mean values of the cycles, and information on the maximum and minimum of the signal.
3. *Load severity in terms of pseudo damage or equivalent load.* When comparing many loads, or when characterizing a population of customers, it is practical (and often more or less necessary) to characterize the severity of the load by a single number. This load severity should reflect the damage potential of the load. Therefore, it is based on the Palmgren-Miner damage accumulation rule together with a simple artificial Wöhler curve, where only the Wöhler exponent is used. The most straightforward severity measure is the pseudo damage, sometimes also called the duty value, which is simply the damage calculated from the artificial Wöhler curve. An alternative is the equivalent amplitude, which has the advantage of having the same unit as the load, thus being more easily interpreted. Further, it is also possible to calculate an equivalent mileage. Details on load severity are found in Section 3.1.12.

Figure 3.26 gives an overview of the most important amplitude-based counting methods and their relation to each other. The connection between the methods indicates that the result of a certain method can be derived from the result of another method without going back to the time signal. It can be seen that the two most important one-dimensional methods, level crossing and range-pair, can be derived from the rainflow count, which again highlights the importance of the rainflow matrix.

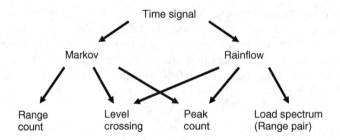

Figure 3.26 Overview some counting methods and their relations

3.2 Frequency-based Methods

Durability of a metal component is a local phenomenon and as such dominated by local quantities such as stress-strain histories. The stresses and strains depend on the forces acting on the component. These component forces in turn (either measured or calculated) depend on both the excitation and the properties of the complete system. Here, a system might be a structure like an axle or a cabin, a subassembly like a suspension, or a full vehicle. The physical behaviour of such systems is further described in Chapter 5. In this section, we use some of these notions but the main ideas do not depend on technical details of that chapter.

At the local stress and strain level, the amplitude domain is most important. But as soon as we start processing non-local quantities such as section forces or even outer excitations, we need to take into account all the properties of the loads which have an influence on the local stress amplitudes. In a rigorous sense, this means that we have to look at the full time signals as the local response of a complex (non-linear) system might depend on "everything". Clearly, the amplitude domain of the outer loads is important, but also other properties such as the so-called frequency behaviour or spectral properties have an impact on the local stress amplitudes.

As an illustration let us consider a non-stiff system, that is a system with eigenfrequencies in the range of the excitation spectrum. An example is the frame of a truck which is excited by road irregularities. In the simplest case (a linear system), the response of such a system (for example, a local stress) to a sinusoidal excitation is also a sinusoidal function. If the input signal consists of a certain composition of different sinusoids, the same is true for the response. Frequencies not contained in the input signal are also not contained in the response. The linear system does not *create* new frequencies, but rather amplifies or reduces the input frequencies. In addition, the phase of the response is shifted due to damping properties of the system. As we will see in Section 3.2.4, the system can be represented by a matrix of complex functions in the frequency domain, which describes the amplification/reduction and phase shifting of the input frequencies. Using that system matrix, we can study the transfer from outer loads to the local response. In order to do this, we need to estimate or calculate the system matrix and analyse the input in the frequency domain.

In this section we will not go into all the details of the so-called spectral analysis, but rather attempt to concentrate on the basic ideas, explain the most important methods and give some hints on how to use these algorithms. The first sub-section gives a brief introduction to the PSD function (referring to the appendix and to Chapter 6). In the second sub-section, the most important PSD estimation method based on the periodogram is introduced. The third sub-section concentrates on the frequency-based system analysis. The section closes with a sub-section on Lalanne's fatigue damage spectrum, which is a special method for combining the amplitude-based reasoning in Section 3.1 with the frequency-based reasoning, and with some remarks on additional methods and the relation to stochastic processes.

In order to explain the most important methods in this area, we will use some notions of harmonic analysis. The reader who is not familiar with that is referred to the corresponding appendix (see Appendix C.1), where some of the technical details and more references concerning Fourier analysis can be found.

3.2.1 The PSD Function and the Periodogram

The frequency content of a signal $x(t)$ is often represented by the power spectral density $\hat{P}_x(f)$ (PSD). For a non-periodic signal $x(t) \in L_2$ (see Appendix C.1) this can be defined (up to normalization) by the absolute value of the Fourier transform squared, that is

$$\hat{P}_x(f) = c \cdot |\hat{x}(f)|^2, \tag{3.37}$$

where c is a constant (see Appendix C.1).

For periodic signals it can be defined (up to normalization) by the coefficients c_k of the Fourier series

$$\hat{P}_x(f_j) = c \cdot |c_j|^2, \; f_j = \frac{j}{N\Delta t}, \; j = 0, \ldots N-1, \tag{3.38}$$

where N is the number of samples and Δt is the sampling rate (see Appendix C.2). This representation is often called the spectral decomposition of the signal.

For stationary stochastic processes (see Chapter 6) there is another definition of the PSD. It states that the PSD function is the Fourier transform of the auto-correlation function $R_x(t)$, that is

$$R_x(t) = \mathbf{E}[x(\tau) \cdot x(t+\tau)] - (\mathbf{E}[x(\tau)])^2 \tag{3.39}$$

$$\hat{P}_x(f) = \hat{R}_x(f), \tag{3.40}$$

where $\mathbf{E}[\cdot]$ denotes the expected value which is calculated by averaging over many representations of the stochastic process $x(t)$. The Wiener-Khinchin theorem (see Oppenheim and Schafer [179] or Orfanidis [181]) gives the relation of this definition with the definitions given in Equations (3.37) and (3.38).

Based on the discrete Fourier transformation and its coefficients $\hat{x}_j$ (see Appendix C.4), we introduce the so-called *periodogram*

$$I_x(f_0) = \frac{1}{N^2}|\hat{x}_0|^2 \tag{3.41}$$

$$I_x(f_j) = \frac{1}{N^2}(|\hat{x}_j|^2 + |\hat{x}_{N-j}|^2) \text{ for } j = 1, \ldots, \frac{N}{2} - 1 \tag{3.42}$$

$$I_x(f_{N/2}) = \frac{1}{N^2}|\hat{x}_{N/2}|^2 \tag{3.43}$$

which plays a central role in the estimation of the PSD function. In contrast to the analytical definitions of the spectral functions given above, the periodogram is applied to observed signals. In Equation (3.41) we consider only frequencies $f_j = \frac{j}{N\Delta t}$ for $j = 1, \ldots N/2 - 1$, where we assume for the sake of simplicity that N is an even number (most often a power of 2 for algorithmic reasons). This is due to the fact that in case of real signals $x(t)$, we have $\hat{x}_j = \hat{x}^*_{N-j}$, where z^* denotes the complex conjugate of z. Therefore the length of the spectral vectors (PSD, FRF, ...) is always half the length of the input block we use for the estimation procedures.

We also introduce the *cross-periodogram*

$$I_{xy}(f_0) = \frac{1}{N^2}\hat{x}_0 \cdot \hat{y}^*_0 \tag{3.44}$$

$$I_{xy}(f_j) = \frac{1}{N^2}(\hat{x}_j \cdot \hat{y}^*_j + \hat{x}_{N-j} \cdot \hat{y}^*_{N-j}) \text{ for } j = 1, \ldots, \frac{N}{2} - 1 \tag{3.45}$$

$$I_{xy}(f_{N/2}) = \frac{1}{N^2}\hat{x}_{N/2} \cdot \hat{y}^*_{N/2} \tag{3.46}$$

for two signals x, y which will be used later for the FRF-based system analysis.

Since the discrete Fourier transformation approximates the coefficients c_j of the Fourier series (see Appendix C.4), the periodogram can be interpreted as an approximation of the PSD function as defined in Equation (3.38). As we will see in the following, this estimator has some drawbacks which need to be corrected. This is the subject of the next sub-section.

Remark 3.1 *If one is interested in the amplitudes of harmonic functions (sinusoidals) resulting from the decomposition of a signal (see Appendix C.4), it is rather natural to calculate*

$$A_x(f_0) = \frac{1}{N}|\hat{x}_0| \tag{3.47}$$

$$A_x(f_j) = \frac{1}{N}(|\hat{x}_j| + |\hat{x}_{N-j}|) \text{ for } j = 1, \ldots, \frac{N}{2} - 1 \tag{3.48}$$

$$A_x(f_{N/2}) = \frac{1}{N}|\hat{x}_{N/2}| \tag{3.49}$$

instead of the periodogram in Equation (3.41). Since $A_x(f_j)$ approximates the magnitude of the corresponding Fourier coefficient c_j of the signal $x(t)$, the set of coefficients might be called the amplitude spectrum.

3.2.2 Estimating the Spectrum Based on the Periodogram

3.2.2.1 Block Averaging

The formula for the discrete approximation of the auto-correlation function and its Fourier transform in the theory of stationary stochastic processes looks very similar to the periodogram in Equation (3.41). In fact it can be shown that the periodogram is an estimator of the PSD (see Brockwell and Davis [40]). However, the variance of this estimator is large and does not depend on the number of samples used for the calculation. As a consequence, the periodogram of the discretized signal is usually very noisy. Increasing the sample size N does not reduce the variance. To improve this situation, the time signal is divided into K (overlapping) sections/patches. The PSD is computed for and averaged over all sections. This reduces the variance by the factor $1/K$ and the periodogram becomes smoother. A deeper understanding and justification of the average procedure is based on the fact that in Equation (3.39), the expected value needs to be calculated. For a stationary process the block averaging approximates that value under the assumption that all blocks are (generally) statistically equivalent. In "real-life" load analysis, this is not always true. Thus averaging over a long time signal might be misleading. A more detailed explanation, however, is beyond the scope of this section. An example of the blocking is given below.

3.2.2.2 Leakage

The calculation of the periodogram is based on a block of the data of a certain length. Blocking can be interpreted as the multiplication of the entire signal with a rectangular window function $y(t) = w(t) \cdot x(t)$ which is 0 outside the block and 1 inside. In the frequency domain the multiplication becomes a convolution $\hat{y}(f) = \int_{-\infty}^{\infty} \hat{w}(f - s) \cdot \hat{x}(s) ds$, where $\hat{w}(f)$ is the Fourier transform of the window function w. This formula shows that the "correct" frequencies $\hat{x}(f)$ become distorted by other frequencies according to the weighting given by $\hat{w}(f)$. This is the so-called leakage effect, which widens the peaks of the PSD. Since windowing (application of a window function) cannot be avoided due to the finite length of the measurements and due to the block averaging procedure, the leakage effect needs to be corrected. To this end, certain window functions $w(t)$ have been developed in such a way that their Fourier transform $\hat{w}(f)$ have a narrow peak at 0 and decay rapidly with increasing f. Many window functions exist (Parzen, Bartlett, Hann, Hamming, Welch, ...), but their different properties are mostly negligible in practice (see, e.g. Press *et al.* [191] or Brockwell and Davis [40]).

3.2.2.3 An Example

In order to illustrate the estimation concepts we introduce a simple example. It is based on two single input single output (SISO) systems as follows:

$$m \cdot \ddot{y} + d \cdot \dot{y} + k \cdot y = c \cdot u \tag{3.50}$$

$$m \cdot \ddot{y} + d \cdot \dot{y} + k \cdot y \cdot |y| = c \cdot u, \tag{3.51}$$

where $\dot{y}$ and $\ddot{y}$ denote the first and second time derivative, respectively. The first one is a linear second order system representing a linear spring/damper mechanism. In the second example, the damping is still linear, but the stiffness term is now non-linear, as can be seen in Figure 3.27. The parameters are chosen such that the linear system has an eigenfrequency of 5 Hz and the damping ratio $\frac{d}{2\sqrt{m \cdot k}} = 5\%$.

The systems are excited by an input function $u(t)$ of the form

$$u(t) = 2 \cdot \sin(2\pi \cdot f_1 \cdot t) + \sin(2\pi \cdot f_2 \cdot t) + \epsilon(t) \tag{3.52}$$

where $f_1 = 10$ Hz, $f_2 = 13$ Hz and $\epsilon(t)$ is a 30 Hz low pass filtered Gaussian white noise process with zero mean and variance 4. Thus the input function has frequency peaks at 10 Hz and 13 Hz. In Figure 3.28 a segment of the input and the response of the systems in time domain are shown.

The spectral analysis described in the following has been performed based on signals of length $T = 40$ s with sampling frequency $f_s = 500$ Hz (sampling rate $\Delta t = \frac{1}{f_s} = 0.002$ s),

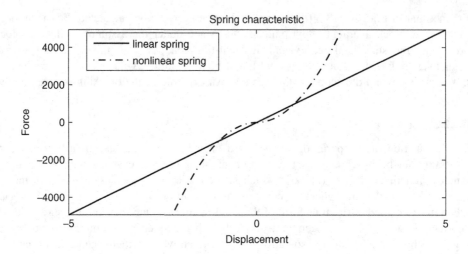

Figure 3.27   Linear and non-linear spring characteristics

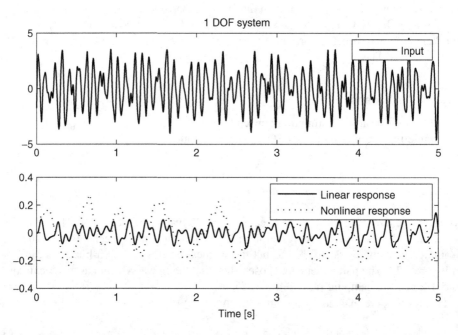

Figure 3.28   A segment of the input and the response of the linear and non-linear system

resulting in a Nyquist frequency $f_N = 250$ Hz and $N = 20000$ samples (see Appendix C.3). Using, for example, a block length of $M = 1024$ and an overlapping of 50%, this results in 38 blocks for the periodogram averaging. This is a rather large number of blocks leading to a well-smoothed spectral estimation.

3.2.2.4 Practical Aspects

The following table gives an overview of the frequency resolution $\Delta f = \frac{1}{M \cdot \Delta t}$ and the number of blocks $K \simeq \frac{2N}{M} - 1$ as a function of the block length M.

M	$\Delta f\,[Hz]$	K
512	0.98	77
1024	0.49	38
2048	0.24	19
4096	0.12	9
8192	0.06	4

In Figure 3.29 the PSD as calculated using the techniques described above is shown. Both input frequencies can be seen clearly in the input PSD as well as in the output PSDs. The output PSDs, however, contain another peak at 5 Hz in case of the linear system and at approximately 2.2 Hz in case of the non-linear system. Since the noise term of the input contains any frequencies up to 30 Hz, the eigenfrequencies of the systems lead to an amplification of the corresponding input and the peak in the response spectrum (see also Section 3.2.4).

In Figure 3.30 the effect of different window functions is demonstrated using the response of the linear system. It can be seen that the peaks of the rectangular window are

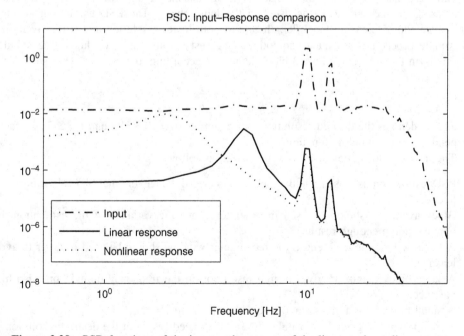

Figure 3.29 PSD functions of the input and response of the linear and non-linear system

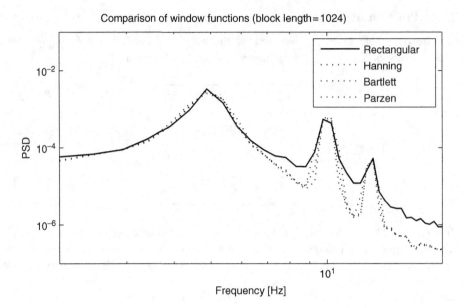

Figure 3.30 Comparison of different window function to handle leakage effects

somewhat wider due to leakage, whereas the other window functions essentially give similar results.

In Figure 3.31 the effect of different block lengths is demonstrated, again using the linear system response. Increasing the block length means increasing frequency resolution and decreasing number of blocks as listed in the table above. The PSD estimation becomes smoother with decreasing block length due to both lower resolution and better averaging. A simple procedure to balance smoothness against resolution is to choose the smallest possible (in terms of smoothness) block length M according to

$$M \geq c_w \cdot \frac{f_s}{f_b}, \tag{3.53}$$

where f_b denotes the desired resolution (for example, $f_b = 0.5$ Hz) and $c_w \simeq 2$ is a factor depending on the window function.

The investigations above can be summarized as follows:

- PSD estimation is based on the periodogram and hence on the FFT algorithm (see Section C.4).
- Block averaging should be used in order to reduce the variance of the estimation and obtain smooth enough results.
- A window function different to the rectangular window should be used in order to avoid leakage.
- A large block length M gives a high resolution in the frequency domain but also high variance.
- A small block length gives smooth results but low resolution.
- A compromise is given by formula (3.53) which needs as input the desired resolution f_b and the window factor c_w.

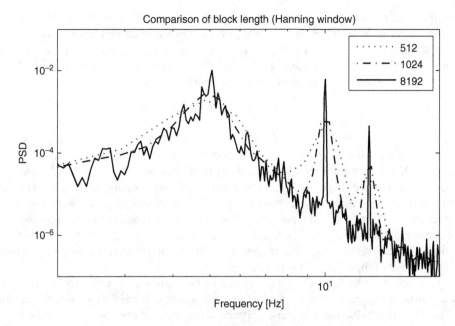

Figure 3.31   Effect of block length on the resolution of the PSD function

These rules should be taken as a guideline in the spectral analysis. For better comparability it is advisable to choose the parameters window type and block length M once and keep them fixed during subsequent analysis. More details can be found, for example, in Oppenheim and Schafer [179], Press *et al.* [191] or Orfanidis [181].

3.2.3   Spectrogram or Waterfall Diagram

Of course, a large number of refinements or specific applications of the methods described above exist. One such application is the "local Fourier analysis". In that case, the signal is allowed to represent a non-stationary stochastic process with varying spectral properties (e.g. varying PSD function). The PSD function is calculated for a certain segment of the signal only and the segment is moved along the time axis. The result is a series of PSD functions, each one valid for a certain point in time (i.e. the centre of the corresponding segment) and it is plotted in a so-called waterfall diagram or spectrogram, where the horizontal axis represents time, the vertical axis represents the frequency, and the colour of the plot represents the value of the PSD at the corresponding time slice and frequency.

3.2.4   Frequency-based System Analysis

3.2.4.1   Transfer Function: Frequency Response Function

The cross-correlation function

$$R_{yu}(t) = \mathbf{E}[u(\tau) \cdot y(t + \tau)] - \mathbf{E}[u(\tau)]\mathbf{E}[y(\tau)] \tag{3.54}$$

describes the relation between two time signals $u(t)$ and $y(t)$ of the same length and sampling rate. Its value gives an indication of the (stochastic) dependency of the sampling points separated by the time lag t. If $y = u$, this is the auto-correlation function as introduced in Equation (3.39). Similar to the fact that the PSD function can be defined as the Fourier transform of the auto-correlation function, the Fourier transform

$$\hat{P}_{yu}(f) = \hat{R}_{yu}(f) \tag{3.55}$$

of the cross-correlation function will play an important role in the system analysis. It can be used to analyse the dependence of the output functions y on the input functions u of an unknown (linear) system S (black box system) frequency-wise.

As was mentioned at the beginning of Section 3.2, the output signal of a linear time-invariant system consists of a certain composition of sinusoids which are given by the spectral decomposition of the input signal. Frequencies not contained in the input signal are not contained in the response. The linear system does not *create* new frequencies, but rather amplifies or reduces the input frequencies. In addition, the phase of the response is shifted due to damping properties of the system.

The transformation from the input spectra to the response spectra can be represented by a matrix of complex-valued functions in the frequency domain which exactly describes the amplification/reduction and phase shifting of the input frequencies. This is the transfer function or frequency response function (FRF). Each matrix element corresponds to a pair of one input and one output signal. In the linear case, the FRF completely describes the system, whereas in the case of a non-linear system, the FRF is a linear approximation of the system around a certain set point and only valid for small perturbations of the set point.

In simple cases, the FRF can be calculated explicitly from the model equations (see below for an example). In more complicated situations it needs to be estimated from input/response pairs (u, y). The derivation of the formulas for the estimation starts with a linear time invariant multiple input multiple output system (MIMO). Let $u^{(n)}(t)$, $n = 1, \ldots, p$ be the vector of input functions and $y^{(m)}(t)$, $m = 1, \ldots, q$ the response vector. Then the system is represented by the convolution of a system matrix A with the input u

$$y^{(m)}(t) = \sum_{n=1}^{p} \int_{-\infty}^{\infty} A_{m,n}(\tau) \cdot u^{(n)}(t - \tau) \, d\tau, \qquad m = 1, \ldots, q. \tag{3.56}$$

Calculating the cross-correlation functions $R_{y^{(m)}u^{(n)}}(t)$, $R_{u^{(n)}u^{(k)}}(t)$ and their Fourier transforms $\hat{P}_{y^{(m)}u^{(n)}}(f)$, $\hat{P}_{u^{(n)}u^{(k)}}(f)$ using Equation (3.56), we get the matrix equation

$$\hat{P}_{yu}(f) = \hat{A}(f) \cdot \hat{P}_{uu}(f) \tag{3.57}$$

where $\hat{A}(f)$ is a $q \times p$ matrix, $\hat{P}_{uu}(f)$ is a $p \times p$ matrix, and p is the number of input and q is the number of response signals. Multiplying this equation with the inverse of the auto-covariance matrix we get the estimation formula

$$\hat{A}(f) = \hat{P}_{yu}(f) \cdot (\hat{P}_{uu}(f))^{-1}. \tag{3.58}$$

The latter operation requires that $\hat{P}_{uu}$ is invertible for all frequencies, which means that the input functions $u^{(n)}(t)$ must not be fully correlated and need to contain all frequencies.

The evaluation of the formula can be done, no matter where the data come from. However, the interpretation of $\hat{A}$ as a system representation is valid only if $y(t)$ is the response of a linear system with respect to $u(t)$.

The calculation of $\hat{P}_{yu}(f)$ and $\hat{P}_{uu}(f)$ is usually done based on the cross-periodogram introduced in Section 3.2.1. As for the ordinary periodogram, the windowing and blocking strategies are used to reduce leakage and variance. The best input data for estimating the FRF are uncorrelated white noise (constant PSD over the frequency range) or coloured noise (non-constant PSD) on all signals, which guarantees that the input signals are independent and all frequencies are considered. In this case the matrix $\hat{P}_{uu}$ is a diagonal matrix.

3.2.4.2 Applications of the FRF

The FRF is used to analyse the frequency behaviour of systems. This can be useful when comparing systems on test tracks, on test rigs, or in simulations. One such example is reported in Schmudde [211]. Here, three systems are compared to each other. The first system is the 'real' system, in this case the suspension of a passenger car during driving on a test track. The second system is the same suspension on a test rig. Note that this is another system due to the specific mounting of the suspension on the test rig, which is close to but not identical to the mounting in the car. The third system is a numerical model (MBS) of the suspension. As input for all three signals, measured wheel forces are considered. The output consists of several channels measured during the test drive and the rig test. For all three systems, the FRF can be estimated using the corresponding data (measured on the test track, measured on the rig, result of the numerical simulation). In that specific case it has been found that the FRF of the test rig system differs from the others with respect to the steering force, which was one of the output signals. The reason was that on the test rig, the corresponding mount point was too stiff compared to the situation in the car. More details can be found in Schmudde [211].

Besides the system analysis, the drive file iteration (iterative learning control, see, for example, Moore [164] or Cuyper [67]) is an important application of transfer functions. It computes the proper input signals for given output data (target). Here, an initial guess for the input u_0, for example, a static equilibrium or the drive files for a similar model, is required. The system can be simulated/driven with white noise added to the initial guess $u = u_0 + u_{noise}$. This ensures that the matrix $\hat{P}_{uu}$ used in the FRF estimation is invertible, see Equation (3.58). With the measured output signals y, the FRF of the linearized system can be identified. By multiplying the residual (output of the undisturbed system - target) in the frequency domain with the inverse system matrix, one gets an input correction. Iterating this procedure improves the input functions. Some more details of the method are described in Section 5.4.2.

The general theory of dynamical systems and system identification is also discussed in, for example, Ljung [146] and the references therein. Practical aspects of the drive file iteration can be found in de Cuyper [67].

3.2.4.3 A Simple Example

We complete this sub-section with the continuation of the example from Section 3.2.2. As we saw in Figure 3.29, the PSD of the response of the systems has a peak which is not contained in the input PSD. This peak is due to the system which in the linear case amplifies the frequencies around 5 Hz contained in the noisy part of the input. In the non-linear case, this statement is no longer true in a rigorous sense.

Since we have a single input single output system (SISO), the FRF matrix of the system reduces to a single transfer function which (for the linear system) is given explicitly by

$$FRF(f) = \frac{c}{m \cdot (i2\pi f)^2 + d \cdot (i2\pi f) + k}. \tag{3.59}$$

The FRFs estimated from the linear and non-linear response and the calculated FRF are shown in Figure 3.32. The FRF estimated from the linear response data cannot be distinguished from the calculated one.

The data available ($N = 40000$, $M = 4096 \Rightarrow K = 9$, $\Delta f = 0.12$ has been used) is sufficient for a reasonable estimation. Consequently the FRF estimated from the linear response is very accurate and cannot be distinguished from the theoretically calculated FRF in Figure 3.32. There is a peak in the magnitude and a phase transition from 0–180 degrees at the eigenfrequency 5 Hz. Clearly, this peak corresponds to the 5 Hz peak in the response PSD.

The FRF estimated from the non-linear response, Equation (3.58), resembles that of a linear system with a smaller eigenfrequency and a higher damping. In fact, if the parameters in Equation (3.50) are changed according to $m' = m, c' = c, d' = 1.2 \cdot d, k' = 0.2 \cdot k$ the calculated transfer function of the modified system is close to the estimated one from the non-linear response as shown in Figure 3.33.

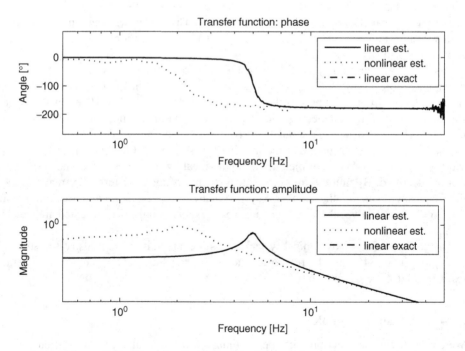

Figure 3.32   Transfer function of the linear and non-linear system

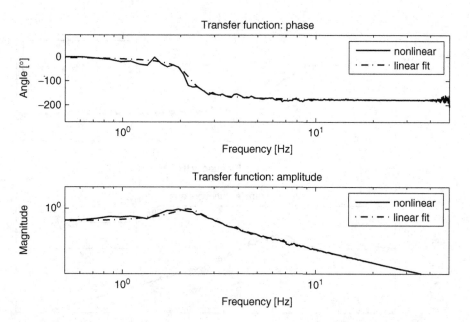

Figure 3.33   Estimated transfer function of the non-linear response and a fitted linear system

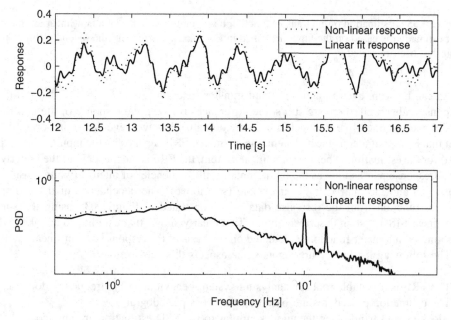

Figure 3.34   Comparison of the non-linear response and the response of the fitted linear model (time domain at the top and frequency domain below)

But this similarity must not be overestimated. If we run the fitted linear model and compare that response to the response of the non-linear system we obtain the results shown in Figure 3.34.

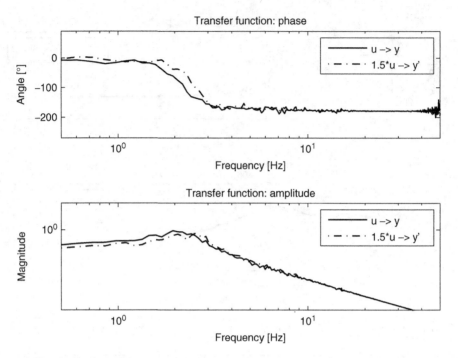

Figure 3.35 Estimation of the transfer function based on the response of a non-linear system. The first estimation is based on the input described in Eq. (3.52), whereas for the second estimation the input has been re-scaled

As can be seen, even though the fitted transfer function resembles that of the non-linear response rather well, there are still some differences between the response of the non-linear model and the fitted linear model in the time domain. However, even more important is the fact that in case of a non-linear system, the estimated FRF depends on the input/response pair used for the estimation, whereas for a linear system, the FRF is independent of the underlying input/response pair as long as the input contains all frequencies of interest (white noise, pink noise ...) and the matrix $\hat{P}_{uu}$ is invertible (see above). The dependence of the estimated transfer function from the response data is demonstrated in Figure 3.35 using the simple non-linear SISO system described above. This clearly shows that care has to be taken when working with transfer functions calculated on the basis of the response of non-linear systems.

The following list briefly highlights some aspects this sub-section:

- The FRF is a suitable tool for analysing system behaviour in the frequency domain.
- The estimation procedures are based on the cross-periodogram.
- Blocking and windowing techniques similar to the PSD estimation are applied.
- Care must be taken when dealing with FRFs estimated on the basis of a non-linear system response as this FRF represents the system only for input signals close enough to the one used during estimation.

3.2.5 Extreme Response and Fatigue Damage Spectrum

The aim of these methods is to analyse the response of a hypothetical mass spring system as a function of its eigenfrequency. To this end, the signal to be analysed is filtered using a transfer function with a pronounced peak at a certain frequency and the damage is then calculated on the resulting output. By varying the location of the peak in the transfer function, the eigenfrequency of the hypothetical mass spring system is modified and the damage becomes a function of the frequency. Thus, the result contains the damage potential of the input signal when applied to a system with a single eigenfrequency.

The process more specifically is as follows: Consider the single degree of freedom system

$$\ddot{y} + 2 \cdot \zeta \cdot \omega_0 \cdot \dot{y} + \omega_0^2 \cdot y = u, \tag{3.60}$$

$$\omega_0 = 2\pi \cdot f_0, \tag{3.61}$$

$$u = u(t), t \in [0, T], \tag{3.62}$$

where f_0 is the eigenfrequency of the (undamped) system and ζ is the damping. Given a signal $u(t)$ to be analysed, we can calculate the response $y(t)$ of the 1-DOF system and derive a quantity of interest, say, S. If this is done for varying system parameters ζ and f_0, S becomes a function of these parameters. If we fix the damping parameter, for example, $\zeta = 5\%$, then $S = S(f_0)$, and we can plot S over f_0.

If S is the maximum absolute value of the response $y(t)$, then S is called the extreme response spectrum (ERS). If S is the pseudo damage calculated on the response $y(t)$, then S is called Lalanne's fatigue damage spectrum (FDS). Historically, the extreme response spectrum was introduced first, mainly in military and earthquake applications. Based on that, the FDS was proposed by Lalanne later.

From Equation (3.60) it is clear that the response $y(t)$ is to be interpreted as a displacement. Thus ERS is the maximum displacement of the mass of the system. If S is the maximum absolute value of $\ddot{y}(t)$, the result is called the shock response spectrum (SRS) as it plots the maximum acceleration of the mass.

As mentioned in the introduction to the frequency-based methods, this approach combines aspects of the amplitude-based methods (represented by calculating the maximum value or the pseudo damage) with aspects of the frequency domain methods (represented by the FRF of the simple 1-DOF system). The purpose of Lalanne's method is to analyse or test a system where the frequency response is unknown. A typical example is the vibration of electronic equipment, where it is practically impossible to identify the frequency response for all the individual electrical components. For more details, see Lalanne [140, 141]. The idea of ERS/SRS goes back to Biot, see Biot [28].

Example 3.21 (ERS and FDS)
As an example, the ERS and FDS (for Wöhler exponent $\beta = 5$) have been calculated with $\zeta = 5\%$ for the input function $u(t)$ defined in Equation (3.52). The results are plotted in Figure 3.36. The peaks in the PSD of $u(t)$ at 10 Hz and 13 Hz are reflected in the FDS, showing that this signal has a large damage potential for systems with eigenfrequencies at 10 Hz or 13 Hz.   □

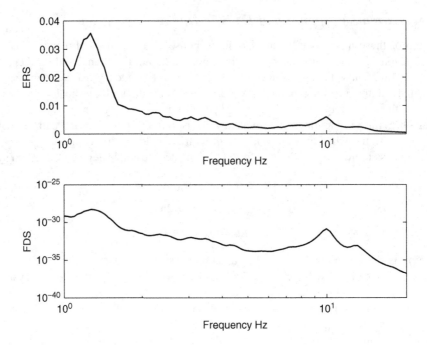

Figure 3.36   Example for ERS (top) and FDS (bottom)

Remark 3.2 *Lalanne's damage spectrum does not predict the damage that would occur if the signal was applied to a more complicated system with multiple eigenfrequencies. The mass spring system acts like a frequency filter. In that sense, the output at different frequencies could be interpreted as a decomposition of the input signal (similar to a harmonic analysis) with respect to frequency. However, the damage spectrum must not be interpreted as a decomposition of the total damage potential in the sense that the integral of the spectrum gives the total damage calculated on the basis of the input signal.*

3.2.6   Wavelet Analysis

A totally different method but still in the range of the frequency-based methods is the Wavelet analysis. Similar to Fourier analysis, it decomposes the signal into a set of base functions. In the case of Fourier analysis, the base functions are the (non-local) harmonic functions $\varphi_k(t) = e^{i2\pi \cdot k \cdot t}$. Using these functions for the decomposition is very natural in the sense that they are special solutions of linear dynamic models of the system, and the superposition principle ensures that a specific solution can be obtained by a linear combination of these base functions.

The Wavelet base functions have no direct relation to linear dynamic systems. They aim at a decomposition of a signal on levels of different coarseness. On each level, there are base functions which are non-zero in a segment of a certain length only called the support of the function. The support of the function is moved along the time axis and the signal is approximated by a combination of the base functions on that level. The remaining error is then approximated by a combination of similar base functions on a refined level with a more localized support. This process is repeated a couple of times until a certain quality or a predefined limit of levels is reached.

The Wavelet analysis can be used, for example, for the analysis of road profiles (see Öijer and Edlund [175, 176]) with respect to potholes. In contrast to Fourier analysis, which is very commonly used, the Wavelet analysis is applied only rarely in load analysis of vehicles. Therefore, we will not go into more detail but refer to Percival and Walden [185] or Press *et al.* [191] for an introduction to Wavelets.

3.2.7   Relation Between Amplitude and Frequency-based Methods

Frequently, load signals are interpreted as stochastic processes. In many situations (usually not in the truck industry but often in the off-shore, wind turbine, or aircraft industry), loads are specified by PSD functions. Consequently, one would like to reconstruct the stochastic process from the PSD function in order to use it either for test rig excitation or for simulation purposes.

Under certain assumptions on the underlying (unknown) stochastic process (typically a stationary Gaussian process is assumed), one can derive formulas which relate amplitude-based notions to frequency-based notions. The simplest example is Rice's formula, which relates the level crossing intensity to the moments of the PSD function.

A more elaborate connection between the amplitude domain and the frequency domain is given by Dirlik's method (Dirlik [73]) under certain assumptions on the underlying signal (stationary Gaussian process, see Chapter 6 for a sketch of the corresponding theory). It calculates an approximation of the distribution of rainflow cycles for the given PSD function. The approximation of the distribution of rainflow cycles from spectral properties is further discussed in Rychlik [200], Bishop and Sherratt [29], Zhao and Baker [248], and in Chapter 6.

If available, it might be tempting to apply convenient 'frequency analysis' instruments (especially when extensive frequency domain techniques are applied any way to utilize iterative learning control methods to prepare test rig signals). However, in cases where time signals are available, for example, from load signal measurements, it is strongly recommended that time history data are analysed for durability purposes, because the assumptions of the PSD-based methods are often not fulfilled for truck applications. This could lead to considerable mismatches between the rainflow content estimated on the basis of the PSD and the 'real' rainflow content.

3.2.8   More Examples and Summary

We conclude the section on frequency-based methods with some more examples on PSD calculation and the parameters to be chosen therein, as well as some hints about the application of the methods.

Since the PSD is probably the most common function in this context, two simple signals are used to illustrate some of the properties of the estimation procedure. The first is a rectangular signal, and the second is the sum of two sine signals with 5 and 7 Hz. Both signals are periodic, such that there is a Fourier series representation and in these specific cases, the coefficients can be calculated exactly. In the first case, we have coefficients at integer frequencies $f_k = 2k - 1$ and $c_k \sim \frac{1}{2k-1}$. In the second case, only the coefficients at $f = 5$ Hz and $f = 7$ Hz are non-zero. Figure 3.37 shows the time signal and corresponding PSD estimations, where the procedure explained in Section 3.2.2 has been used.

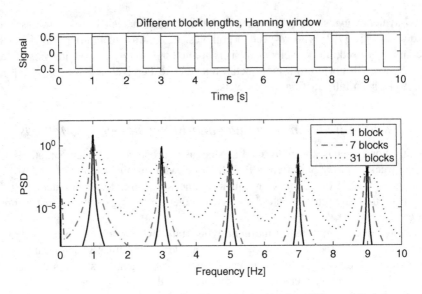

Figure 3.37 Rectangular signal and PSDs calculated using different block lengths

Windowing (Hanning window), averaging (different number of blocks and different block lengths), and overlapping (half block length) have been applied. The results reflect the theoretical result (contributions at $f_k = k$ decaying like $\frac{1}{k}$) and show clearly that the resolution decreases with decreasing block length. Since the signal is perfectly periodic without noise, all PSD estimations are smooth, independent of the amount of averaging.

In Figure 3.38 similar results are shown for the combination of two sinusoidal functions. The same remarks as for the rectangular signals apply.

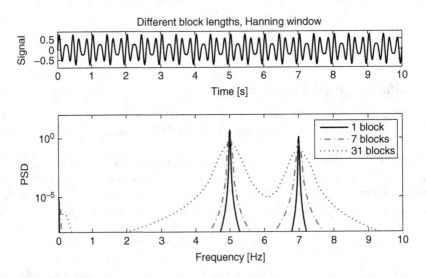

Figure 3.38 A combination of two sinusoids and PSDs calculated using different block lengths

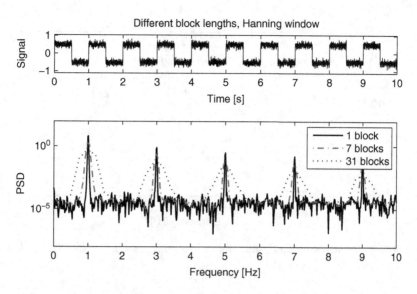

Figure 3.39 Rectangular signal + noise and PSDs calculated using different block lengths

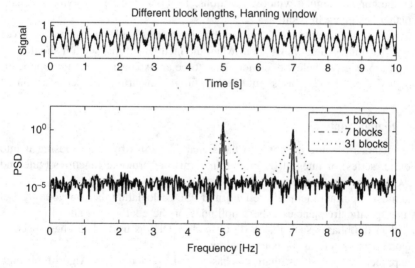

Figure 3.40 A combination of two sinusoids + noise and PSDs calculated using different block lengths

If we add some noise to the signals, we get the results shown in Figures 3.39 and 3.40. Again we see how the spectral resolution depends on the block length. In addition, we now see that the averaging procedure smooths out the large variances we have in the PSD estimation especially when only one block is used. For the sine combination example, we show in addition how the window type influences the results (see Figure 3.41).

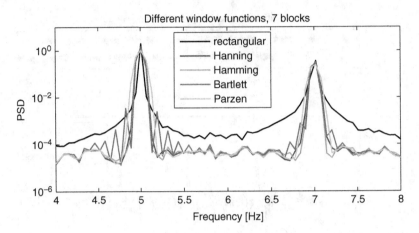

Figure 3.41   PSD functions for a combination of two sinusoids + noise, calculated using different window functions

Especially for the second peak at 7 Hz, the width produced by the rectangular window is much larger than for the other window functions, whereas there is no significant difference between the non-rectangular window functions. This is in accordance with the example and the explanations given in Section 3.2.2.

We conclude the section on frequency-based analysis with an attempt to give a short summary and some general hints about when the methods of this section should be applied. The list of topics given below does not claim to be an exhaustive range of applications, but it highlights some of them and should be helpful and inspiring in other situations.

3.2.8.1   Signal Analysis

1. The methods provide insight into the signal properties by decomposing it into a sum (Fourier series) or integral (Fourier transform) of harmonic functions (sine and cosine waves) at specific frequencies.
2. The most important methods are DFT and PSD, calculating the contribution coefficients at the specific frequencies. The amplitudes of the coefficients are often more important than the phases such that PSD instead of DFT is used. The phase becomes more important in system analysis (see below).
3. The procedure for PSD estimation is based on the periodogram. Algorithmic and statistical effects due to working with limited and sampled (discrete) data make it necessary to refine the basic estimation algorithm by averaging and windowing techniques. Hints on how to choose the corresponding parameters like block length and type of window have been given in Section 3.2.2. The two simple examples above illustrate these effects.
4. PSD calculation can be used to check data collected in measurement campaigns: Is there much noise in the data above a certain frequency threshold? Are there peaks in the spectrum which cannot be explained by physical reasoning (no system behaviour)?
5. Local loads like strains or forces acting on a stiff component need not be analysed in the spectral domain as a matter of routine.

6. Loads which act on systems capable of vibration are potentially subject to spectral analysis. Whether you should perform a spectral analysis depends on what questions have to be answered and what tasks have to be accomplished based on the data:

   (a) Does a special event (for instance, emergency braking) excite certain eigenfrequencies of a structure? PSD calculation shows the excitation potential of the load.

   (b) Compare public road measurements with test track measurements or highway with city driving: Spectral analysis is important as eigenfrequencies that are not excited on the public road might be excited on the test track and vice versa. This might lead to fatigue behaviour on the test track which is not typical of customer usage.

   (c) Apply some of the loads to a numerical model: Spectral analysis is important as the model needs to be capable of handling the load frequencies. It need not be capable of handling very high frequencies which are not excited.

   (d) Use some of the loads as targets for a rig test: Apply the methods to check for usability of data on a test rig. Does the frequency range fit the specifications of the rig?

3.2.8.2 Data-based System Analysis

1. The methods provide insight into the system properties with respect to the response to variable excitation by decomposition techniques.
2. The most important method is the transfer function or frequency response function (FRF) analysis, calculating how the harmonic fractions of the input are transferred by the system to the harmonic fractions of the output.
3. The procedure for FRF estimation is based on the periodogram and the cross-periodogram. The remarks given above on details of the PSD estimation procedure apply here as well.
4. Stiff systems or structures whose eigenfrequencies are well above the highest excitation frequency need not undergo an FRF analysis as a matter of routine. Their response to an excitation can be estimated by quasi-static calculations.
5. FRF analysis of systems or subsystems is helpful in

   (a) the verification of numerical models: Does the numerical model of an air spring fit the real behaviour? In which frequency ranges?

   (b) the verification of a rig configuration (see the corresponding example given in Section 3.2.4): Is the load path of a suspension on the rig the same as the load path of the suspension in the vehicle?

   (c) the drive file iteration process for test rigs. In fact, it is a central methodology in that context. There are usually software tools for handling this process, but often the automatically computed FRF representation is manually edited by the test engineer to enable or speed up convergence of the iteration process. This requires a thorough understanding of the method.

3.3 Multi-input Loads

The preceding sections in this chapter mainly focus on load analysis of one input signal. In most applications, however, there are many loads that act on a system or on a component. This section describes some of the most important methods for dealing with this situation and shows how to extend one-input methods to the situation of multi-input loads.

The first Section, 3.3.1, considers the load path from a multi-input load signal acting on a component to the local stress histories and uses this relation to motivate the specific analysis for multi-input loads. In essence, the idea is to perform a rainflow analysis on the components of the local stress tensor history and map this back to the outer loads (RP method). The next Section, 3.3.2, explains the RP method in more detail. The following sections describe some more methods for dealing with multi-input load signals, such as pseudo damage, equivalent multi-input loads, phase plots, and correlation.

3.3.1   From Outer Loads to Local Loads

3.3.1.1   Stiff Components

In Section 3.1 it was shown that there is good reason to apply the rainflow method on outer loads. The choice of method is supported by the theory of fatigue, at least in the case of stiff metallic components where the relation between local stress $\sigma(x, t)$ at point x and external load $L(t)$, which is to be analysed, is of the form $\sigma(x, t) = c(x) \cdot L(t)$. In general, neither is the external load one-dimensional, nor is the local stress uniaxial. However, if we stick to stiff components, the relation between external multi-input loads $L_k(t)$ and the local stress tensor $\sigma_{ij}(x, t)$ can be written as a linear combination of the external loads

$$\sigma_{ij}(x, t) = c_{ij}^{(1)}(x) \cdot L_1(t) + c_{ij}^{(2)}(x) \cdot L_2(t) + \cdots + c_{ij}^{(n)}(x) \cdot L_n(t), \qquad (3.63)$$

where n is the number of external loads acting on a component, and x denotes a specific location. This formula presumes validity of linear elastic behaviour as well as stiffness of the component. Here, stiffness is not only a property of the component. It means that the highest frequencies contained in the load signals are much smaller than the lowest eigenfrequencies of the component. In this case, the coefficients $c_{ij}^{(k)}(x)$ can be obtained from linear static FE analysis (see Chapter 5).

3.3.1.2   Critical Plane Approach

In the presence of a locally multi-axial stress state, the notion of hysteresis cycles given above is no longer valid. There are many ways of approaching the extension of the uniaxial case. However, a detailed description will not be given here as it falls outside the scope of this guide. Among these approaches the critical plane approach is the most important in our context and gives good reasons for selecting the multi-axial rainflow approach.

At the surface of a component, the stress is usually a two-dimensional tensor given by the tensile/compressive stresses σ_x, σ_y and the shear stress τ. A crack which may arise due to the load will have a certain orientation (critical plane), given by some angle α. If we now consider the stress normal to the plane, we can write it as a linear combination of the two-dimensional stress tensor components. The same is true for the shear stress along this plane. Combining this with the formula given above, we can write

$$\sigma_p(x, \alpha, t) = c_p^{(1)}(x, \alpha) \cdot L_1(t) + \cdots + c_p^{(n)}(x, \alpha) \cdot L_n(t), \qquad (3.64)$$

where the index p indicates a certain stress component of interest. This extends the simple formula $\sigma(t) = c \cdot L(t)$ to the multi-input case and expresses the stress components of interest as linear combinations of the external loads.

3.3.1.3 Non-linear Behaviour

The approach sketched above utilizes a linear relationship between loads and stresses. Often the biggest part of a component is loaded in a range where this is justified. The hot spots where the stresses are no longer in a linear regime are typically small. For this situation, there are a couple of correction methods (the most famous example is Neuber's rule, see Neuber [170]) which offer a transformation from the stresses calculated by linear theory (sometimes called pseudo stresses) to real stresses.

3.3.1.4 Non-stiff Components

If the component is assumed to exhibit vibrational behaviour, such that the eigenmodes may be excited by the loads, the component can be mathematically represented using a combination of static modes and normal modes, the latter describing the vibrational dynamic behaviour, the former describing the static stiffness properties.

This is often used during multibody simulation of a complex system such as a full vehicle, if some of the sub-structures (the cabin, for example) are excited within their frequency range. See, for example, Craig and Bampton [60], Craig and Kurdila [61], ADAMS [2], Schwertassek and Wallrapp [212] or Dombrowski [75]. The Craig Bampton approach is the one most often used nowadays. Its application is automated and straightforward with the exception of the modal damping which is harder to derive and often chosen by experience (see Lion [144]). However, the damping in the surrounding elements of the multibody model (shock absorbers, joints, rubber bearings, etc.) is typically more important than structural damping.

The local stress tensor histories can again be written in the form given above, where the coefficients c represent the modes and the L_k are the so-called modal participation factors (MPF). They can be calculated using a combination of finite element and multibody system analysis (see Chapter 5).

The MPFs are not measurable quantities as they arise from a purely mathematical model describing the structure. Thus they are available only if a CAE process including FE and MBS simulation is established. Nevertheless, analysing this kind of data with the methods described in this guide can be very helpful.

As an example of such an application, consider the case where a modification (for instance, a new adjustment of an air spring) comes into a late phase of the development process. The tests for the durability release of the component are already finished. The incoming modification does not affect the component directly but it changes the load path from the road excitation to the component. In this situation, comparing the MPFs acting on the component before and after the modification may help to decide whether the whole release process should be repeated or not. The remark on linear or pseudo stresses and real stresses applies here too.

3.3.1.5 Summary

Since the emphasis here is on load analysis and not on FE analysis, we will not go into more detail of stress calculation, but use the two formulas (3.63) and (3.64) above as a foundation and justification for the multiaxial rainflow method, which is explained in some detail in Section 3.3.2.

3.3.2   The RP Method

As has been explained in the preceding subsection, quantities of the form

$$L_p(t) = c_1 \cdot L_1(t) + c_2 \cdot L_2(t) + \cdots + c_n \cdot L_n(t) \tag{3.65}$$

as in Equation (3.63) or in Equation (3.64), where $L_k(t)$ is the external load acting on a structure, contain important information about the 'damage potential' of the load. If a specific structure and its critical points are well known, the coefficients of this formula can be calculated via FE modelling and a detailed analysis can be made. In this case, one would calculate the local stress histories and apply load analysis methods thereon.

The emphasis of the methods to be described here, however, is to offer a possibility to analyse multi-input loads without reference to a specific structure the load acts on. This can be achieved, for example, by analysing all possible linear combinations of the form $c_1 \cdot L_1(t) + \cdots + c_n \cdot L_n(t)$ or a representative set of such combinations.

Example 3.22 (Two Sinusoidal Loads)
As an illustrative example, consider the simplest case of 2 sinusoidal loads $L_1(t) = \sin(t)$, $L_2(t, \theta) = \sin(t + \theta)$ acting on a component. Since we do not want to use any information about the component the load is acting on, we cannot restrict the analysis to a certain fixed combination of the loads. Instead we have to check the rainflow content of all possible combinations, for example, the combination $(1, 1) : \sigma_1(t, \theta) = L_1(t) + L_2(t, \theta)$ or the combination $(1, -1) : \sigma_2(t, \theta) = L_1(t) - L_2(t, \theta)$.

If $\theta = 0$ we have $\sigma_1(t, 0) = 2 \cdot \sin(t)$ and $\sigma_2(t, 0) = 0$. In that case we have an amplitude of $\Delta\sigma_1 = 2$ at $(1, 1)$ and of $\Delta\sigma_2 = 0$ at $(1, -1)$. If $\theta = \frac{\pi}{2}$, we have $\sigma_1(t, \frac{\pi}{2}) = \sin(t) + \cos(t)$ and $\sigma_2(t, \frac{\pi}{2}) = \sin(t) - \cos(t)$. In the latter case we have an amplitude of $\Delta\sigma_1 = \sqrt{2}$ at $(1, 1)$ and at $(1, -1)$.

If we restrict the analysis to the input loads $L_1(t), L_2(t, \theta)$, different phase-shifts θ do not make a difference. But looking at the combination $(1, 1)$, we observe a difference in the amplitude of 2 if $\theta = 0$ and $\sqrt{2}$ if $\theta = \frac{\pi}{2}$. This is shown in Figure 3.42. For the combination $(1, -1)$, the difference is even larger: 0 if $\theta = 0$ and $\sqrt{2}$ if $\theta = \frac{\pi}{2}$.   □

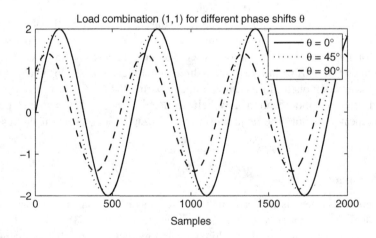

Figure 3.42   The $(1, 1)$-combination of two sinusoidal loads for different phase angles θ

To summarize the simple example, the traditional approach (rainflow analysis of both loads) does not distinguish between different phase-shifts θ although the impact on the stress at a certain location might be totally different. In contrast, the analysis of the additional combinations $(1, 1)$ and $(1, -1)$ reveals a difference between both loads.

Of course, the amount of additional information increases with increasing number of combinations. Since the number of all possible combinations is infinite, the idea is to construct a representative set of combinations and analyse these using the uniaxial rainflow methods described in Section 3.1.3. This approach will be called the Rainflow Projection (RP) approach in the following (see Beste *et al.* [25] or Dressler *et al.* [79]).

As counting proportional time signals leads to rainflow matrices with the same shape but just different ranges, the coefficients of the combinations can be normalized such that $c_1^2 + \cdots + c_n^2 = 1$ and $c_i > 0$, without losing any information. Nevertheless, the number of combinations increases rapidly with increasing number of load signals (either physical loads or modal participation factors). Each combination can be interpreted as a unit vector in n-dimensional space where n is the number of loads (RP direction).

The following algorithm might be used for the selection of combinations. Let n denote the dimension of the space we are working in, that is the number of loads to be analysed. Let k denote a parameter defining the discretization level. Then a uniform grid on the boundary of the unit cube $[-1, 1]^n$ is generated. The number of points on each axis is given by k. For the last axis only positive values are taken. This results in N points, where

$$N(n, k) = \frac{k+1}{2} \cdot k^{n-1} - \frac{k-1}{2} \cdot (k-2)^{n-1}. \tag{3.66}$$

The number $N(n, k)$ of directions increases rapidly with n and k, such that k is usually small. The following table contains the number $N(n, k)$ for some values of n and k:

n	$N(n, 3)$	$N(n, 5)$
2	5	9
3	17	57
4	53	321
5	161	1713
6	485	8889

The points on the boundary of the cube are projected onto the unit circle giving the normalized direction vectors c. This algorithm is illustrated in Figure 3.43 for $k = 7$.

The RP approach as an extension of the uniaxial rainflow counting method fits, by definition, well into the framework of fatigue. As has become clear in the preceding sections, the approach concentrates on the rainflow content of a hypothetical local stress tensor component and it is therefore more or less an amplitude-based method.

3.3.3 Plotting Pseudo Damage and Examples

Since the result of the RP counting is an entire set of rainflow matrices, simply plotting all matrices might be impractical. However, a first overview of the data and a good basis for

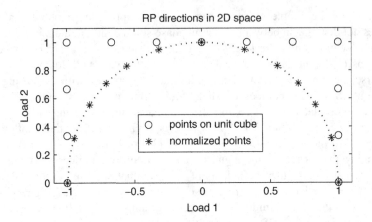

Figure 3.43   The directions for the RP approach in 2D space with $k = 7$

comparison of several signals are obtained by calculating a pseudo damage number on the basis of the rainflow matrices using a hypothetical Wöhler curve and then plot the results in a bar diagram for each direction. In the case of three loads the coefficients for all directions lie on the surface of the unit sphere. Thus, can also plot the results in the form of a contour plot, showing the "northern" half of the unit sphere. The position within that plot indicates the coefficients of the RP direction and the colour indicates the pseudo damage. This type of plot may be called load influence sphere.

Example 3.23 (RP Analysis Of Wheel Forces Of a Truck)
The following example uses the truck measurements in the city, on the highway, and on the country road which have been used already in Section 3.1.4 (see Figure 3.12). The three wheel forces at the right front axle have been analysed. Prior to the RP analysis, the forces have been normalized to maximum absolute value 1, such that the magnitude of the different force directions become comparable. This, to some extent, reflects the fact that the suspension is designed for different load magnitudes in different directions. In Figure 3.44, the load influence spheres for the three road types are shown. As can be seen, the city and highway data have similar characteristics, whereas the country road data behaves different. This can be seen again in Figure 3.45. The pseudo damage on the country road is much greater than in the city in directions with a large longitudinal coefficient.   □

Example 3.24 (RP Analysis Of Wheel Forces Of a Passenger Car)
We apply the RP method to the wheel forces of a passenger car, measured on a race track and a test track. In Figure 3.46, the load influence plot of one of the tracks is shown on the right-hand side. The centre of the sphere belongs to direction $(0, 0, 1)$. The red spot in the plot indicates that the combination $(0, 0.71, 0.71)$ is such that it contains the highest damage potential. The components F_y and F_z seem to be correlated. As shown on the left-hand side, the bar diagram can easily be used to compare the two different measurements. Apparently, both loads have similar properties with respect to the RP results. It should be kept in mind that these plots are independent of any specific component. Thus the

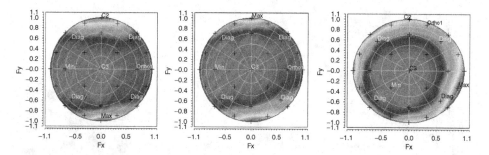

Figure 3.44   Load influence spheres for the city, highway and country road measurements. The longitudinal force is on the horizontal axis (x), the lateral force on the vertical axis (y), and the vertical force is on the axis perpendicular to the paper (z). While the scaled lateral force is most damaging in the city and on the highway, a combination of longitudinal and lateral force is most damaging on the country road. As we are only interested in the relation between the signals, the legend containing the absolute pseudo damage numbers is omitted

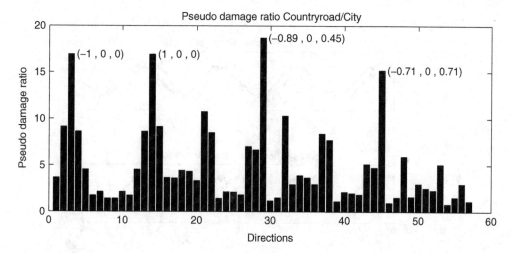

Figure 3.45   Pseudo damage ratio of country road over city per km ($\beta = 5$) for 57 RP directions. The order of the load signals is longitudinal, lateral, vertical. The direction coefficients are shown for some directions

combination (0, 0.71, 0.71) with the highest damage potential might not be relevant when the loads are applied to a certain structure. For more details on this example, see Bremer *et al.* [36].   □

Example 3.25 (RP Analysis Of a Synthetic Signal)
Another example is given by the load signals plotted in Figure 3.47. As the first signal is the same in both cases (upper and lower plot) and the second signal differs only in its phase delay θ (neglecting differences in the noise), ordinary rainflow counting gives similar

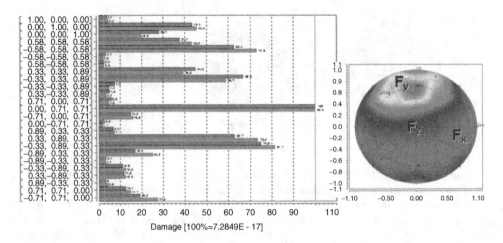

Figure 3.46   Bar diagram of two different multi-input loads and the representation of one of them as a load influence sphere

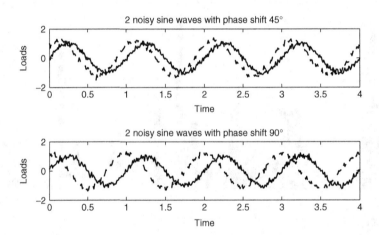

Figure 3.47   Example for two-dimensional RP counting: Two pairs of noisy sine waves with a phase delay of 45 (top) and 90 (bottom). In both cases, the amplitude of the first signal is 1 and the amplitude of the second signal is 1.2

results in both cases. Thus, the same is true for the level crossing as well as the range-pair counting. However, applying RP counting reveals the essential difference between both pairs (the phase-shift). This is illustrated in Figure 3.48.

The pseudo damage values for the original signals (direction $[1, 0]$ =1st signal, direction $[0, 1]$ =2nd signal) are the same, while the values for the combinations $[0.71, 0.71]$ differ by a factor of 3 approximately. This means that, if a component is loaded with these signals $L_1(t)$, $L_2(t)$, and at a hotspot x of the component, we have $\sigma(x, t) = c_{ij}^1(x)L_1(t) + c_{ij}^2(x)L_2(t)$ with $c_{ij}^1(x) \approx c_{ij}^2(x)$, the fatigue life will also differ by a factor of 3.   □

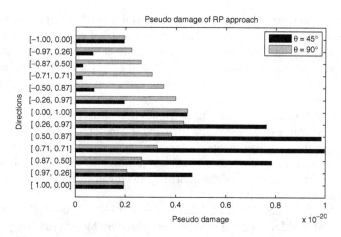

Figure 3.48 Pseudo damage (slope of S-N-curve $= 5$) of the different directions of the 2-dimensional RP approach applied to the signals in Figure 3.47

3.3.4 Equivalent Multi-input Loads

The aim is to find a simpler multi-input load that is equivalent to the measured load in terms of damage. The first task is to specify the characteristics of the multi-input equivalent load which is not as easy as for the one input case, see Section 3.1.12. The most straightforward extension is to generalize the constant amplitude equivalent load for the one-input case and treat each load signal individually. For the two-input case, the equivalent load with N_0 cycles can then be expressed as

$$L_{eq,1}(t) = A_1 \sin(2\pi t), \quad L_{eq,2}(t) = A_2 \sin(2\pi t), \qquad 0 \le t \le N_0. \qquad (3.67)$$

Thus, the two amplitudes A_1 and A_2 represent the equivalent load. However, the drawback is that the dependencies between the input loads cannot be captured. The next generalization is to introduce a phase-shift between the equivalent constant amplitude loads, which for the two-input case becomes

$$L_{eq,1}(t) = A_1 \sin(2\pi t), \quad L_{eq,2}(t) = A_2 \sin(2\pi t + \theta), \qquad 0 \le t \le N_0, \qquad (3.68)$$

resulting in three parameters defining the equivalent load: the two amplitudes A_1 and A_2, and the phase angle θ. More complicated definitions of equivalent loads are possible by introducing even more parameters. However, by increasing the complexity of the equivalent load, the interpretation also becomes harder, especially when the severity of different load signals is to be compared. Thus, the choice of a proper complexity of the multi-input equivalent load depends on its intended use, for example, comparing the severity of load signals, or producing an equivalent test sequence. The construction and definition of multi-input loads is the topic of Genet [99].

The idea is that the equivalent load should be damage-equivalent to the measured load for any linear combination of the load components. In general, this is not possible to obtain, as there is a damage equivalence requirement for each individual linear combination of load components, but only a few parameters defining the equivalent load. One solution is to select the same number of linear combinations as the number of parameters defining the equivalent load, and then solve the corresponding system of equations. For the case of two inputs and two parameters, Equation (3.67), the two linear combinations would be the two load components themselves, that is, $c_1 = (1, 0)$ and $c_2 = (0, 1)$. By introducing the phase angle, see Equation (3.68), the third linear combination can, for example, be the sum of the two load components, $c_3 = (1/\sqrt{2}, 1/\sqrt{2})$.

Another solution to approximate the damage equivalence is to minimize the distance between the equivalent pseudo damage $d_{eq}(c)$ and the measured pseudo damage $d_{life}(c)$ for all linear combinations of load inputs. In practice, this can be implemented by choosing a sufficiently large number of linear combinations, resulting in a system with more equations than unknowns. The parameters of the equivalent load can then be estimated by using, for example, the least-squares method.

The methodology is developed in detail in Genet [99] for Basquin's damage criterion, Basquin [13], and for Morel's multiaxial damage criterion, Morel [165, 166], which is based on Dang Van's criterion (Dang Van [63]). The equivalent load can be defined as a constant amplitude load, as above, or some more complicated load signals. It can also be defined as a certain class of random processes, for example, narrow band Gaussian loads or Markov loads, which are the two cases treated in Genet [99].

Example 3.26 (A Two-Input Sinusoidal Equivalent Load)
We finish off with a simple example of a two-input sinusoidal equivalent load; Genet [99, Example 7.3.2]. The measured reference load together with the resulting equivalent load is presented in Figure 3.49. Here the reference load is a measurement of one lap of the test track, which is represented by $N_0 = 500$ equivalent cycles. The equivalent load is defined by two amplitudes and one phase angle, as in Equation (3.68). Using the damage exponent of $\beta = 8$, the two amplitudes are estimated at $A_1 = 1.68$ and $A_2 = 2.34$, and the phase angle becomes $\theta = 84^{\circ}$. □

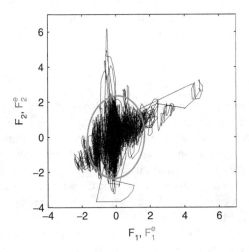

Figure 3.49 The measured reference load together with the resulting equivalent sinusoidal load; Genet [99, Figure 7.1]

3.3.5 Phase Plots and Correlation Matrices for Multi-input Loads

A good method for investigating the correlation between input loads is the concept of phase plots. This method, however, does not focus on fatigue aspects. Instead of plotting each input signal $L_i(t)$ as a function of time t, one load $L_i(t)$ is plotted as a function of another load $L_j(t)$. Using figures of that type, it can easily be seen whether the signals are correlated. If the samples $(L_i(t_k), L_j(t_k))$, $k = 1, 2, \ldots$ lie close to a straight line, then the signals are strongly correlated. If the samples cover a ball-shaped area, there is no correlation between the signals. These plots give a visual impression of the correlation of signals.

Example 3.27 (Two Sinusoids with Different Phase Angles)
Figure 3.50 shows the phase plot of 2 sinusoids with different phase angles θ. It can be seen that the signals are perfectly correlated for $\theta = 0$ and uncorrelated for $\theta = \pi/2$. The case $\theta = \pi/4$ is somewhere in between.   □

Example 3.28 (Phase Plot of Truck Signals)
Figure 3.51 shows two phase plots of a truck measurement during a city drive. To the left, the lateral force is plotted over the vertical force. It can be seen that there is a certain positive correlation between both forces. This is induced by curves which increase/decrease both the vertical and the lateral force at the same time. To the right, longitudinal force over braking/accelerating torque is plotted. Here we see a strong, negative correlation which is induced by braking and accelerating events. At small values of the torque, there is a certain 'noise' in the longitudinal force coming from the road roughness.   □

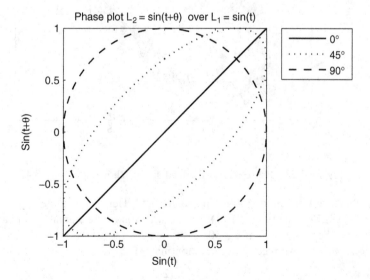

Figure 3.50 Phase plot of two sinusoids with different phase-shifts θ

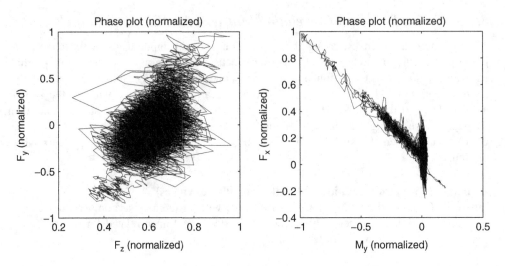

Figure 3.51 Phase plot of truck wheel forces during a city drive. To the left F_y over F_z, to the right F_x over M_y

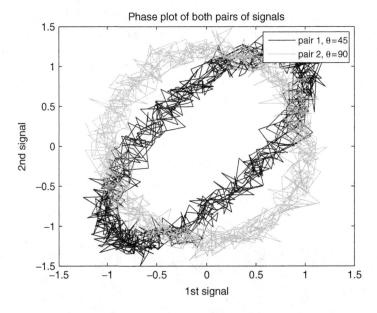

Figure 3.52 Phase plot of the load signals shown in Figure 3.47

Another phase plot example is given in Figure 3.52. Here we use the signals introduced in the previous section (see Figure 3.47). By construction of the signals, the phase plot very much resembles the plots in Figure 3.50.

In addition, one can estimate the correlation matrix

$$corr(L_i, L_j) = \frac{cov(L_i, L_j)}{\sqrt{cov(L_i, L_i) \cdot cov(L_j, L_j)}} \tag{3.69}$$

$$cov(L_i, L_j) = \frac{1}{N} \cdot \sum_{k=1}^{N} (L_i(t_k) - \bar{L}_i) \cdot (L_j(t_k) - \bar{L}_j) \qquad (3.70)$$

for $i, j = 1, \ldots, n$, where n is the number of input signals, N is the number of samples and $\bar{L}_i$ denotes the mean value of $L_i(t)$. The correlation of two signals is a number (correlation coefficient) between -1 and 1, where 1 means that the mean value-corrected signals are proportional to each other with a positive constant, -1 means that the proportionality constant is negative. A small absolute correlation coefficient means that the signals are not or only weakly correlated. In Figure 3.53 the correlation matrix (absolute values) is shown for the same data as above. The strong (negative) correlation between longitudinal force and braking torque results in a correlation coefficient of about -0.9. The correlation between lateral and vertical force is about 0.4.

3.3.5.1 Correlation Matrices in the Frequency Domain

The idea of the correlation or covariance matrices in the time domain as explained above can be transformed into the frequency domain using the Fourier transformation. In Section 3.2.4 this approach has already been formulated for continuous signals and used for the FRF-based system analysis. The corresponding formulas for sampled data are described in the following.

$$R_{L_i L_j}(\tau) = \frac{1}{N} \cdot \sum_{l=1}^{N} (L_i(t_l + \tau) - \bar{L}_i) \cdot (L_j(t_l) - \bar{L}_j) \qquad (3.71)$$

$$\gamma_{L_i L_j}(\tau) = \frac{R_{L_i L_j}(\tau_k)}{\sqrt{R_{L_i L_i}(0) \cdot R_{L_j L_j}(0)}} \qquad (3.72)$$

The function γ_{L_i, L_j} describes the relation between the time signals $L_i(t)$ and $L_j(t)$. Its value gives an indication of the dependency of the sampling points separated by the time lag τ.

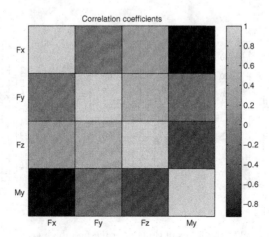

Figure 3.53 Correlation coefficients of truck wheel forces during a city drive

The Fourier transform $\hat{\gamma}_{L_i, L_j}(f)$ maps this relation into the frequency domain, such that the correlation between the signals can be studied 'frequency-wise'. As already mentioned, it is mainly applied in the context of system analysis or system identification. It is also used for modelling random processes, especially for characterizing Gaussian vector processes.

3.3.6 Multi-input Time at Level Counting

The multi-input time at level count has already been mentioned in Section 3.1.13 for rotating components. If we discretize the range of each signal as usual, the n-dimensional cube of values of the load signals (n is the number of loads) is divided into a large number of cells. Now we can count how many samples lie in a certain cell. The result is an n-dimensional histogram which can be mapped to the corresponding time at level count by multiplying with the sampling rate Δt. For two input signals, we obtain an ordinary matrix that can be easily plotted. For more signals we get an 'n-dimensional' matrix. The latter is usually visualized by plotting certain projections. As an example, consider a 3D gear box signal consisting of the gear index, the torque, and the rotational speed. The resulting 3D time at level count will be plotted as series of 2D matrices, each representing one gear.

An example of such a counting result is shown in Figure 3.54 for the first three gears of a passenger car driving on a public road.

3.3.7 Biaxiality Plots

Depending on the loads acting on a component, the orientation of the plane stresses $\sigma_X, \sigma_Y, \sigma_{XY}$ at the surface of a structure may vary over time. To analyse this, the principal stresses σ_1, σ_2 and the angle φ are introduced, where σ_1 denotes the eigenvalue of the stress tensor with the largest magnitude, the angle φ defines the orientation of the corresponding eigenvector and σ_2 is the second eigenvalue. The biaxiality ratio is defined as the quotient $\frac{\sigma_2}{\sigma_1}$. A biaxiality of 0 corresponds to a uniaxial stress, a value of -1 corresponds to a pure shear stress, and a value of 1 corresponds to a pure biaxial state.

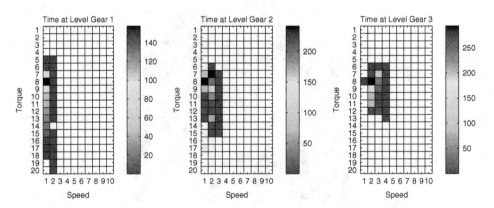

Figure 3.54 Time at level counting result (in [s]) for the first three gears of a passenger car during a public road drive. Since we are only interested in the principle of that kind of plots, the magnitude of torque and speed is omitted

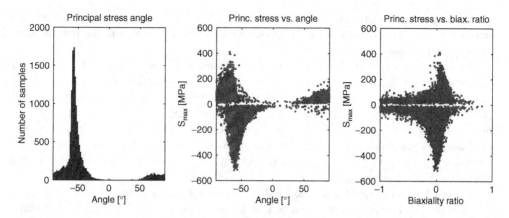

Figure 3.55 Plots analysing the local stress state

Figure 3.55 shows plots of these quantities resulting from the stresses at a trailer axle during a rough road drive. The left figure shows a histogram of the principal stress angles during the rough road drive with a sharp peak at about $-60°$. Thus, most of the time, the orientation of the stress tensors is the same. The second plot in the middle reconfirms this observation. Large stresses mainly occur in the same direction. The plot on the right shows the principal stresses over the biaxiality ratios. Here we see that we have large stresses only in situations, where the local stress state is uniaxial (biaxiality ratio = 0).

Since we are mainly concerned with outer load analysis and the biaxiality ratio is a local stress analysis option, we do not go into more detail here.

3.3.8 The Wang-Brown Multi-axial Cycle Counting Method

An interesting multi-axial cycle counting method, developed for low cycle fatigue, is presented in Wang and Brown [239, 240]. The counting procedure is rather complicated, but the principle is to count hysteresis cycles making use of the von Mises stress. The method allows the critical plane to change with every reversal. An interesting observation is that for uniaxial loads it becomes equivalent to the rainflow cycle counting method.

3.4 Summary

This chapter has described the fundamental methods for characterizing the durability properties of load signals. The first two sections treat the case of one-dimensional load signals, focusing first on amplitude-based methods and then on frequency-based methods. The third section introduces the case of multi-input loads, that is, many load signals are acting simultaneously on the system.

The section on amplitude-based methods, presented in Section 3.1, starts with the preprocessing of load signals by extracting turning points and optionally removing small amplitude cycles through rainflow filtering. The rainflow and range-pair counting are described in great detail, but also Markov counting, level crossing counting, and other counting methods are discussed. The severity of a load often needs to be described by a single number, and the severity measures of pseudo damage, equivalent amplitude, and equivalent mileage are introduced. Finally, methods for rotating components are discussed and the time at level counting is presented.

The frequency-based methods, presented in Section 3.2, mainly deal with the spectral representation of the load, which means that the frequency content of the load signal is studied. Concepts such as Power Spectral Density (PSD), Frequency Response Function (FRF), and correlation function are introduced in connection with system analysis. For non-stationary loads with varying spectral properties, the combination of frequency and time analysis can studied by using a spectrogram or Wavelet analysis. For systems with unknown frequency response, the Extreme Response Spectrum (ERS) or Fatigue Damage Spectrum (FDS) can be useful.

Methods for multi-input loads, discussed in Section 3.3, cover amplitude- and frequency-based methods. The main conclusion is that for stiff components the so-called Rainflow Projection (RP) method is recommended, while for non-stiff components and systems the frequency behaviour needs to be studied in addition. Most of the presented material is more or less related to the RP method, for example, pseudo damage, load influence sphere and multi-input equivalent loads. Other methods for presenting multi-input loads are phase plots, multi-input time at level counting, and biaxiality plots.

4

Load Editing and Generation of Time Signals

4.1 Introduction

In this chapter, we will deal with methods to modify loads which have been either measured directly or derived from measurements, for instance, section forces calculated by a multibody simulation based on measured wheel forces (see Chapter 5). The reasons for modifying these loads are manifold. We distinguish between the following aspects:

1. **Correction:** The first reason is simply that there are errors in the data, originating, for example, from problems with the measurement equipment or a numerical calculation procedure. These errors have to be eliminated prior to analysis, otherwise erroneous conclusions might follow. Methods for checking and correcting such problems are described in detail in Section 4.2.
2. **Acceleration:** The second reason is to shorten the load signals in order to speed up subsequent tasks such as rig testing or numerical simulations. In this case, the impact of the load on the system should not be changed. Section 4.3 is mainly (but not completely) dedicated to that task.
3. **Adjustment:** The third reason is to edit the load such that the impact on a system changes in a predefined way. A simple task in this context is to scale a load (because we want to obtain a rough estimate of the effect of a higher payload) or to apply a frequency filter to eliminate high frequent noise. A more complex example is extrapolating the load to a longer period or to a more severe event. Such methods are described in Section 4.4.

Finally, Section 4.5 is concerned with methods for the generation of load time signals based on some "incomplete" information such as a rainflow matrix. This means that the information we start with does not uniquely define a load signal, but there is a whole set of possible signals, from which we have to select an appropriate one. Figure 4.1 illustrates that different time signals might be mapped to the same rainflow matrices or load spectra, and different rainflow matrices or load spectra might be mapped to the same pseudo damage number.

Guide to Load Analysis for Durability in Vehicle Engineering, First Edition. Edited by P. Johannesson and M. Speckert.
© 2014 Fraunhofer-Chalmers Research Centre for Industrial Mathematics.

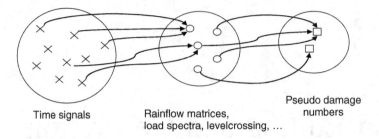

Time signals   Rainflow matrices,   Pseudo damage
                              load spectra, levelcrossing, ...   numbers

Figure 4.1   Equivalence classes of time signals and counting results

This is due to the loss of information inherent in all counting methods. It is important
to control what kind of information is neglected by a certain method and whether this is
acceptable in view of the system the load signal acts on. In order to assess the editing or
generation methods, a structured view of the properties of loads and their corresponding
impact on a system is essential.

4.1.1   Essential Load Properties

The properties which are of importance for the assessment in our context depend on the
type of quantity we are studying. For a local stress or strain or for a single force acting
on a stiff component, the most important properties are related to the amplitude domain,
in particular, the rainflow domain. This has been discussed in Chapter 1 and Chapter 3,
especially in Section 3.1 and Section 3.1.3 (rate-independent methods).

For a load acting on a non-stiff system capable of vibration, this is no longer sufficient,
because the response of such systems is not rate-independent. Analysis in the frequency
domain, that is, Fourier analysis (also known as spectral analysis), is appropriate for linear
vibrating systems. The most important methods are PSD and transfer function calculation.
These types of analysis were introduced in Section 3.2.

Things become more complex if multiple loads are considered. In that case the interaction
of all load signals needs to be considered because the impact of the loads on a structure
depends not only on the rainflow or frequency domain of each of the signals, but also on
their correlation. Methods to handle this were described in Section 3.3. In general, all three
aspects, namely the amplitude domain, the frequency domain, and the "correlation domain"
have to be considered.

In view of these three types of load properties, we introduce some measures in the
following Section, 4.1.2, which can be used for comparing loads. These measures can then
be applied to assess the effects of the load editing procedures described later in this chapter.

4.1.2   Criteria for Equivalence

We want to use the structure of load properties formulated above to deduce criteria for
assessing equivalence and differences of load signals. In general, the severity of a load
is not a property of the load signal alone, but it also depends on the system the loads
are applied to, for example, the configuration of the truck. Accordingly, all criteria will

have their limited range of validity and the application of the methods needs engineering judgement.

4.1.2.1 Rainflow Domain

Comparing two different load signals in the rainflow domain of course means comparing the rainflow matrices. As the rainflow matrix is a 2D data structure, we cannot simply define an order relation which tells us that one matrix is "bigger" than another. A bin-by-bin comparison is both impractical and much too detailed. Thus, we have to concentrate on a more condensed data structure. In the simplest case, we can use the methods described in Section 3.1.12 and calculate the pseudo damage numbers (or equivalent load amplitude or equivalent mileage) based on some synthetic Wöhler curve for both matrices and compare the resulting scalar values.

The main drawback of this simple procedure is that equivalence in the pseudo damage numbers does not necessarily mean equivalence with respect to fatigue of specific components. The result of the comparison depends on the slope of the Wöhler curve and implicitly presumes the validity of the Palmgren-Miner rule. Of course, more refined pseudo damage calculations could be made to take, for example, mean stress effects into account. In fact, this can help if we analyse "local loads" where we know more about the structure the load is acting on. But we often need a general comparison procedure (see again Section 3.1.12) that can be used for large sets of data containing different types of signals, for example, for wheel forces or cabin accelerations. It should be applicable without specifying the structure it acts upon. In such situations, the limitations of the simple pseudo damage numbers, while we are aware of them, are usually accepted.

Bearing all the amplitude-based analysis methods (Section 3.1) in mind, we see that more refined measures like the difference of two range-pair histograms or load spectra are obvious. But these measures do not give a clear statement such as "signal A is x times more severe than signal B" and they need engineering judgement. A compromise between a single pseudo damage number and a load spectrum is to define a few clusters in the rainflow matrix (e.g. small, medium and large cycles or such with a high mean load), calculate the pseudo damage numbers for each cluster and make comparisons on the basis of these few scalar numbers. We will come back to that point later in the course of rainflow extrapolation and optimum track mixing.

4.1.2.2 Frequency Domain

It is less straightforward to develop measures for comparing signals in the frequency domain. The main reason for this is that assessing the potential impact of a load on a component or system in the frequency domain depends heavily on the component or system and not primarily on the load signal itself. Although this is also true for the amplitude domain, it becomes even more important for the frequency domain.

The simplest thing that can be done in this context is to compare the PSD functions in certain frequency bands. This gives an answer to the question whether one signal contains more potential to excite vibrations in that frequency range than another. It does not say anything about the sensitivity of the system to that frequency range. If a transfer function of the system is known, then the product of the Fourier transform of the signal and the transfer function gives the Fourier transform of some response which can be analysed in addition.

Lalanne's fatigue damage spectrum is an approach trying to combine both the amplitude and frequency domain. Here, the pseudo damage of the response of a simple 1-DOF system induced by the load signal is analysed as a function of the eigenfrequency of the system. See Section 3.2.5 for a description of the method.

4.1.2.3 Correlation Domain

If we deal with multiple load signals acting on a component, the correlation of the signals becomes important. The basis for measures of load equivalence is defined in Section 3.3, where methods for the analysis of multi-input loads have been described. An extension of uniaxial amplitude domain methods is given by the RP method (see Beste *et al.* [25] or Dressler *et al.* [79]). Essentially, this method tries to cover the correlation by rainflow-analysing not only the given input signals, but also combinations of these signals that depend on their correlation. Thus the pseudo damage-based measures for the comparison can be extended and applied to multi-input loads. However, in that case we always have to deal with more than just one single number as we have at least one value for each input signal and for each additional combination.

Of course, additional methods such as phase plots or multi-input time-at-level can also be used. Finally, we refer to the work of Genet [99], where a systematic study of equivalence of loads based on two local damage criteria is presented. Some details of the method are given in Section 3.3.

4.2 Data Inspection and Correction

4.2.1 Examples and Inspection of Data

Frequently, measured data are incorrect in the sense that the data show some deviation from what was intended to measure, mostly due to problems during the measurement process or properties of the equipment. Besides statistical aspects (noise, etc.), which are discussed in a later section, there are essentially three types of disturbances, namely, offsets, drifts, and spikes. If present, these types of errors are such that they have to be corrected in order to get valuable results during later analysis.

We begin with some illustrating examples. Figure 4.2 shows a measured signal, representing the stress at the boogie of a train travelling from Oslo to Kristiansand, see Section 2.1.4. One can easily detect an offset and a more careful look also shows drifts during the segments of low activity.

Another example is shown in Figure 4.3, a strain signal measured at the arm of an excavator during work. The first plot represents the total data. Here, sections of different activity can be seen but no outstanding peaks can be identified. The second plot zooms into a specific segment, which shows a sequence of peaks. The third line zooms into another segment showing a single peak.

A more careful analysis of the data reveals that the sequence of peaks comes from an impact event, whereas the single peak in the third plot is erroneous (spike). However, the first peak of the impact sequence can hardly be distinguished from the peak in the third

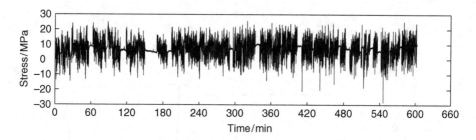

Figure 4.2 Drift and offset problems in a stress-signal calculated from a strain measurement

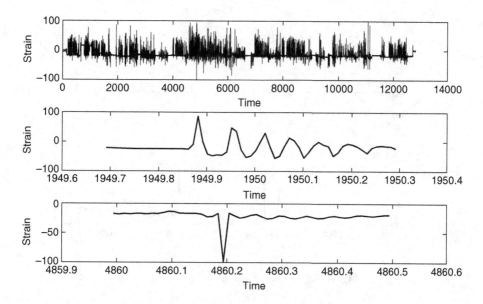

Figure 4.3 Possible spikes in a strain signal

plot due to the small sampling rate, which cannot accurately resolve the impact signal. This example shows that automated detection and deletion of such spikes are hard to achieve in general.

Figures 4.4 and 4.5 show a rather slowly changing cylinder stroke signal and an oil flow measurement. In the first case two different erroneous segments can easily be identified, whereas in the second case the peaks between 60 and 61 seconds are suspicious and need further attention.

Simple statistical measures such as extreme values, mean, standard deviation, RMS, etc., may help to check the data, but often, as some of the examples show, a visual inspection of the data is required to find possible problems and decide what to do. Unfortunately there are no reliable procedures which can be applied to correct the data in a fully automated way. Thus, this early part of the analysis is often very time-consuming. The next section gives some hints on how to find and correct problems.

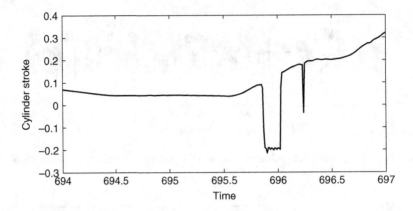

Figure 4.4   Possible spikes in a stroke signal

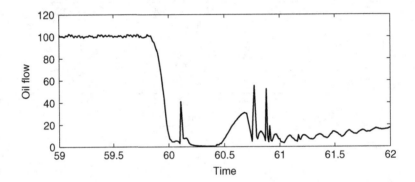

Figure 4.5   Possible spikes in an oil flow signal

4.2.2   *Detection and Correction*

For drifts and offsets there are a couple of robust methods available for correcting the data. As illustrated in Figure 4.2 simple offsets are easily found and corrected by calculating the mean and subtracting it from the signal.

A somewhat more sophisticated method is to fit a smooth curve y to the data x (for example, a linear function) and subtract it from the data to get the corrected signal $\hat{x} = x - y$. This eliminates offsets as well as linear drifts. For more complicated deviations, a piecewise linear fit could be used. Another approach is to apply frame-based averaging according to

$$y_i = \frac{1}{m+1} \sum_{k=i-m/2}^{i+m/2} x_k, \qquad \hat{x} = x - y, \qquad (4.1)$$

where m is called the window or frame length and controls the degree of smoothing. This parameter has to be adapted to the properties of the signal disturbances. If, for example, the drift comes from a rather slowly varying temperature (as is the case in Figure 4.2), a large window length m may be used.

Alternatively, high pass filters can be used (see Section 3.2) to eliminate drifts and offsets. Here the cutting frequency has to be adapted to the data and to the type of disturbance. A small cutting frequency eliminates an offset and slow variations, corresponding to a broad window in the frame-based approach. A large cutting frequency also eliminates higher frequent variations, corresponding to a narrow window in the frame-based approach.

Even though such methods could be applied automatically, the engineer first has to decide whether a drift or an offset has to be corrected at all or whether the suspicious segment is a consequence of a special event or property of the signal (like an offset of the vertical wheel load of a truck due to the weight or the result of loading/unloading the vehicle, e.g. a tipper truck).

Spikes are more difficult to find and eliminate. A spike might be restricted to a single point in time (see Figure 4.3) or to a certain sequence of points (see Figure 4.4). The incorrect values might well be separated from the correct physical range of values (very large or very small, see Figure 4.6) or might even lie within the correct range. In the former case, a simple threshold approach is suitable for finding the problematic sections, in the latter case more sophisticated methods are needed.

One possible approach is to fit a smooth curve y to the data x (for example, using a piecewise linear approximation based on moving averages similar to the drift elimination in formula (4.1)), subtract this from the data to isolate the rough part $r = x - y$, calculate the local variances v, and search for sections where this variance is above a certain threshold h:

$$r_i = x_i - y_i, \quad \text{with} \quad y_i = \frac{1}{m+1} \sum_{k=i-m/2}^{i+m/2} x_k, \tag{4.2}$$

$$v_i = \frac{1}{n+1} \sum_{k=i-n/2}^{i+n/2} (r_k - \bar{r}_k)^2, \quad \text{where} \quad \bar{r}_k = \frac{1}{n+1} \sum_{k=i-n/2}^{i+n/2} r_k \tag{4.3}$$

$$\text{eliminate or correct sample } i \text{ if } v_i > h. \tag{4.4}$$

This kind of approach is called a variance filter or statistical filter. A good choice of the parameters m, n, h used in this process (block sizes for moving averages and threshold), however, depends on both the type of data to be analysed and the type of problem that caused the incorrect values. As an example, consider the stroke signal in Figure 4.4. In contrast to the strain channels and due to the fact that the stroke is a low frequency channel, it may be corrected by a two-step variance filter. The first step is a filter with

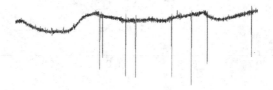

Figure 4.6   Some obvious spikes in a signal

block length $m \simeq 30$ for the mean and block length $n \simeq 10$ for the variance calculation. Spikes are detected and corrected if $v > 0.01$. This filter eliminates broader spikes (several points). The second step uses a filter with block length $m \simeq 10$ for the mean and block length $n \simeq 3$ for the variance calculation. Spikes are detected and corrected if $v > 0.0001$. This kind of filter eliminates the narrow spikes.

Instead of using average values (mean values) of some signal segment during the drift/offset correction or spike detection, it is also possible to use quantiles to check for irregular samples. In image processing the median filter is a well-known tool, which calculates the median value for a certain window of the input signal. The window is moved along the signal and the sequence of corresponding median values is interpreted as the "clean" signal. As the median of a set of samples is robust with respect to single spikes, the median filter is able to follow a step signal exactly.

In Figure 4.7 an example of the behaviour of the median filter is given. It can be seen that if the block length is too small, a spike might not be fully removed. If the block length increases, the spikes are removed, but there is also a reduction of the amplitudes in areas where apparently there is no spike. It is essentially this property (this is also true for moving average filters), that leads us to the two-step approach mentioned above, that is (a) calculate the deviation between the measured and a smoothed signal and (b) use this deviation to detect spikes. This procedure is conservative in the sense that a correction of the signal is made only if the deviation is significant. In all other cases, the signal remains untouched, whereas a moving average filter would round off the step.

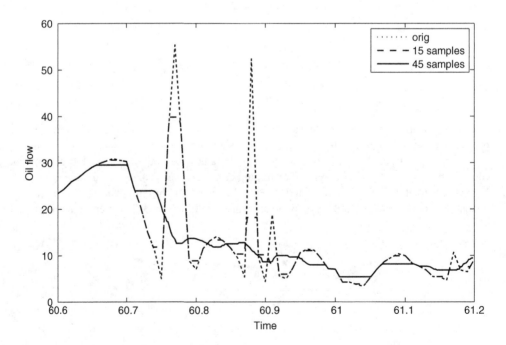

Figure 4.7   The oil flow signal and the results of median filters with different block lengths

4.2.2.1 Recommendations

Measurement campaigns in the vehicle industry often result in several hundred files, each containing between 50 and 200 channels. The length of the files range from a few kilometres to about 100 km. Consequently, a huge amount of data has to be checked and corrected if necessary before further analysis. As explained above, this can hardly be achieved in a fully automated way. However, a semi-automatic approach could be the following:

1. Calculate simple statistical values (min, max, mean, RMS, etc.) and check if these are reasonable.
2. Pick a few measurements including the suspicious ones with respect to the simple statistics and check the data using a suitable display tool.
3. If there are problems, apply the methods described above and tune the parameters to get reasonable results.
4. Insofar as the tuned methods gave good results for the test measurements, apply them to the entire data set.
5. Repeat the first step for the corrected data.

Of course this is not a recipe for solving all the problems, but it should at least solve most of them. For the remaining issues, there is no way round a separate treatment unless the corresponding channels and measurements can be skipped in a further analysis.

4.3 Load Editing in the Time Domain

This section deals with the manipulation of loads represented as time signals. The methods described below apply to measured data as well as results from a numerical simulation (e.g. multibody simulation).

In addition to what we present in the following, there is a large number of simple editing procedures which are important in practice. Examples are

- cut off unimportant sections (e.g. the driveway to the test track)
- fade in and fade out (improve stability of test rigs or numerical simulation)
- split data into several parts or concatenate several parts
- smooth transitions
- interpolation.

These basic methods are not specific to durability applications and will not be described in this guide.

The following presentation of the more durability-specific editing algorithms takes into account the three aspects of amplitude, frequency, and correlation domain, which have been discussed above.

4.3.1 Amplitude-based Editing of Time Signals

We begin the description of methods for editing time signals by emphasizing the amplitude domain. The corresponding methods preserve or modify specific properties in the amplitude domain and do not consider the frequency domain. The basic justification for these methods was given in detail in Section 3.1.3. These arguments apply here too. As a consequence, amplitude-based editing methods are applicable in case of loads acting on stiff structures or systems or on local stress or strain data.

4.3.1.1   1D Signals

For a single load signal the concept of rate independency can be applied directly leading to the most basic editing algorithms, namely the min-max filter and the hysteresis filter. Mainly for reasons of completeness, the simple **min-max-filter** is mentioned first. It checks for local extremes (minima and maxima) and deletes all points in monotonic sections of the signal. An example is given in Figure 4.8.

As can be seen, the method preserves all changes in load direction, no matter how small they are. The amount of data reduction of course depends on the signal, its sampling rate, and the amount of measurement noise. In this example (sampling rate 300 Hz), the amount of data reduces by a factor of 6. For very smooth signals this factor might be much more significant. This method is often used as a pre-processing step for more refined methods (see the hysteresis filter below), but it is valuable also, for example, for speeding up a quasi-static multibody simulation.

The **hysteresis filter** (also called the rainflow filter) has already been described in the discussion of rate independency, hysteresis models, and stress-strain paths. It eliminates hysteresis cycles smaller than a user-specified threshold and replaces them with a smooth signal. It works similar to the rainflow counting procedure and can easily be implemented as an on-line procedure. The basic hysteresis filter only checks the amplitude of hysteresis cycles. It can be extended to a more refined approach taking into account, for example, mean load effects or other damage calculation issues, see the PhD thesis by Hack [109] for an algorithmic description including mean load consideration. These extensions are not described in detail in this guide, because they require more detailed knowledge about damage calculation procedures, and their applicability to the specific situation needs to be justified. Figures 4.9 and 4.10 show some results of the basic hysteresis filter process.

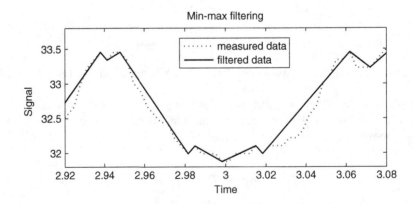

Figure 4.8   A measured vertical wheel force and the result of the min-max filter

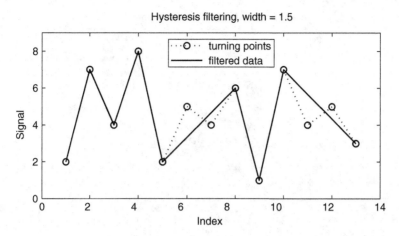

Figure 4.9 Result of the hysteresis filter applied to a simple example of discretized turning points (the vertical axis shows the bin numbers). The threshold (width of the filter) has been set to 1.5 times the width of a bin

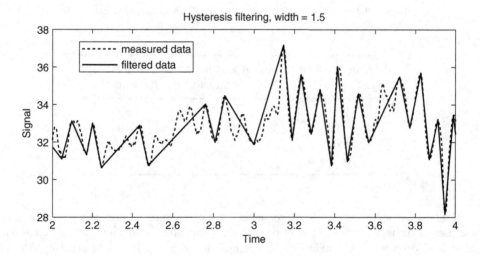

Figure 4.10 Part of the result of the hysteresis filter applied to vertical wheel force measurement

Example 4.1 (How Much Damage Gets Lost During Filtering?)

Of course, the amount of data reduction depends on the width of the filter as well as the properties of the signal (sampling rate, type of measurement, etc.). In case of the vertical force signal shown in Figure 4.10 (RFM is shown in Figure 4.11), the factor between original length and filtered length for different thresholds defined relatively to the maximum signal amplitude is given in the following table:

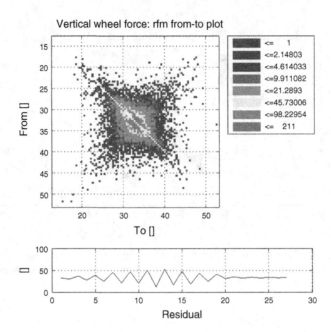

Figure 4.11   RFM and RES of the vertical wheel force measurement

threshold in % of largest cycle	signal length reduction factor	relative damage ($\beta=3$) in %
0.1	6	100
1.0	16	100
5.0	30	100
10	52	99

The third column of this table shows the amount of damage left after the filter process. In the pseudo damage calculation, a slope of 3, no mean stress correction, and no endurance limit have been used. The reduction in damage for increasing threshold naturally depends on the shape of the load spectrum of the signal and will be different for other signals.   □

Example 4.2 (How to Choose The Filter Level?)
Plots of the type shown in Figure 4.12 can be used to find the threshold, given the amount of damage reduction which can be accepted.

Figure 4.12 shows the cumulative damage per amplitude of the vertical force signal. In order to determine the maximum filter level (threshold) so that not too much damage is lost by the filter process, first decide on an acceptable amount of loss of damage, for instance, 10%. Then find the intersection of the curve with the 90% vertical line and read

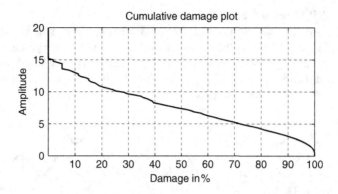

Figure 4.12   Plot of cumulative damage ($\beta=3$) per amplitude for the vertical wheel force

off the corresponding amplitude value. This is the maximum threshold you are looking for. The calculation naturally depends on the parameters of the damage calculation used for plotting the curve (slope = 3 in our example). The example will be continued below after describing the test pre-processing.   □

4.3.1.2   Multidimensional Loads

A basic requirement for all multidimensional editing methods is to preserve the correlation of the signals. If points are inserted or deleted in one of the signals, then the same number of points need to be inserted or deleted in the remaining channels.

The simplest multidimensional method is an extension of the min-max filter, which is often called **peak slicing**. For all samples, it checks if, at that point in time, there is a turning point in at least one of the signals. If not, that sample can be deleted from all signals. Otherwise that sample is kept in all signals. By construction, this method preserves the correlation and rainflow content of the individual signals. However, the examples given in Section 3.3 show that this algorithm might delete points which are not turning points in the given signals but turning points in some combination. Thus, the algorithm changes the multidimensional rainflow content (RP counting result).

This drawback can easily be avoided by a straightforward application of the methods proposed in Section 3.3: calculate additional load combinations (RP combinations or directions) of the form $L_p(t) = c_{p,1} \cdot L_1(t) + c_{p,2} \cdot L_2(t) + \cdots + c_{p,n} \cdot L_n(t)$, where the $L_i(t)$ denote the given load signals, and apply the peak slicing method to the set of signals and additional signal combinations. In doing so, we preserve the RP content of the signals while deleting all those monotonic parts which do not contribute to a hysteresis cycle in any of the signals or RP combinations.

This is a kind of min-max filter taking the RP concept into account. The next step now is to apply a hysteresis filter to all signals and the RP combinations and delete only those samples that do not contribute to a hysteresis cycle greater than a predefined threshold in any of the signals or RP combinations. This is the "RP extension" of the uniaxial hysteresis filter to the multidimensional domain, also called **RP filter**.

Example 4.3 (RP Filter for a 2D Noisy Sine Wave)
We take up the simple example used in Section 3.3.3 (two noisy sine waves with a phase delay of $45°$). Figure 4.13 shows the signals before and after the filter process. It can be seen that the length of the signals reduces to about 25% of the original length for the RP filter with additional directions and to 10% of the original length for the simultaneous rainflow filter without additional directions. This is the expected behaviour.

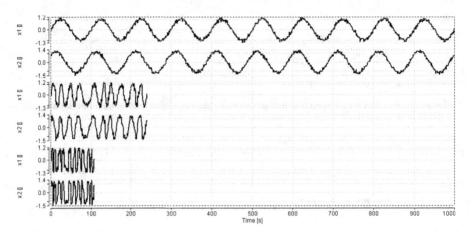

Figure 4.13 2D example for RP filter: The two top signals are the original time series. The next two signals are the results of the RP filter (12 additional directions) with a hysteresis filter width of 10% of the maximum amplitude. The last two signals are the result of the rainflow filter without taking additional linear combinations into account

To assess the quality of the shortened signals we have to compare the pseudo damage of the additional directions of both results. In Figure 4.14 the ratio of the pseudo damage for the shortened signals is shown. By definition of the filter algorithm, there is no difference in the signal itself (directions $[1, 0]$ and $[0, 1]$). However, there are directions where the damage of the simultaneous rainflow filter signal is between $1/3$ and $1/2$ of the damage of the RP-filtered signal.   □

Although inspired by the stress superposition formula of linear elastic analysis, the RP approach of load analysis is independent of the properties of the loaded component. Thus, the same is true for the RP-filter algorithm. This is achieved by the general procedure for calculating the coefficients $c_{p,i}$ as described in Section 3.3. However, it is easily possible to adapt the method to a specific application, if the FE results of the structure of interest are given. In that case, we can use a small number of combinations only, which are derived from the stresses of the corresponding static unit load cases at specific spots of interest. The potential for additional reduction of the load signal is large. In Jung *et al.* [130], an application to a passenger car knuckle is described. The procedure can be used to speed up numerical fatigue life estimations but also for deriving target loads for a rig test. Especially in the latter case, the corresponding time and cost savings can be considerable.

The RP filter does not take into account the frequency properties of the signal. Thus it should not be applied to loads acting on a structure capable of vibration. But it can be of value for a fatigue life calculation of such components if an FE-based modal superposition

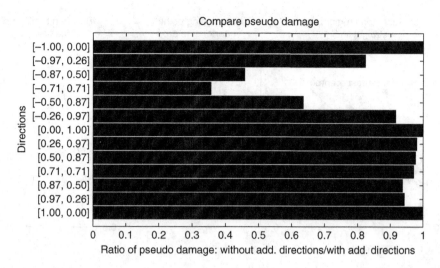

Figure 4.14 Ratio of the pseudo damage when filtering without and with additional directions

approach (see Section 5.3.2) has been taken. In this situation, we can apply it to the modal participation factors (MPF), which are the result of a modal transient FE analysis. This is justified because the dynamic behaviour of the component is already reflected by the MPF signals and the local stresses are just a linear combination of the form $\sigma(x, t) = \sigma_1(x) \cdot MPF_1(t) + \sigma_2(x) \cdot MPF_2(t) + \cdots + \sigma_m(x) \cdot MPF_m(t)$. Here, m denotes the number of modes to be considered and $\sigma_i(x)$ denotes the i:th stress mode shape.

4.3.1.3 Test Pre-processing

Once we have reduced a time signal to a sequence of turning points that still contains the correct rainflow cycles, we might want to use it as a target for a rig test. For instance, it could represent a force to be applied to a control arm in a uniaxial rig test. Since we are interested in an accelerated test, we do not want to use the original time scale of the signal before the filter process. Instead we turn the sequence of turning points into a new time signal simply by assigning a new sampling rate Δt to it.

Now we can check the usability of this signal at the test rig. We assume that there are limitations of the test rig concerning maximum allowed gradients m_{rig} and maximum frequency. The local gradient (slope) of the signal x is given by $m = \frac{|x(t_{i+1})-x(t_i)|}{\Delta t}$. If $m > m_{rig}$, then a certain number n of additional points is inserted between $x(t_{i+1})$ and $x(t_i)$ as sketched in Figure 4.15.

There are a variety of ways to calculate the additional points. Here, we follow the method implemented in LMS [147]. If both points are part of a monotone region (to the left of the figure), then the additional points are inserted along a straight line between $x(t_{i+1})$ and $x(t_i)$. If one of the points is a turning point (in the middle of the figure), the additional points are interpolated such that the stretched arc follows a "quarter sine wave". If both points are turning points, then the interpolation follows a "half sine wave". The number n of points to

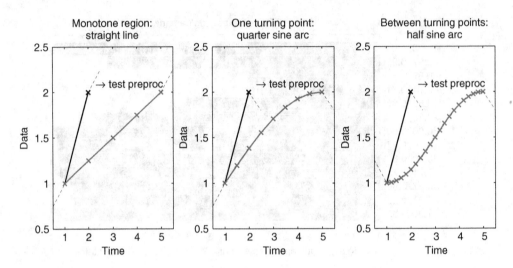

Figure 4.15   Sketch of the test pre-processing algorithm to reduce the slope

be inserted is calculated such that the slope after the interpolation is below the given limit m_{rig} and n is as small as possible. This is achieved by using the following formulas:

$$\text{monotone part: } n \geq \frac{|x(t_{i+1}) - x(t_i)|}{m_{rig}\Delta t} - 1 \tag{4.5}$$

$$\text{non monotone part: } n \geq \frac{|x(t_{i+1}) - x(t_i)|\pi}{2m_{rig}\Delta t} - 1 \tag{4.6}$$

The procedure is performed for all points $x(t_i), x(t_{i+1})$ of the signal. It ensures that the resulting signal satisfies the slope limit. By construction of the method, the rainflow content of the signal is untouched as it only modifies the rate of the signal, not the turning points.

Next, we have to extend test pre-processing to multiple input loads. Formulas (4.5 and 4.6) are applied to all signals in parallel to calculate the number of additional points for each signal. The largest of these numbers defines the number of points to be inserted for all signals between t_i and t_{i+1}. This simple approach ensures that the correlation of the multiple time signals is maintained during the test pre-processing. Although the RP content of the multiple signals has not been taken into account explicitly, there are almost no differences in the RP result before and after the pre-processing.

Example 4.4 (Test Pre-processing of a Short Demo Example)
Figure 4.16 shows a simple example containing nine discretized turning points.

Two different values for m_{rig} have been used to illustrate how the method works. The first two points of the signal are 2 and 7. The sampling rate is $\Delta t = 1$ and the limits for the slope are 3.5 and 2.5. Putting the numbers in the second formula (4.5) gives $n \geq \frac{|2-7|\pi}{2\cdot3.5\cdot1} - 1 = 1.24$ resp. $n \geq \frac{|2-7|\pi}{2\cdot2.5\cdot1} - 1 = 2.14$. Thus we insert 2 and 3 new points respectively in the first range of the signal. The calculation for the remaining ranges is similar.   □

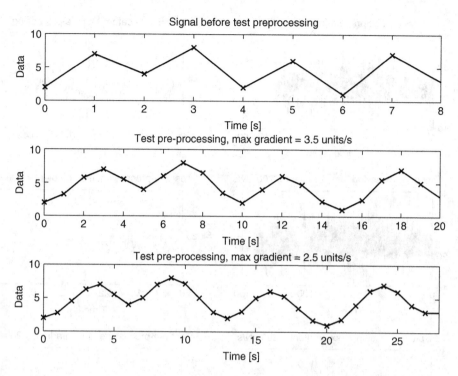

Figure 4.16 Application of the test pre-processing procedure to the simple example shown in the first plot. The second and third plots show the result of the method for maximum gradient of 3.5 units/s and 2.5 units/s, respectively

Example 4.5 (Test Pre-processing of a Vertical Wheel Force)

Figure 4.17 shows the result of the method when applied to the vertical wheel force measurement, which we have already used above to illustrate the hysteresis filter. Again we use two different values for m_{rig}.

The plots on the left display the entire signal in order to show the total length of the results. To the right, a detail at the beginning of the signal is plotted to give an impression of the interpolation algorithm. If the maximum gradient is limited to 700 kN/s, the test pre-processing algorithm increases the signal length by a factor of 3 approximately. If only 200 kN/s is allowed, the length increases by a factor of 8. The following table summarizes the combined effect of the hysteresis filter (threshold = 10% of largest cycle) and the preceding test pre-processing with allowed gradients of 200 kN/s and 700 kN/s, respectively.

	signal length reduction factor	relative damage $(\beta = 3)$ in %
TP signal after hysteresis filter	52	99
TP signal + gradient limit 700 kN/s	18	99
TP signal + gradient limit 200 kN/s	6	99

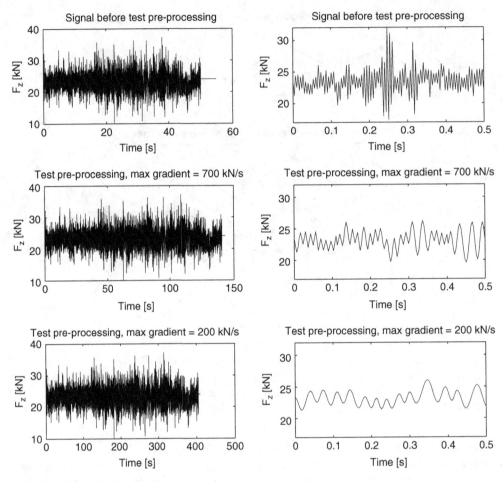

Figure 4.17 Application of the test pre-processing procedure to the vertical wheel force measurement shown in the top plots. To the left is the whole time range, to the right a detail. The second and third rows show the result of the method for a maximum gradient of 700 kN/s and 200 kN/s, respectively

Of course, these numbers depend on the data the method is applied to and will in general differ from the examples shown here.   □

While test pre-processing is usually required for filtered data, it may occasionally be necessary even if one is working with the original measurements. This depends on the severity of the measured signal and the capabilities of the rig. In Bremer *et al.* [36], an application of the RP filter and the test pre-processing is described. In our example we have a bound on the gradient of 200 kN/s and 700 kN/s, respectively. The maximum gradient of the measured signal, however, is about 900 kN/s, that is the rig cannot follow the signal exactly. Instead, sharp peaks would probably be smoothed by the rig leading to a reduction of the largest cycles. Thus it is important to check this, prior to the application of the signal on the rig.

In Klätschke and Schütz [135] there is a description of another method called "Simultanverfahren", which combines a multi-input hysteresis filter with a test rig adjustment similar to the one described here. Essentially, the method is equivalent to a successive application of the RP filter and test pre-processing.

4.3.1.4 Extrapolation of a Turning Points Signal

A measured signal often represents only a very short part of the design life. When performing variable amplitude tests, it is customary to use a measured load history, and repeat this load block until failure. This has the drawback that only the cycles in the measured signal will appear in the extrapolation, even though other cycles are also possible. This can be particularly critical for the most damaging large amplitude cycles. A methodology presented in Johannesson [121] is to repeat the measured load block, but modify the highest maxima and lowest minima in each block. The random regeneration of each block is based on statistical extreme value theory. An example showing the principle of the method is presented in Figure 4.18, where three repetitions of a measured block are compared to three extrapolated blocks. Note that only the maxima above a high load level u_{max} and the minima below a low load level u_{min} are randomly regenerated. Also note that the extrapolation is performed on the turning points of the signal, and not on the originally sampled signal. This also gives the opportunity to remove irrelevant small cycles before the extrapolation.

The theoretical base for the method is the so-called Peak Over Threshold (POT) technique in statistical extreme value theory, where the extreme excesses over a high level u are modelled, see Appendix B.3 for details. We then have the approximation for the excess $Z = Max - u \in Exp(m)$, with cumulative distribution function

$$F(z) = 1 - \exp(-z/m), \qquad m = \text{"mean excesses over } u\text{"} \qquad (4.7)$$

where the mean is estimated from the observed excesses $z_i = Max_i - u, i = 1, \ldots, n$, as

$$m = \frac{1}{n} \sum_{i=1}^{n} z_i. \qquad (4.8)$$

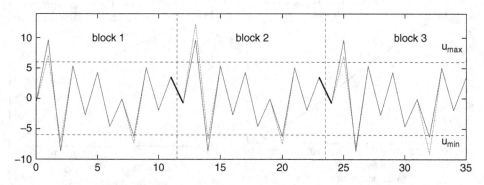

Figure 4.18 Three repetitions of a measured block (solid line), compared to three extrapolated blocks (dotted line). The horizontal dashed lines represent the threshold levels, $u_{min} = -6$ and $u_{max} = 6$, where the extrapolation starts

4.3.1.5 Algorithm

The generation of a k-fold extrapolated signal (sequence of turning points) can be performed in the following stages.

1. Start with a time signal.
2. Extract the turning points of the time signal (small cycles should be removed by using a rainflow filter).
3. Choose threshold levels u_{min} and u_{max} for the POT extrapolation. The choice of levels is discussed in Johannesson [121]. Extract the excesses under u_{min} and the excesses over u_{max}.
4. Estimate the mean excesses m_{min} and m_{max} under and over the thresholds u_{min} and u_{max}, respectively.
5. Generate an extrapolated load block by simulating independent excesses as exponential random numbers. Replace each observed excess under u_{min} by a simulated exponential number with mean m_{min}, and each observed excess over u_{max} by a simulated exponential number with mean m_{max}.
6. Repeat step 5 until k extrapolated load blocks have been generated.
7. The k-fold extrapolated signal is obtained by putting the generated load blocks after each other.

From the k-fold extrapolated signal, we can perform a rainflow count to obtain the k-fold extrapolated rainflow matrix or load spectrum, see Figure 4.19 for examples.

4.3.2 Frequency-based Editing of Time Signals

Frequency-based editing means modifying a signal such that its spectral content (e.g. PSD) changes in a certain predefined way. Often, the goal is to eliminate certain sections (bands) of the spectrum of the signal. Examples are low-pass filtering to eliminate high frequent

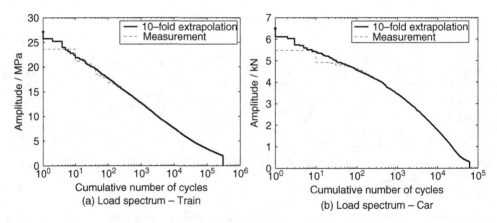

Figure 4.19 Load spectra of ten-fold extrapolated signals, compared to 10 repetitions of the measured signal

noise or high-pass filtering to get rid of a drift. The main idea is to transform the signal into the frequency domain using the continuous Fourier transform or the discrete FFT algorithm, perform the manipulation on the "spectrum" and apply the back transformation to return to the time domain. Some methods actually work like that (Fourier filter), others (FIR or IIR filters) are developed in the frequency domain but implemented in the time domain. They do not perform a Fourier transformation. Instead they use the samples $x_k, x_{k-1}, x_{k-2}, \ldots$ of the signal and the samples $y_k, y_{k-1}, y_{k-2}, \ldots$ of the filtered signal in order to calculate the next sample y_{k+1}. This enables the construction of on-line algorithms.

As the amplitude domain is not explicitly considered during the development of the methods, the rainflow content will not be preserved exactly. However, it is often not hard to predict the impact of filter algorithms to the amplitude domain. A low-pass filter will eliminate small hysteresis cycles and decrease the amplitude of the remaining cycles somewhat, a high-pass filter eliminates offsets and drifts and thus modifies the mean load of cycles. Anyway, the rainflow content after the filter process can and should be compared to the content before filtering.

Most of the frequency-based editing methods do not change the length or duration of a signal. This is in contrast to most of the amplitude-based methods.

Frequency-based methods, digital filters, and related topics are covered in a large number of textbooks. We recommend Oppenheim and Schafer [179], which covers the subject from an electrical engineering point of view, Priestley [192] deals with the mathematical theory, and Kay [134] takes a middle ground between mathematics and engineering applications.

4.3.2.1 Digital Sampling and Signal Pre-processing

As a digital signal is recorded at discrete time points, it contains only a limited amount of information about the frequencies of the original, analog signal. This notion is made precise by the Nyquist-Shannon Sampling Theorem discussed in Appendix C.3. The theorem states that a digital signal permits a unique reconstruction of the original, provided that the sampling frequency is greater than twice the bandwidth of the signal. Note that for the purpose of pre-processing, we only deal with the case of low-pass filtered signals where the bandwidth is the maximal frequency present in the signal before it is digitized.

According to the sampling theorem, a digital signal with sampling rate f_s only contains unambiguous information about frequencies below the so-called *Nyquist-frequency* $f_N = \frac{f_s}{2}$. Thus, the frequency-based analysis of a digital signal requires that this rate be sufficiently high:

- The phenomena of interest should occur at frequencies below the Nyquist-frequency. For example, if we are interested in the behaviour of a mechanical structure where the largest eigenfrequency of interest is 100 Hz, the loads acting on the structure need to be recorded with a rate exceeding 200 Hz.
- Before digital recording, the signal should be processed by an analog low-pass filter with a cutoff frequency below f_N (the so-called *anti-aliasing filter*) to ensure that the presence of higher frequencies does not distort the results.

If frequencies equal to or higher than the Nyquist-frequency are present in the signal prior to digital sampling, they cannot be distinguished from frequencies below f_N after sampling, an effect known as *aliasing*:

Example 4.6 (Aliasing and Reconstruction)

Consider the simple analog signals

$$x_1(t) = \sin(4 \times 2\pi t) \quad x_2(t) = -\sin(6 \times 2\pi t) \quad x_3(t) = 0.5(x_1(t) + x_2(t)). \quad (4.9)$$

If t is measured in seconds, the first sine curve has a frequency of $f_1 = 4$ Hz, and the second $f_2 = 6$ Hz. Now assume that we sample from these signals with $f_s = 10$ Hz ($\Delta t = 0.1$ s). The Nyquist-frequency is $f_N = 5$ Hz, so according to the sampling theorem, only x_1 can be reconstructed uniquely.

Figure 4.20 shows the signals in the interval from 0.5 to 2.5 seconds. All three continuous-time signals have the same representation in discrete-time as the three curves coincide at the sampling points. Thus, all signals will lead to the same continuous reconstruction. As we know that x_1 can be restored perfectly, all three signals will in fact be reconstructed as x_1. If we use an analog low-pass filter with a cut-off frequency of f_N before recording the digital sample, we obtain better results in terms of the low frequencies:

Original signal	Analog Filter	Digital signal	Reconstruction
$x_1(t)$	none	$x_1(t_k)$	$x_1(t)$
$x_2(t)$	none	$x_1(t_k)$	$x_1(t)$
$x_3(t)$	none	$x_1(t_k)$	$x_1(t)$
$x_1(t)$	low-pass (5 Hz)	$x_1(t_k)$	$x_1(t)$
$x_2(t)$	low-pass (5 Hz)	0	0
$x_3(t)$	low-pass (5 Hz)	$0.5x_1(t_k)$	$0.5x_1(t)$

Note how frequencies equal to or greater than the Nyquist-frequency appear as lower frequencies after discretization. For example, $f_2 = 6$ Hz is indistinguishable from its *alias* $f_1 = 4$Hz. In fact, the signals

$$-\sin(16 \times 2\pi t), -\sin(26 \times 2\pi t), \ldots, \sin(14 \times 2\pi t), \sin(24 \times 2\pi t), \ldots \quad (4.10)$$

all appear as x_1. Depending on the actual frequency, the sine may be phase shifted w.r.t. x_1 (thus the minus signs), but the apparent frequency is always 4 Hz. By applying the anti-aliasing filter, we lose all information about high frequencies, but the contribution of frequencies below f_N can be identified correctly. □

For the remainder of this section, we will assume that the available signal satisfies the conditions of the Nyquist-Shannon Sampling Theorem, meaning that it is a digital recording of an analog signal that was treated by an anti-aliasing filter.

4.3.2.2 Transformation of Sampling Rate

Some applications require that a digital signal sampled with a frequency f_1 be transformed to a time series with a different rate f_2 but containing the same information. This might be due

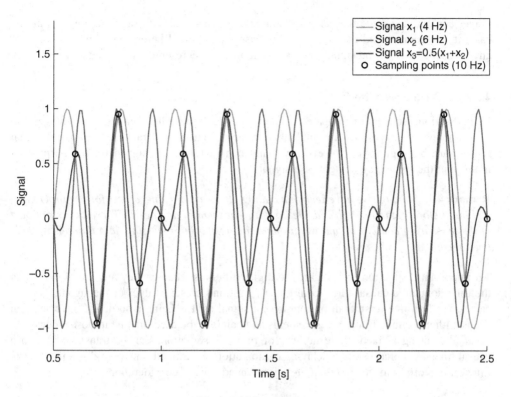

Figure 4.20 Aliasing

to the need to synchronize it with other measurements, or as a means of data compression. If $f_1 > f_2$, we speak of downsampling, otherwise of upsampling. During downsampling, the number of data points is reduced, otherwise it increases.

If the signal with frequency f_1 satisfies the conditions of the Nyquist-Shannon Theorem, the original continuous time signal can be reconstructed as shown in Appendix C.3 as

$$x(t) = \sum_{k=-\infty}^{\infty} x_k \frac{\sin(\pi(t - t_k)f_1)}{\pi(t - t_k)f_1} \tag{4.11}$$

with time points $t_k = \frac{k}{f_1}$ and discrete measurements $x_k = x(t_k)$. Using this relationship, the transformation to a signal $\tilde{x}_k$ with sampling rate f_2 is straightforward:

$$\tilde{t}_k := \frac{k}{f_2} \qquad \tilde{x}_k := x(\tilde{t}_k) = \sum_{l=-\infty}^{\infty} x_l \frac{\sin(\pi(\tilde{t}_k - t_l)f_1)}{\pi(\tilde{t}_k - t_l)f_1}. \tag{4.12}$$

If signals need to be synchronized, it may be necessary to shift the origin as well, which is done simply by adding a constant offset to each $\tilde{t}_k$.

For upsampling, applying Equation (4.12) is completely sufficient. However, downsampling becomes problematic if the Nyquist-frequency $\frac{f_2}{2}$ for the new signal is less than or equal to twice the actual bandwidth of the signal. If this happens, the resampled signal no longer contains the same frequency information as the signal with sampling rate f_1. In this case, it is recommended that a digital low-pass filter is used before resampling to remove any frequencies from the signal that can no longer be represented at the reduced rate f_2.

4.3.2.3 Frequency Filtering

In this section, we want to compare different types of digital filters and their advantages and disadvantages. In principle, a digital filter is any function that transforms a digital signal x_k into another signal y_k. However, we only consider filters designed for the purpose of modifying the frequency content of a signal.

Remark 4.1 *Although we have seen that analog filters are often necessary for pre-processing a signal before it can be digitized, we do not treat them in any detail. These filters are part of the measuring equipment and we are more interested in processing data that has already been recorded.*

An important distinction with regards to digital filters is whether they are causal or not. In the time domain, a causal filter requires only past and present values of the signal. Thus, it can be used on-line to smooth or compress a signal as it is being recorded. Conversely, an acausal filter requires future measurements and can only be used off-line in post-processing. As a filter using a fixed, finite number of future observations can be transformed into a causal filter by a phase shift – delaying computations by a fixed time period – only acausal filters that work with the entire (finite) signal need special consideration.

Fourier filter

The Fourier filter is an example of an acausal filter. It works by transforming the complete signal into the frequency domain using the discrete Fourier transform (DFT), setting all $\hat{x}_k$ belonging to frequencies outside the desired range to 0, and transforming the reduced spectrum back via IDFT (see Appendix C.4 for a description of DFT/IDFT). The major advantage of this filter is that there is no phase shift, that is, the phase information for all frequencies of interest is retained.

Theoretically, the Fourier filter is also an ideal filter, for example, a band-pass filter acting on the Fourier series representation has an abrupt cut-off at the specified bandwidth and does not distort the amplitudes of lower frequencies. However, this is only correct if the series coefficients are known exactly. Using DFT, we assume that the signal is periodic and was observed for an integer multiple of its period. Unfortunately, any violation of this assumption leads to the leakage phenomenon discussed in Section 3.2.2.

Example 4.7 (Leakage)

Figure 4.21 shows parts of two periodograms calculated via DFT from the signal

$$x(t) = \sin(2\pi t) \tag{4.13}$$

which is simply a sine wave with a frequency of 1 Hz. The sampling rate was $f_s = 10$ Hz in both cases, so the Nyquist-frequency $f_N = 5$ Hz is sufficiently high to prevent aliasing. However, the sampling interval was taken as either 10 s or 10.2 s, and only the former is an integer multiple of the true period of 1 s.

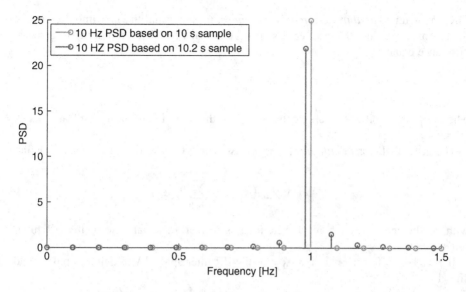

Figure 4.21 Leakage

As can be seen in the graph, the periodogram based on 10 s of data attributes all of the signal energy to the proper frequency of 1 Hz. This is not true for the second data set, where 1 Hz is not even one of the Fourier frequencies and the energy is distributed among frequencies in the neighbourhood of 1 Hz. □

One way to interpret leakage is to note that the periodic signal obtained by joining infinitely many copies of the analog original has a jump discontinuity and phase shift at multiples of the observation period. And since DFT yields a series expansion for the discretized version of this infinite continuation, it introduces other frequencies than those present in the original signal. The windowing techniques discussed in Section 3.2.2 are one way to reduce leakage as they smooth the signal before applying DFT, such that the values at the beginning and end coincide.

In practice, most time series are not really periodic, either due to noise or occasional shifts in the behaviour of the system. And even if there is a dominant periodic component, the period length may not be known exactly. Thus, we also consider other filter designs that are less sensitive to leakage:

FIR and IIR filters

FIR and IIR filters are causal designs implemented in the time domain, meaning that they work directly with the signal. They can, however, be specified in terms of their effect on the PSD. Both filter types are linear, time-invariant transformations of the original signal, but they are distinguished by their response to the unit impulse

$$x_k = \begin{cases} 1 & \text{if } k = 0 \\ 0 & \text{else} \end{cases} \tag{4.14}$$

FIR stands for *finite-duration impulse response*, meaning that the unit impulse affects only finitely many values of the filtered signal. A causal FIR filter of order q is given by the difference equation

$$y_k = \sum_{j=0}^{q} b_j x_{k-j} \tag{4.15}$$

where b_j are arbitrary coefficients. If x_k is the unit impulse, it is clear that $y_k = 0$ for $k > q$.

The general IIR or *infinite-duration impulse response* filter of order (p, q) is defined as

$$y_k + \sum_{j=1}^{p} a_j y_{k-j} = \sum_{j=0}^{q} b_j x_{k-j} \tag{4.16}$$

with coefficients a_j and b_j. The FIR filter is in fact a special case of the IIR filter with order $p = 0$.

If $p = 1$, $q = 0$, $|a_1| < 1$, and $b_0 = 1$, the IIR filter applied to the unit impulse yields the signal

$$y_k = -a_1 y_{k-1} + x_k \iff y_k = \begin{cases} 0 & \text{if } k < 0 \\ (-a_1)^k & \text{else} \end{cases} \tag{4.17}$$

Thus, the impulse affects infinitely many future values y_k, although its impact decays at an exponential rate. Note that for $|a_1| > 1$, the filter would continue to amplify the impulse over time, leading to instable behaviour. Thus, we need some notion of stability to design useful IIR filters:

A filter is said to be *BIBO stable (BIBO = 'bounded input, bounded output')*, if and only if it transforms every bounded signal into another bounded signal. A signal x_k is bounded if $|x_k| \leq C$ for some positive constant C and all time instants k. In the case of an IIR filter, an equivalent condition for BIBO stability is that all (complex) roots of the polynomial

$$a(z) = 1 + \sum_{j=1}^{p} a_j z^j \tag{4.18}$$

lie outside the unit circle, that is, their absolute value is greater than 1. Note that FIR filters (as well as Fourier filters) are always BIBO stable.

Remark 4.2 *FIR and IIR filters are closely related to ARMA models for time series (see Section 4.5.2). To be precise, a causal, stationary ARMA(p, q) process is equivalent to a BIBO stable IIR filter of order (p, q) applied to a white noise signal, while an FIR filter yields an MA(q) process.*

The effect of an FIR or IIR filter in the frequency domain can be seen by taking the DFT of Equation (4.16) and solving for $\hat{y}_k$:

$$\hat{y}_k = \hat{x}_k H\left(e^{-\frac{2\pi i k}{N}}\right) \qquad H(z) = \frac{\sum_{j=0}^{q} b_j z^j}{1 + \sum_{j=1}^{p} a_j z^j}. \tag{4.19}$$

This is due to the fact that the DFT of the time-shifted signal is

$$x_{k-j} \xrightarrow{\text{DFT}} e^{-\frac{2\pi ijk}{N}} \hat{x}_k. \tag{4.20}$$

As the PSD of the filtered signal y_k evaluated at the kth Fourier frequency $f_k = \frac{k}{N} f_s$ is proportional to the absolute value of the Fourier transform $\hat{y}_k$ squared, we obtain

$$\hat{P}_y(f_k) = \hat{P}_x(f_k) \left| H\left(e^{-\frac{2\pi ik}{N}}\right) \right|^2. \tag{4.21}$$

If we are given a desired gain (amplification factor) for the amplitudes of all frequencies f, we can design an appropriate FIR or IIR filter by choosing the order (p, q) and coefficients a_j, b_j of the continuous function

$$H(f) = H\left(e^{-\frac{2\pi if}{f_s}}\right), \quad f \in [0, f_N) \tag{4.22}$$

such that $|H(f)|$ approximates that gain. Here, we use the fact that the amplitude for a single frequency in the signal is proportional to the square root of energy. Note that $|H(f)|^2$ is the frequency response function (FRF) of the IIR filter, as the latter is a special case of a linear time-invariant system (see Section 3.2.4).

Example 4.8 (Designing a Low-Pass IIR Filter)

An ideal low-pass filter with cut-off frequency $f_c < f_N$ would require

$$|H(f)| = \begin{cases} 1 & f \in [0, f_c] \\ 0 & f > f_c \end{cases} \tag{4.23}$$

This relationship cannot be satisfied exactly by a function of the form (4.22), as H is continuous and can only be constant on an interval if it is constant overall. In particular, if H is not zero everywhere, it has only finitely many zeros.

In practice, one has to relax the constraints on the filter somewhat by prescribing tolerances, for example

- $|H(f)| \in [0.95, 1.05]$ if $f \in [0, f_c]$.
- $|H(f)| \leq 0.01$ if $f \geq 1.2 f_c$.
- $|H(f)|$ is strictly decreasing in $(f_c, 1.2 f_c)]$.

This filter has a passband of frequencies $[0, f_c]$ with amplitudes distorted by no more than 5%, a stopband of frequencies $[1.2 f_c, f_N)$ with amplitudes reduced to at most 1%, and a transition zone $(f_c, 1.2 f_c)$ of undesirable frequencies that are not fully removed by the filter. □

As the example shows, it is usually impossible to fit an IIR filter exactly to specifications. However, the approximation can always be improved by increasing the number of coefficients. There are various algorithms for filter design intended to optimize certain properties of the filter given a fixed order (p, q). For example, Butterworth-type filters are intended for minimal distortion of the amplitudes in the passband, while Chebyshev-type filters have faster roll-off, which means they decay more quickly in transition.

We will not discuss the details of filter design here, as these are rather involved, and the methods are part of most software packages for signal processing. It is enough to keep in mind that the choice of filter should be adapted to the task at hand, as there is a trade-off between signal quality and computational effort.

However, we need to address another important point here: FIR and IIR filters generally change the phase for each frequency in the signal. This can be problematic if the data is later to be used for other purposes than a purely frequency-based analysis. For example, the rainflow-count of the filtered signal can differ markedly from the original. A simple trick to turn an arbitrary FIR or IIR filter into one with no phase-shift is this:

> Take the filtered signal, reverse it, filter it again with the same filter, and reverse the output.

This way, the phase-shift introduced by the filter will cancel itself. Unfortunately, the resulting algorithm is no longer causal. The IIR filter requires the entire (finite) signal to work, whereas the FIR filter uses q future values of the forward-filtered signal. Thus, an FIR filter can still be used in an on-line context by introducing a delay, while an IIR filter cannot.

Remark 4.3 *It is possible to derive IIR and FIR filters with no phase-shift by imposing additional conditions on the coefficients a_j and b_j, but it can be shown that the resulting filters are by necessity acausal and of the same type as above.*

The properties of FIR and IIR filters can be summarized as follows:

Stability FIR filters are always BIBO stable, while IIR filters require certain conditions (see Equation (4.18)).

Phase shift Neither filter type can be realized as a causal filter without a phase shift, but the zero-phase FIR filter only requires a finite delay.

Filter order An FIR filter tends to require more coefficients than an IIR filter of the same quality. As the impact of past observations on a stable IIR filter decays exponentially fast, it is in fact possible to approximate an IIR filter with a high-order FIR filter to any desired precision.

Error propagation Since the FIR filter only uses a finite number of past observations, measurement errors do not propagate to infinity as they do for an IIR filter. However, if the latter is stable, the impact of any single error will become negligible over time.

Eliminating offset, drift, and noise After discussing common types of frequency filters, we can now look at their application in removing certain kinds of errors from the data:

The *offset* of a signal is its mean amplitude, or equivalently the energy contributed by a frequency of 0 in the Fourier series representation. While the magnitude of the offset is of interest in many applications, it is usually desirable to separate the signal into a base level (the offset) and fluctuations around that level. For example, for load data, one wants to distinguish between static and dynamic stress. During postprocessing, removing the offset from a signal is very easy: just calculate the mean and subtract it from all observations. However, in an on-line setting, it may not be possible to determine the offset of a highly dynamical process accurately until many observations are available. In this situation, a high-pass filter can be used to remove the offset.

Similar to the offset, the *drift* of a signal is the contribution of low frequencies. Often, these are errors introduced by the measurement device, e.g. integrated noise or a reaction to gradual changes in the environment (ambient temperature, etc.). Even if the drift is indeed a part of the observed process, its frequencies are usually too low to be of interest. For load data, this would be the case if they are far below the first eigenfrequency of the structure. Given that the contribution of low frequencies to the signal is also hard to estimate accurately from a short sample, it is common practice to remove any frequencies well below the range of interest using a high-pass filter.

The term *noise* refers to measurement errors that are truly random in nature. A common assumption is that of *white noise*, where the energy is distributed evenly across the entire frequency spectrum. The anti-aliasing filter will take care of very high frequency noise before the signal is digitized, and a high-pass filter for drift also takes care of slowly varying noise. If it is clear in advance that the interesting frequencies for the analysis are well below f_N, a digital low-pass can be applied to the data, which also permits further compression via downsampling. Unfortunately, noise occurring at the same frequencies as the process under observation cannot be removed by filtering. In this case, the sensor needs to be sufficiently accurate.

The overall recommendation is to use a digital band-pass filter to remove some of the measurement errors. This requires information both about the precision of the sensor as well as the properties of the structure under observation. And this knowledge should be used already when planning the measurement campaign. Otherwise, the recorded data may be insufficient for the analysis.

Example 4.9 (Digital Band-pass Filter)
Consider the signal shown in Figure 4.22:

$$x_k = 10 + 0.01t_k + \sin(2\pi t_k) + \epsilon_k, \quad k \in \{0, \dots, 999\}. \qquad (4.24)$$

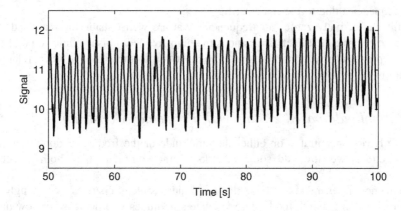

Figure 4.22   Signal with offset, drift, and noise

The time grid is $t_k = k \times 0.1s$ and ϵ_t is uncorrelated Gaussian white noise with standard deviation $\sigma = 0.2$. Thus, we have a sine wave with an offset, distorted by a linear drift and additional measurement errors across the frequency spectrum.

Assume that we are only interested in frequencies between 0.5 and 2.0 Hz. Figure 4.23 compares the pure sine wave to the last 10 seconds of the signal after applying an appropriate band-pass filter (= combination of a Butterworth high-pass and low-pass filter, each of order 10). As can be seen, the filter has removed the offset and drift, but not all of the noise.

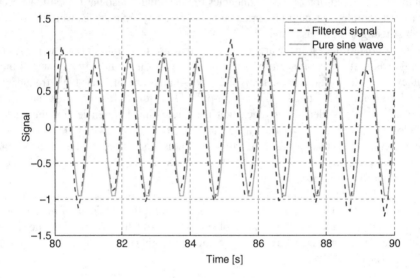

Figure 4.23   Reconstructed signal (part)

Figure 4.24 shows the periodograms for the noisy signal, filtered signal, and pure sine wave from 0 Hz to 2.5 Hz. The y-axis has been truncated, as the dominant frequency of 1 Hz registers as a very large peak in all three graphs. There is no leakage as the length of the sampling interval is an integer multiple of the period length 1 s. Note that the original signal has a second peak at 0 Hz due to the offset, as well as comparatively large energies at low frequencies due to the linear drift.

The filter effectively zeroes all frequencies that are well outside the passband, which is sufficient to remove the offset, drift, and some high-frequent noise. While some of the noise inside the passband is amplified, its total contribution to the signal is small relative to the contribution of the sine wave.   □

4.3.3 Amplitude-based Editing with Frequency Constraints

While we have concentrated on either the amplitude or the frequency domain in the latter two subsections, we now add some remarks on the combination of both aspects during editing.

If frequency filtering is applied, the amplitude content changes accordingly without explicit control. In general, the hysteresis cycle amplitudes will decrease by low pass filtering because of the smoothing effect. To some extent, this is a desired effect (elimination of

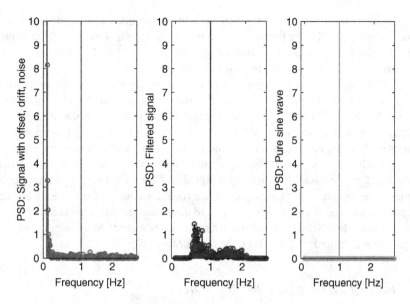

Figure 4.24   Power spectral densities

noise), although it is usually difficult to decide what is a good cut-off frequency. The degree of alteration of the amplitude domain needs to be controlled implicitly by checking these properties after the frequency filtering. If necessary, the filter parameters (cut-off frequency, etc.) have to be modified accordingly.

If the test pre-processing method (see Section 4.3.1) is applied as the final step of the amplitude-based editing process, only the maximum signal frequency, not the shape of the frequency spectrum, can be controlled by assigning a suitable sampling rate Δt. If we want to have more control over the frequency properties of the resulting signal, we have to reduce the amount of compression induced by the min-max or hysteresis filter algorithms (amplitude filter).

A rather simple extension of an amplitude filter is given by the following procedure:

- Run the amplitude filter, but only mark (do not delete) samples in the signal (1 means delete the sample, 0 means keep the sample).
- Partition the signal into segments such that each either contains only samples marked for deletion or samples marked for keeping.
- Compress the signal according to one of the following rules:
  1. Delete the samples corresponding to a deletion segment only if it is long enough (minimum length of deleted segments).
  2. If a sample is marked to be kept, also mark a certain neighbourhood to be kept (minimum length of kept segments).

The reduction achieved by the compression is controlled by the two parameters: the minimum length of deletion segments or the minimum length of kept segments. The parameters can be expressed in number of samples or in seconds. If these length parameters are set to a single sample, then the algorithm is identical to the amplitude filter we started with and we have the highest possible compression. If the minimum length is increased, the compression reduces considerably and the algorithm behaves more and more like a manual deletion of segments based on a visual inspection by an engineer.

The method can be enhanced even further by controlling the frequency properties (PSD) of the segments to be potentially deleted. For the calculation of the PSD of a segment, a certain neighbourhood is included if the segment is too short. Then the segment is deleted only if the amount of energy of the PSD within a predefined frequency band is not too large. By that procedure, we can make sure that no energy is lost within a certain frequency band.

The approach can easily be extended to a multi-input setting. In order to do this, one has to calculate the digital channel for each input signal and combine these by an 'and-operation': the value of the combined digital channel is one (deletion) only if the value for all individual digital channels is one. The remaining steps are then based on the combined digital channel. The extension of the PSD-based control is also straightforward. An implementation of these techniques can be found in LMS Tecware (see LMS [147]).

4.3.4   Editing of Time Signals: Summary

Various methods for editing time signals have been presented. The goal of editing in the time domain is either to accelerate rig tests or numerical calculations or to adapt the signal, for example, with respect to its frequency properties. In the following, the most important methods and their applications are briefly summarized.

1. Prior to editing a signal, check whether you need to preserve the rainflow domain, the frequency domain, the correlation domain, or a combination of these three properties. This decision depends on the system or component the loads will be applied to. Select the method or the combination of several methods accordingly.
2. The min-max and the hysteresis filter are rate-independent methods taking the amplitude domain into account. The min-max filter preserves all cycles exactly. The hysteresis filter is also called rainflow filter. It preserves the rainflow domain above the filter level and the deletes all those parts of the signal, contributing to small cycles only. The frequency domain is not taken into account. The shape of the frequency spectra will change accordingly.
3. Finding an appropriate threshold for the hysteresis filter: Calculate the cumulative damage per amplitude curve and determine the threshold by intersecting this curve with the vertical line corresponding to the acceptable loss of damage (see Section 4.3.1). An even stronger acceleration (reduction of the number of samples) can be achieved if you set the threshold such that you end up with, say, 50% of damage and repeat the signal twice in order to compensate for that loss of damage.
4. Multi-input extensions of the hysteresis filter exist. One is called the RP filter and is based on the RP concept described in Section 3.3. It preserves the correlation of the signals and the rainflow domain above the threshold. It does not take the frequency properties into account. It is particularly helpful for the acceleration of rig tests of stiff structures like a knuckle or a control arm. Another method for pre-processing multi-input loads that combines filtering and test pre-processing is the "Simultanverfahren" described in Klätschke and Schütz [135].

5. The frequency filter methods do not reduce the length of the signals. Instead, they are designed to modify the frequency spectra (PSD) of a signal in a well-defined way. Such methods can be helpful for the correction of data (high-pass filters for eliminating offset and drift or low-pass filters for eliminating high frequent noise). These methods modify the amplitude domain of the signals. Typically, low-pass filters reduce the amplitude of large cycles, high-pass filters modify the mean load of the cycles.

6. There are a couple of different frequency filter methods. An important group of methods are the FIR and IIR methods, which can be implemented as on-line procedures. Thus they are suitable for on-board data processing. The most common tasks like low-pass, high-pass, and band-pass filtering are easily achieved with these methods. By construction, they introduce a small phase shift. This can be eliminated by filtering the signal a second time in backward direction. However, this is no longer an on-line algorithm.

7. The loads acting on large structures or subsystems capable of vibration must not be compressed by a purely rate-independent method. There is an extension of the hysteresis or RP filter, which is to some extent capable of preserving the upper part of the frequency domain. It is a compromise between reduction of the length of the signal and preserving the frequency properties as much as possible. The filter level is chosen in the same way as for the pure hysteresis filter methods. For the selection of the remaining parameters several trials should be calculated and the best result with respect to the reduction on the one hand and the properties of the frequency spectra on the other hand should be chosen.

8. Check the maximum gradients and frequencies of the signals prior to the application on a rig. Use the test pre-processing procedure to avoid losing the largest cycles because of limitations of the rig. This might also be helpful for the drive-file iteration.

4.4   Load Editing in the Rainflow Domain

This section deals with methods to process rainflow matrices like re-scaling, superposition, and extrapolation. The idea is to edit the results of the rainflow counting without going back to the time domain. The methods are explained in detail for rainflow matrices, but they can also be applied to one-dimensional counting results like range-pair or level crossing.

4.4.1   Re-scaling

The rainflow matrix of a signal depends on the selected counting range and the number of bins. These parameters can be different for different signals, especially if the counting range is adapted automatically to the range of the data. For the superposition (see Section 4.4.2) and the quantile extrapolation (see Section 4.4.4), rainflow matrices with equal counting range are needed. To avoid repeated rainflow counting, the matrices can be estimated by re-scaling to the new counting parameters. These parameters have to be selected to fit all signals under consideration. The re-scaling algorithm is explained in the following, see LMS [147].

With m_1, M_1, N_1 and m_2, M_2, N_2 respectively, we denote the counting range of the original matrix and the new counting parameters. $\Delta_1 = (M_1 - m_1)/N_1$ and $\Delta_2 = (M_2 - m_2)/N_2$ denote the width of each bin.

To transform the matrix to the new counting range, we first estimate physical values for every cycle. A cycle from the bin (i_1, j_1) corresponds to physical values (x, y) with

$$x \in [m_1 + (i_1 - 1)\Delta_1, m_1 + i_1\Delta_1), \qquad y \in [m_1 + (j_1 - 1)\Delta_1, m_1 + j_1\Delta_1).$$

A random value for x and y is selected from this interval. Then we have to find the bin numbers (i_2, j_2) for this cycle with respect to the new counting parameters, i.e.

$$i_2 \text{ with } x \in [m_2 + (i_2 - 1)\Delta_2, m_2 + i_2\Delta_2), \quad j_2 \text{ with } y \in [m_2 + (j_2 - 1)\Delta_2, m_2 + j_2\Delta_2),$$

and add the cycle to this bin (i_2, j_2). In the same way all the cycles of the original matrix and the values of the residual are transformed. Finally, cycles with amplitudes smaller than the bandwidth of the filter for the original matrix are removed. Then we have to check if the new rainflow matrix and residual are consistent. If not, the residual has to be corrected accordingly.

Example 4.10 (Re-scaling of Rainflow Matrices)
Consider the rainflow matrix on the top of Figure 4.25 with 100 classes. The matrix is rescaled to 60% of its original counting range, again with 100 classes (on the bottom of Figure 4.25).

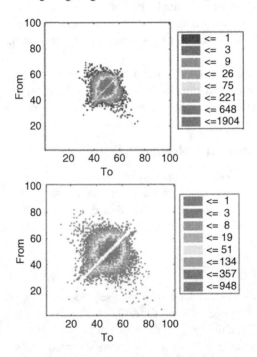

Figure 4.25 Original (top) and matrix rescaled to 60% of the counting range (bottom) with 100 classes

To investigate the effects of the re-scaling on the damage values, we compute the damage of both matrices in Figure 4.25, once with the slope of the Wöhler curve $\beta = 5$ and once with $\beta = 3$. Additionally, we also compute both damage values for the rescaled matrices with 60% of the original counting range but with 50 and 128 classes respectively, as well as for the matrices rescaled to 140% of the original counting range with 50, 100 and 128 classes. The relative counting range, the number of classes, N, and the relative damage values, compared to the damage of the original matrix, are given in the following table:

relative counting range (%)	N	relative damage with $\beta = 5$ (%)	relative damage with $\beta = 3$ (%)
100	100	100	100
60	50	99.0	99.9
60	100	98.6	98.7
60	128	99.5	98.7
140	50	110.2	106.7
140	100	102.1	100.4
140	128	108.3	99.5

We get the largest deviation for both slopes for the increased counting range with only 50 classes. This is no surprise, since one new class of this counting range contains nearly 3 classes of the original counting range, which makes the damage calculation inaccurate compared to the original counting. However, the largest deviation is only 10% in damage, which is not too much and can be accepted in practical applications. □

4.4.2 Superposition

A target for a test rig or test track is often constructed out of a combination of rainflow matrices. We then have n signals with corresponding rainflow matrices $RFM_1, \ldots, RFM_n$ and look for the rainflow matrix of the combination with given repetition factors $w_1, \ldots, w_n$. To achieve this, one could concatenate the time series and compute the rainflow matrix of these new time series. However, for large time series or large repetition factors, this is not feasible. In this case, we can compute the resulting rainflow matrix directly from the rainflow matrices of the individual signals and the repetition factors by rainflow superposition. The algorithm is implemented in LMS [147].

For the superposition, all matrices must have the same counting range. If this is not the case, they have to be rescaled first (see Section 4.4.1).

We can then combine the individual rainflow matrices with the given repetition factors. To account for cycles closing during the second run, we first have to identify them by rainflow counting of twice the residual. We add this new rainflow matrices, multiplied with a factor $(w_i - 1)$ for signal i, as these cycles are closing in the second as well as every following run.

Finally, we have to account for the cycles closing due to the composition of the different signals. They can be computed by rainflow counting the concatenation of all residuals

$RES_1, \ldots, RES_n$. The remaining residual is the one for the new matrix. The following formulas summarize the procedure:

$$RFM_s = w_1 RFM_1 + \ldots + w_n RFM_n$$

$$+ (w_1 - 1)RFM(RES_1, RES_1) + \ldots + (w_n - 1)RFM(RES_n, RES_n)$$

$$+ RFM(RES_1, RES_2, \ldots, RES_n), \tag{4.25}$$

$$RES_s = RES(RES_1, RES_2, \ldots, RES_n). \tag{4.26}$$

There may be a slight difference between the result of this algorithm and the rainflow count of the concatenation of the individual time series. Due to memory effects, the orientation (from lower to upper triangle of the rainflow matrix) of some cycles may be changed by the rainflow superposition. However, this makes no difference in the range-pair or level-crossing histograms and, moreover, does not affect the pseudo-damage calculation used in our context.

Example 4.11 (Superposition of Rainflow Matrices)

We consider two short time series (see Figure 4.26) that we want to combine with a factor $w_1 = 2$ and $w_2 = 1$. By rainflow counting (see Section 3.1.3), we get the cycles of each of the two time series. The first one has the cycles $(3, 2)$, $(4, 1)$, $(3, 4)$ and the residual $(1, 5, 2, 4, 3)$. The second time series gives the cycles $(3, 2)$, $(3, 2)$, $(3, 4)$, $(5, 2)$ with the residual $(2, 3, 1, 5)$. Rainflow counting of twice the residual of the first series gives the cycles $(2, 4)$ and $(5, 1)$, the concatenation of the residuals of series 1 and 2 the cycles $(2, 4)$, $(2, 3)$ and $(5, 1)$.

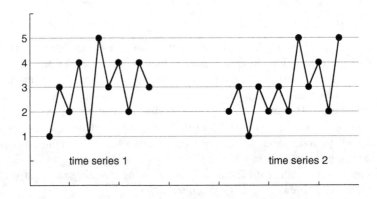

Figure 4.26   Two (short) time series

Now, we combine twice the first and once the second time series (see Figure 4.27). The resulting rainflow matrix and residual can be computed according to Equations 4.25 and 4.26,

$$RFM_s = 2RFM_1 + RFM_2 + RFM(RES_1, RES_1) + RFM(RES_1, RES_2),$$

$$RES_s = RES(RES_1, RES_2),$$

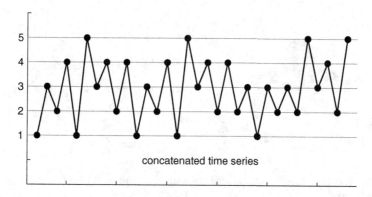

Figure 4.27   Concatenation of twice the first and once the second time series

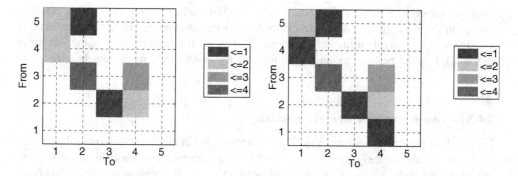

Figure 4.28   Resulting matrix from rainflow superposition (left) and from rainflow counting of concatenated time series (right). The orientation of cycle $(1, 4)$ is changed to $(4, 1)$ in the rainflow superposition

which gives the rainflow matrix on the left in Figure 4.28. Alternatively, we can do rainflow counting of the concatenated time series of Figure 4.27. The result is shown in Figure 4.28 on the right. The only difference between the two rainflow matrices is that the orientation of cycle $(1, 4)$ is changed to $(4, 1)$ in the rainflow superposition. The residual $(1, 5)$ is the same in both approaches.   □

4.4.3   Extrapolation on Length or Test Duration

Because of cost and time limitations, the length of a measurement is restricted. However, we are interested in loads for a long period. In the rainflow domain, we want to know what the rainflow matrix and therefore also the damage would look like if we extended a measurement from e.g. 20 to 200 kilometres, or from 2 to 200 laps on the test track.

Just multiplying the whole matrix with a constant extrapolation factor does not take into account that even the same driver on the same road will not generate the same rainflow matrix for a repetition of the measurement. He does not drive with exactly the same speed,

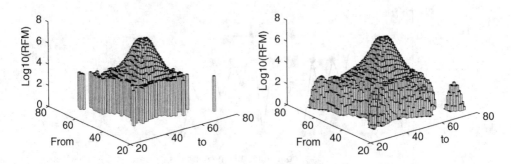

Figure 4.29 Rainflow matrix extrapolated without (left) and with (right) smoothing, logarithmic scale on the z-axis

does not brake with exactly the same force, etc. With this simple approach we get large entries in those rainflow cells where cycles have been observed in the short measurement, and zeros in the neighbouring cells without observation (see Figure 4.29 on the left for an example). It is not realistic to get such a form of the matrix for a measurement over a long period.

4.4.3.1 Non-parametric Kernel Smoothing

Especially if we are not only interested in the damage, but in the extrapolated rainflow matrix itself, we should not only multiply each entry with a given factor. Instead, some smoothing of the rainflow matrix should be used prior to the multiplication by the extrapolation factor.

The reason for this smoothing is the following. The time series of the measurement can be considered a realization of a stochastic process (see Chapter 6). This process defines a distribution of cycles, that is, a two-dimensional distribution of the entries in the rainflow matrix. The longer we measure, the better we know this distribution. If we have only a short measurement, we have to estimate the distribution from the observed cycles.

As there are many different forms of rainflow matrices, a parametric estimation is very difficult. Usually non-parametric density estimators for this distribution are used. In each entry of the matrix a frequency of occurrence (probability) is determined by a weighted averaging over neighbouring cells:

$$p(i, j) = \sum_{k,l} w(i - k, j - l) RFM(k, l). \qquad (4.27)$$

The bandwidth of the kernel w may be constant for all entries of the matrix or may be adapted to the data. The underlying idea of adaptivity is that the amount of smoothing is determined by the entries in the matrix: many entries $\rightarrow$ little smoothing, few entries $\rightarrow$ much smoothing, as few observations indicate little knowledge about the underlying distribution.

There are several possibilities for the choice of the kernel functions w. In, for example, the MATLAB Toolbox WAFO (see WAFO Group [237, 238]), circular Gaussian kernels are used with the normalization

$$\sum_{i,j} w(i - k, j - l) = 1, \qquad (4.28)$$

in Equation (4.27), that is, each entry of the original rainflow matrix is distributed to the other cells with weights from Gaussian circular kernels. If the width of the kernels is adapted to the data, the number of cycles in each cell of the rainflow matrix determines how fast the corresponding kernel decays. Equation (4.28) is still valid for adaptive bandwidth, and the number of cycles in the original and the smoothed matrix is therefore always the same. See also Johannesson [121] for a detailed description of this smoothing.

In LMS Tecware (see LMS [147]) elliptic Gaussian kernels with the normalization

$$\sum_{k,l} w(i - k, j - l) = 1 \qquad (4.29)$$

are used. In the region where these weights are small, they are set to zero, that is, the support of the kernels does not range over all cells of the matrix. In contrast to the method considered above, not each entry of the original matrix is distributed to the other cells, but the entries of the smoothed matrix are collected from neighbouring cells. If the bandwidth is constant, this is exactly the same. But in the case of adaptive bandwidth, the latter approach may lead to a different number of cycles in the smoothed matrix compared to the original matrix.

For adaptive bandwidth (here the length of the half axis of the elliptic kernels), we start with a small bandwidth for each entry and consider the number of cycles contained in the support of the kernel. If the number is small, we expand the kernel until enough cycles are contained.

However, also other than Gaussian kernels can be used and physical limits for the smoothing (e.g. no negative forces allowed, etc.) can be taken into account. For more details on non-parametric density estimation for rainflow matrices, see, for example, Dressler *et al.* [81], Socie [216], Socie and Pompetzki [217], and again Johannesson [121] and WAFO Group [237, 238].

From the smoothed distribution, we can then randomly select cycles for the extrapolated matrix. Figures 4.30 and 4.31 both show an example of a rainflow matrix. It is extrapolated using the WAFO toolbox with an extrapolation factor 1000 and with different smoothing parameters. Figure 4.30 shows at the bottom the resulting matrix for smoothing with the default constant bandwidth. At the bottom of Figure 4.31 the bandwidth for the smoothing is also constant, but larger. Figure 4.32 shows the result of smoothing with adaptive bandwidth. While the difference of the adaptive and non-adaptive smoothing is small near the diagonal, it is much greater in the large amplitude areas.

Such an extrapolation will give good results only if we extrapolate to longer measurements with the same characteristic. For example, extrapolation from a measurement in the city to a longer measurement with different road types is impossible, as events in the long measurement (driving with high speed, etc.) are not contained in the original time signal.

4.4.3.2 Extreme Value Extrapolation of Rainflow Cycles

As the number of observed cycles with large amplitude is usually small, the density estimation is critical in this region of a rainflow matrix. On the other hand, these large cycles

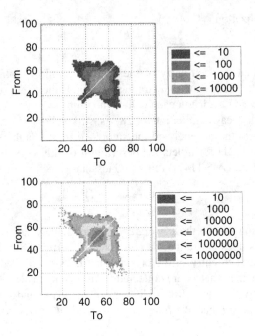

Figure 4.30 Example of smoothing of a RFM with constant bandwidth in the WAFO Toolbox. Original matrix (top), smoothing with default (small) bandwidth (bottom)

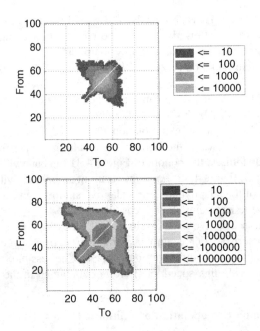

Figure 4.31 Example of smoothing of a RFM with constant bandwidth in the WAFO Toolbox. Original matrix (top), smoothing with larger bandwidth (bottom)

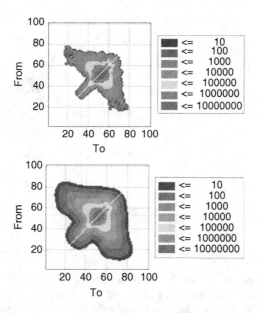

Figure 4.32 Example of smoothing of a RFM with adaptive bandwidth in the WAFO Toolbox. On the bottom plot a larger variation in the bandwidth is allowed

are especially important due to their high damage contribution. This region of the rainflow matrix can be estimated using extreme value theory, see Johannesson and Thomas [123]. To obtain the result for the extrapolated rainflow matrix, the equivalence between counting rainflow cycles and counting crossings of intervals is used. The result is a simple approximate expression for the cumulative rainflow matrix

$$\mu^+(u, v) \approx \frac{\mu^+(u)\mu^+(v)}{\mu^+(u) + \mu^+(v)} \tag{4.30}$$

depending only on the level upcrossing intensity $\mu^+(u)$, which is a function of the level u. The approximation is supposed to be good when u is a low level, and v a high level. To demonstrate the accuracy of the approximation, we have defined a Markov load for which it is possible to compute the exact density of rainflow cycles, see Chapter 6. The approximation is supposed to be good for large amplitudes, which is confirmed by Figure 4.33a.

For a measured load, the intensity of level upcrossings is not known, but can be estimated from measurements. In Johannesson and Thomas [123], an estimation method is proposed, where the intensity of level upcrossings is extrapolated for high and for low levels, by using the generalized Pareto distribution. Note that it is easier to estimate $\mu^+(u)$, the intensity of level upcrossings, than to estimate directly the density of rainflow cycles. As the formula (4.30) is only valid for large amplitudes, the remaining part has to be estimated by other means, here by means of a kernel smoother. The combination becomes the final estimate of the extrapolated RFM, see Figure 4.33b for an example.

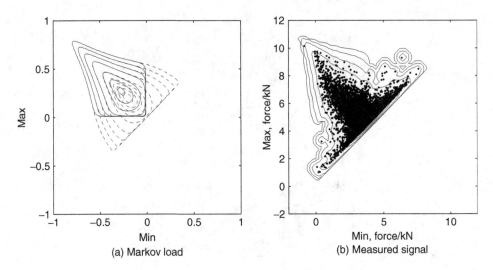

(a) Markov load   (b) Measured signal

Figure 4.33 (a) Iso-lines of the density of rainflow cycles for a Markov load, with the solid lines showing the approximation by Equation (4.30), and the dashed lines the exact one. (b) Iso-lines of the extrapolated RFM, with the dots representing the cycles in the signal. The iso-lines enclose 10%, 30%, 50%, 70%, 90%, 99%, 99.9%, 99.99%, 99.999%, respectively, of the cycles

4.4.3.3 Comparison of Time and Rainflow Domain Methods

The method described previously in Section 4.3.1 is an extrapolation in the time domain, whereas here the extrapolation is in the rainflow cycle domain. The two methods for extrapolation are both based on the theory of statistical extreme values. Consequently, the resulting extrapolated load spectra should be similar. Note that it is also possible to extrapolate directly the range-pair count, using the POT method, see Example 6.5 on page 210.

The time domain method results in an extrapolated time signal where the order of the cycles is preserved, while the rainflow domain method gives an extrapolated rainflow matrix. More precisely, the time domain method simulates extrapolated turning points, and then the extrapolated rainflow matrix can be calculated from this extrapolated signal. Thus, for the time domain method, the result is an N-fold extrapolated load, while for the rainflow domain method the result is a long-run distribution representing $N = \infty$ number of repetitions. In Figure 4.34, the load spectra from 100 repetitions of the original measurements, from 100-fold time domain extrapolations, and from rainflow extrapolations, are shown. Both types of extrapolations give similar results. However, the rainflow domain method is well adopted for really long extrapolations of the load spectrum, whereas the time domain method in those cases will be computationally more demanding. Also note that the maximum load amplitude is extrapolated. The train load measurement has a maximum amplitude of 24 MPa and the extrapolation about 32 MPa. The corresponding maximum values for the car load is 5.5 kN and 7 kN. Consequently, in a fatigue test for the car application, instead of repeating the amplitude 5.5 kN 100 times, the extrapolation applies one amplitude of 7 kN and about 100 different amplitudes in between 5.5 kN and 7 kN. This is our informed guess of what would be the load spectrum if we had a 100 times longer measurement.

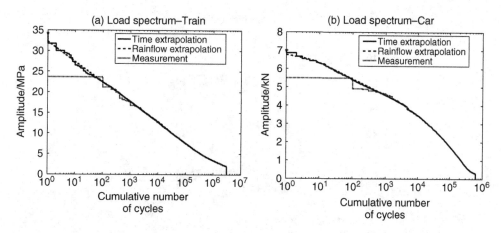

Figure 4.34 Load spectra of hundred-fold extrapolations comparing time and rainflow domain methods

4.4.3.4 Estimation Based on Parametric Models

As an alternative for matrices with a certain shape, specific parametric distributions may be used. In this case, an assumption on the family of distribution is made, and the corresponding parameters are estimated from the measurement. For example, a sum of several 2D-normal distributions could be used. As these methods are, as far as we know, of minor practical importance, we will not go into more detail here.

A completely different approach is to identify a stochastic process from the data. This information can then be used to simulate the process as long as it is given by the desired extrapolation time. From the corresponding time signal, the rainflow matrix or other results can then be derived easily (see Chapter 6).

4.4.3.5 Pseudo Damage of Extrapolated Rainflow Matrices

An important criterion for the quality of the extrapolation procedure is given by the pseudo damage value calculated from the matrix. If we extrapolate by a factor of 100, do we expect the damage of the matrix to increase by the same factor of 100 or should the damage increase somewhat more due to the higher extreme amplitudes we expect after the extrapolation?

To discuss this question, we will assume that we know the 'true' distribution of the cycles from the underlying stochastic process or from an infinitely long measurement. Assume, for example, that the synthetic distribution shown in Figure 4.35 is the distribution of the cycles. The density is such that drawing a cycle with a small amplitude is much more probable than one with a large amplitude.

Then we can simulate the situation of measurements of different length by fixing a certain number of cycles and selecting a sample of this size from the given distribution. As in a real measurement, we then remove the cycles smaller than a given threshold and compute the damage value from the rainflow matrix. We can compare this value to the expectation value of the 'true' distribution (bootstrap method). Since we can select arbitrarily many samples, we get a distribution of damage values for a given number of cycles (duration of

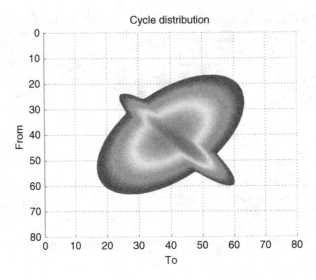

Figure 4.35 Synthetic distribution for the simulation of measurements with a different number of cycles

measurement). If we repeat this procedure for different numbers of cycles, we can study the properties of these distributions and find that the variance of the damage values decreases with increasing number of cycles. In Figure 4.36 we see an example of samples of size 500, 1000, 5000, 10000 and 50000 from the given synthetic distribution.

   The mean of the damage is equal to the expectation value of the underlying 'true' distribution, but the median is smaller. It is therefore more probable to underestimate the damage with a small sample. Overestimating is less probable, but deviations from the expected value are larger in that case. This is a result of the chosen form of the distribution of the cycles. Cycles with small amplitudes are more probable than large ones. This is usually also true in a real measurement. Due to the damage calculation, which amplifies each cycle with a power of the slope of the Wöhler curve, β, this skewness is further increased, the more, the larger β is. The damage in this example is computed with $\beta = 5$. An example of two different rainflow matrices, one from a sample of 1000 cycles, and one of 50000 cycles is given in Figure 4.37.

   The results in Figure 4.36 show that if the number of cycles observed is not too small, then the deviation of the damage value from the expected value is not too large. Of course, this depends on the specific matrix and its shape, but a rainflow extrapolation with a certain amount of smoothing should nevertheless be such that the resulting damage is only slightly above the damage obtained by simply multiplying the initial value with the extrapolation factor.

4.4.4 Extrapolation to Extreme Usage

Assume we have several measurements, for example, from different drivers on the same track. The conditions for all measurements are the same, only one factor, the driver, varies. We now want to find out what is an extreme quantile (often 90%, 95% or 99%) of the load a driver can create on this track. On the level of damage, this is a one-dimensional

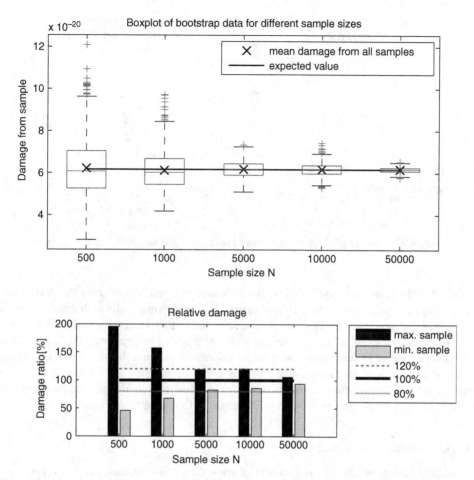

Figure 4.36 Boxplots of the damage distribution for samples of different size drawn from the synthetic distribution and relative damage of the max. and min. sample compared to the expected value

problem since we have a single value representing each observation. We can compute the damage value of each driver, estimate (if possible) a parametric distribution and compute the quantile of this distribution.

However, if we are interested not only in the damage, but also in the shape of the range-pair diagram or the rainflow matrix of this quantile, the problem is no longer one-dimensional. But since the number of cells of a rainflow matrix is usually large (100 classes give 10,000 cells), considering each entry separately is not a feasible extension.

4.4.4.1 Clustering

The quantile extrapolation of rainflow matrices (see Dressler *et al.* [81], Socie and Pompetzki [217]) therefore starts with the definition of several clusters of the matrix to reduce the dimension of the problem.

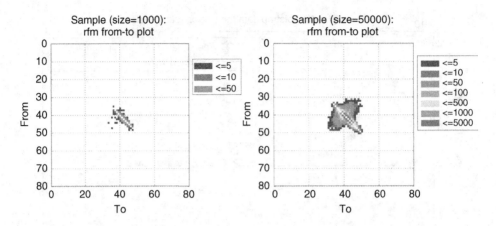

Figure 4.37 Two rainflow matrices from samples of 1000 and 50000 cycles respectively

In LMS Tecware (see LMS [147]), these clusters are built considering the variance (or variability) in each cell of the rainflow matrix over all drivers. Cells with similar variance belong to the same cluster. The idea is that cells with small variance, that is, with more or less the same entries for all drivers, represent the influence of the road and other external conditions. In contrast to this, entries with large variance show the influence of the driver itself.

Another approach for clustering the entries could be building the clusters according to similar size of the amplitudes, that is, split the rainflow matrix along some diagonals. We could also choose another partitioning of the matrix to reduce the size of the problem.

4.4.4.2 Multivariate Estimation

Next, the damage of the rainflow matrix of each driver is computed for each cluster separately. From these values a multivariate parametric distribution is estimated. For, for instance, three clusters we get three damage values for each driver and a three-dimensional distribution is estimated on the basis of these values.

Additionally, a rainflow matrix RFM_{mean} with the mean entry over all drivers in each cell is computed, together with the corresponding cluster damages.

4.4.4.3 Quantile Matrix

As explained above, the total damage value of an extreme quantile can be computed from the 1D distribution of all total damage values. The total damage of the quantile rainflow matrix is defined as this value D_q. The portion of the damage that belongs to each cluster, however, is not yet defined by this procedure and so far we do not have a corresponding rainflow matrix.

To get this, we can now take many samples from the estimated multivariate distribution of the cluster damages and choose one where the cluster damages sum up to D_q. As there are infinitely many combinations of that type, the rainflow matrix belonging to a quantile value is not unique. We then build a matrix with exactly this cluster damages by multiplying each cluster of the mean rainflow matrix RFM_{mean} with an appropriate factor. Instead of simply multiplying the clusters of the mean matrix we can also apply smoothing procedures as discussed above to the clusters to get entries outside the range of the given matrices.

If the measurements contain more than one varying factor (e.g. different drivers and different roads), the procedure described above takes the overall variability into account to compute an extreme quantile. However, it is then not possible to separate the different influences. One method to analyse the influence of different factors on the damage distribution, and to define quantiles and corresponding rainflow matrices is considered in Chapter 8.

Example 4.12 (Quantile Matrix)

We will now consider an example with one varying factor. We have 12 measurements for the same driver, each for several kilometres on different highways, and want to define a 90% quantile road for this driver.

We compute the quantile matrix with LMS Tecware. As a first step, the cluster matrix is defined. We choose the (default) setting of three different clusters that are shown in Figure 4.38. The first, with the entries of smallest variance in black, the second, with medium variance in grey. The cells with the largest variance, together with the cells with value zero for all roads, are the remaining cells of the matrix.

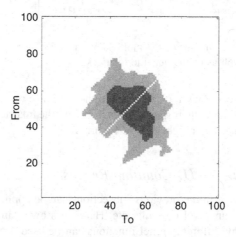

Figure 4.38 Cluster matrix for the quantile extrapolation of the example. The first cluster is shown in black, the second in grey and the third in white

Next, the damage values in all three clusters for the 12 roads are computed with $\beta = 5$. From this 3×12 matrix a three-dimensional distribution is estimated. In Figure 4.39 projections of this distribution together with the damage values of the clusters of the matrices are shown. The lower right plot shows the projection of the distribution to the first and third cluster. The plots on the left and on top show the projection to clusters two and three and to clusters one and two respectively. The upper three plots each show the distribution in one cluster. In all distribution plots, crosses mark the damage values of the original matrices, circles the values of five possible quantile matrices.

As mentioned above, the extrapolated matrices are not unique and differ with respect to the damage values in each of the three clusters. In Figure 4.40 three of the original matrices

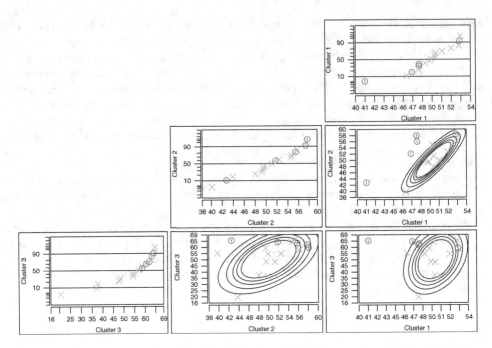

Figure 4.39 Projections of the estimated multivariate distribution together with the cluster damages of the original as well as of the quantile matrices

are shown. As can be seen, they look similar but they are different in the largest amplitudes and also differ somewhat with respect to the form of the matrices. One of the quantile matrices is shown in the lower right plot of Figure 4.40. □

4.4.5 Load Editing for 1D Counting Results

All the methods explained in the previous subsections can be applied similarly to 1D counting results like range-pair, level crossing, etc. Here, it is even simpler than in the case of rainflow matrices. Many different kernel functions can be used, for instance, hat functions, piecewise polynomial functions, or Gaussian kernels. The bandwidth (i.e. the support of the kernel) can be chosen to implement a more or less strong smoothing effect and it can be adapted to the data in the same way as described for rainflow matrices.

Often a mixture of certain basic shapes are used, for example, a sum of two or three Weibull or Rayleigh distributions with a corresponding set of parameters might be used. In Dirlik's method for estimating a range-pair histogram from a PSD function, such an approach is taken (see Section 3.2.7 and Dirlik [73] for more details). However, we will not go into details of these methods.

4.4.6 Summary, Hints and Recommendations

We briefly summarize the methods explained in the present section on editing in the rainflow domain and give some hints on and recommendations for practical applications.

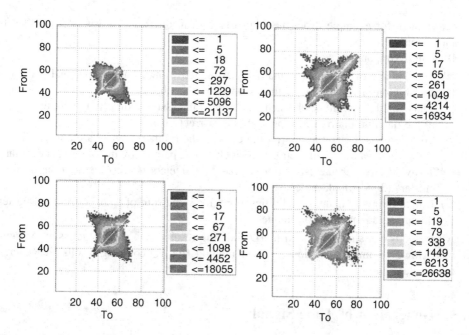

Figure 4.40   Three of the original matrices of the example and a possible quantile matrix for the 90% quantile (lower right plot)

1. Re-scaling: In general, the bins for rainflow counting should be chosen consistently with the signal, that is, not the whole range of the time series should be contained in a few bins. The counting range should at least be adapted to the type of signal. For example, it is good practice to calculate a common range for all vertical wheel forces of one measurement campaign, but a different range for all lateral wheel forces. Then no re-scaling is necessary as long as data from that campaign is analysed. However, if this recipe is not applied or matrices from different campaigns are to be superimposed, the counting ranges for the matrices need to be changed by the re-scaling procedure. Fortunately, the errors introduced by that method are very small in practical applications.
2. Superposition: To compute the rainflow matrix of the concatenation of different signals with possibly high repetition factors, only the rainflow matrices, the residuals, and the rainflow count of twice the residual for each signal are necessary. Together with a rainflow count of the concatenation of all residuals, the resulting rainflow matrix and residual of the superposition can be computed.
3. Extrapolation on length: To extrapolate from a short measurement to a longer one, a rainflow matrix should not only be multiplied with a constant factor. Instead some smoothing or other density estimation should be applied first. This accounts for variances in the repetition of the same measurement. The smoothing parameters should not be chosen too extreme, but should account for the uncertainty expected in the measurement. Short measurement → high uncertainty in the measurement→ strong smoothing; long measurement → lower uncertainty → moderate smoothing. Adequate smoothing parameters can be

determined, for example, from the comparison of a long-term measurement with some short measurements.

4. Quantile Extrapolation:

   (a) The damage of the quantile matrix is estimated from a one-dimensional parametric distribution. The validity of the assumption that the chosen distribution fits to the data should therefore be checked carefully. If the assumption is not valid, the damage of the quantile matrix can represent an incorrect quantile.

   (b) To determine a quantile matrix, the cells of the rainflow matrix are clustered. From the damage values of all matrices in all clusters a parametric distribution is estimated. To avoid errors in the estimation of many parameters for this distribution, only few clusters (3–5) should be chosen.

   (c) Due to infinitely many possibilities to get the same sum of damage from all clusters, the quantile matrix is not unique.

5. All methods can also be applied to 1D counting results such as level crossing or range-pair. However, implementations of these methods for 1D counting results are generally not available.

4.5 Generation of Time Signals

Similar to Section 4.3, we want to derive time signals for validation purposes. In contrast to that section, we will not modify measured time signals to meet certain requirements, but start 'from scratch' and construct time signals with certain properties. There are many possibilities to define these properties, for instance, by prescribing

- amplitude-based counting results like Markov or rainflow matrices, range-pair or level crossing histograms, etc.,
- frequency-based or spectral features such as the PSD function,
- stochastic process parameters, or
- multiaxial properties like correlation matrices or RP counting results.

In principle, we could start with all types of analysis results that we have seen in Chapter 3. While it is often rather simple to derive algorithms for a single time signal, it is much more difficult to deal with multi-input loads. Thus, most of the methods described below are 1D methods, trying to invert the corresponding analysis algorithm used to derive the properties from a time signal. For that reason, these methods are sometimes called reconstruction methods. A different and rather general approach, the optimum track mixing, is described in detail in Section 9.8.

4.5.1 Amplitude- or Cycle-based Generation of Time Signals

This subsection deals with the reconstruction of time signals from amplitude-based histograms. It is often possible to construct a time signal that exactly corresponds to the histogram. This means that, if we apply the counting method to the reconstructed time signal, we get exactly the histogram we started with. This is true for most of the 1D methods dealing with a single signal only. It is no longer true for the multi-input load signals.

A common feature of all those reconstruction methods is that they generate a sequence of turning points rather than a time signal. By assigning a sampling rate Δt, the indices turn into an equidistant time channel. Then we can analyse the reconstructed signal with respect to specific properties such as maximum slope or maximum frequency, and apply test pre-processing methods as described in Section 4.3 if necessary.

4.5.1.1 Markov Reconstruction

We begin the description of the reconstruction methods with the Markov matrix. Since it contains the transitions from one bin to another, we can simply choose an arbitrary starting bin i and randomly select a transition $i \rightarrow j$ from the i:th row of the Markov matrix MM. Then we delete that entry from MM and add the corresponding destination bin j to the time signal: (i, j). Next, we select a transition $j \rightarrow k$ from the j:th row, delete the entry from MM, and add bin k to get the signal (i, j, k). This process is repeated until MM is empty.

In general, this simple algorithm may run into dead ends. There are three reasons for this. One reason is, that the starting bin has not been chosen properly. The starting bin i_S is defined by the following conditions for its row sum $R_{i_S} = \sum_{j=1}^{N} MM(i_s, j)$ and its column sum $C_{i_S} = \sum_{k=1}^{N} MM(k, i_s)$: $R_{i_S} = C_{i_S} + 1$. If there is no such index i_S, an arbitrary starting bin with $R_{i_S} > 0$ is allowed.

Another reason is that the orientation of the first range has not been chosen properly. Let $N_{up} = \sum_{i<j} MM(i, j)$, $N_{down} = \sum_{i>j} MM(i, j)$ denote the number of upward and downward ranges resp. We then have to start with an upward range if $N_{up} = N_{down} + 1$ and with a downward range if $N_{up} = N_{down} - 1$. The starting range is arbitrary if $N_{up} = N_{down}$.

A third reason is that the current point during the reconstruction belongs to an empty row of the Markov matrix while other rows still contain entries. To avoid that, we have to know the destination of the last transition starting from a bin. This can be stored during Markov counting by a straightforward extension of the counting algorithm. Then we can ensure that we reserve that transition during the random selection until there are no other transitions left. Details of the method are given in Krüger *et al.* [139].

The algorithm described so far is a reconstruction algorithm trying to invert the counting. Often it is not necessary to end with a time signal exactly meeting the Markov matrix; interpreting the matrix as a probability $p(i, j) = \frac{MM(i, j)}{\sum_{k,l} MM(k,l)}$ of transitions is enough. In that case, we proceed as before (randomly select a transition $i \rightarrow j$ according to $p(i, j)$, add j to the time signal, randomly select a transition $j \rightarrow k$, add $k \ldots$), but we do not delete the chosen transitions. Instead, we repeat that process as long as we want and end with a time signal whose Markov matrix is a realization of a stochastic process which is defined by the matrix we started with.

4.5.1.2 Rainflow Reconstruction

In spite of the fact that rainflow counting is a non-local procedure suitable for the complex nesting of hysteresis cycles in the stress-strain path, there is a 'perfect' rainflow reconstruction method in the following sense:

1. The time signal constructed from the rainflow matrix and the residual exactly corresponds to the matrix.

2. The reconstruction is randomized such that all signals with that property have the same probability of being reconstructed.
3. The algorithm can be implemented online, meaning that the points of the signal are constructed in the order of appearance in the signal.

The first property is a natural requirement for a reconstruction algorithm. The second one ensures that the algorithm does not introduce a systematic sequence effect, for example, all small cycles at the beginning, all large cycles at the end of the signal. The third property might be interesting for technical reasons (online implementation on a test rig, arbitrarily long signals because of small memory consumption).

The basic idea is to invert the 4-point counting rule described in Section 3.1.3. A cycle (i_2, i_3) is identified by the 4-point algorithm if $min(i_1, i_4) \leq i_2, i_3 \leq max(i_1, i_4)$, where (i_1, i_2, i_3, i_4) is the 4-point stack. Inverting this rule means that, for each cycle (j_1, j_2) in the rainflow matrix, we have to look for a segment (r_k, r_{k+1}) in the residual (or the residual enlarged by some previously inserted cycles), such that $min(r_k, r_{k+1}) \leq j_1, j_2 \leq max(r_k, r_{k+1})$. If we have found such a range, then we can insert the points j_1, j_2 into the segment r_k, r_{k+1} leading to the enlarged segment r_k, j_1, j_2, r_{k+1}. Then the cycle is deleted from the rainflow matrix and we go on with the next cycle until the matrix is empty. The number of points of that signal is twice the number of cycles in the matrix plus the length of the residual.

In order to circumvent dead ends, we have to start with the cycles with the largest amplitudes, as cycles of smaller amplitudes can later be inserted into larger cycles but not vice versa. In general, there are many possibilities to build in a cycle into the (enlarged) residual. Thus we can select one of these randomly. If this is properly done, we fulfil condition 2 mentioned above. The algorithm described so far is not an online algorithm, but it can be extended to an online algorithm by some additional considerations. The details of the complete method are given in Krüger *et al.* [139] and Dressler *et al.* [82].

Example 4.13 (Off-line Reconstruction of a Short Demo Example)
We illustrate the (offline) method using the simple example introduced in Section 3.1.3. In Figure 4.41, we have the residual and the cycles to be reconstructed.

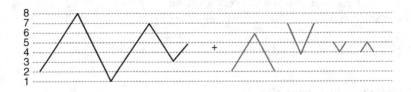

Figure 4.41 The residual (left) and the cycles (right) in the rainflow matrix

We start to insert the largest cycle (2, 6) into the residual. We have to look for a downward range which is large enough to insert the standing cycle (2, 6). As shown in Figure 4.42, there is only one possible position in which to insert that cycle.

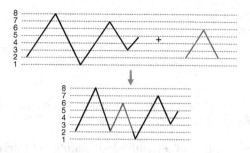

Figure 4.42 Integration of the first cycle into the residual

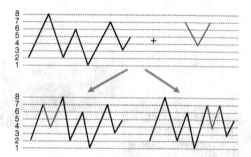

Figure 4.43 Integration of the next cycle into the enlarged residual

The cycle is inserted into the residual at that position and we proceed with the enlarged residual. In Figure 4.43, the next cycle $(7, 4)$ is inserted into the enlarged residual. As shown, there are two possible positions for the insertion.

We choose one of these possibilities (here the first one) and go on with one of the next cycles. Both remaining cycles are of the same size. We proceed with cycle $(5, 4)$ and find four feasible positions for the insertion shown in Figure 4.44.

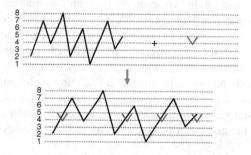

Figure 4.44 Integration of another cycle into the enlarged residual

At first glance, there seems to be a fifth possible position at the upward range $4 \to 8$ of the enlarged residual. But if we insert the cycle $(5, 4)$ here, it would turn into a standing cycle $(4, 5)$ because of the preceding downward range $7 \to 4$. This needs to be checked and treated accordingly, if the cycle to be inserted meets the lower or upper bound of the range where it is inserted.

Again, we have to choose one of the four possible reconstructions and go on with the last remaining cycle $(4, 5)$. The procedure is the same as before and not shown here in detail.   □

The considerations needed to fulfil conditions 2 and 3 mentioned above are more involved and can be found in Krüger *et al.* [139] and Dressler *et al.* [82].

Example 4.14 (Reconstruction of a Synthesized Rainflow Matrix)
We finish the description of the rainflow reconstruction with an example. Consider the time signal shown in Figure 4.45.

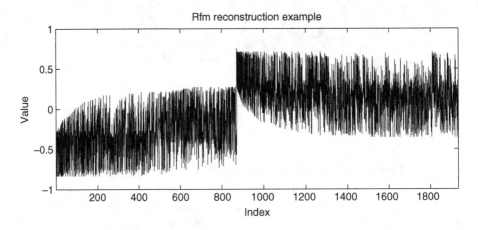

Figure 4.45 A time signal used as an example for the rainflow reconstruction

We are going to rainflow count the signal using 100 bins and a suitable counting range. We count the first half of the signal, then the second half, and finally the entire signal. The results are shown in Figure 4.46. To interpret the rainflow matrices, we look at the time signal and notice that the first half contains cycles with a lower mean load, whereas the second half contains cycles with a higher mean load. This corresponds to what we get in the rainflow matrices. From the viewpoint of the reconstruction algorithm, the cycles with the highest mean load are standing cycles and can thus only be built into the second part of the residual. The cycles with the lowest mean load are hanging cycles and can only be built into the first half of the residual. Of course, the rainflow matrix of this example has been created first and the time signal has been generated by the reconstruction algorithm. Incidentally, this shows again that the algorithm works perfectly in the sense described above.   □

4.5.1.3   Reconstruction from 1D Counting Results

As most of the important 1D counting methods can be derived from either the Markov or the rainflow matrix (see Figure 3.26), we can derive a time signal from a 1D counting result by first deriving a Markov or a rainflow matrix from the 1D counts and then applying the reconstruction methods mentioned above.

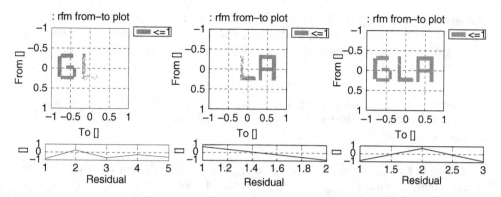

Figure 4.46 Rainflow matrices corresponding to the first half, the second half, and the full signal shown in Figure 4.45

Although there are also algorithms directly constructing time signals from 1D counting results, see e.g. Holm and de Maré [115] and Svensson [226] for reconstruction from level crossing counts, we will not describe these in detail here. The advantage of the two-step procedure

"1D histogram" → "Markov or rainflow matrix" → "time signal"

is that the intermediate matrix representation can be interpreted with respect to certain properties. We are forced to define those properties which are needed for the intermediate matrix representation but are not given by the underlying 1D histogram. This makes non-uniqueness, drawbacks, and assumptions of the algorithms obvious and explicit.

In the following, we consider the two most important histograms, namely range-pair and level crossing in more detail.

The (cumulative) *range-pair histogram* RP_{cum}: As it does not contain information about hanging and standing cycles, we reconstruct a symmetric rainflow matrix. The first step is to find a residual. To this end, the largest amplitude A_{max} is determined from the range-pair histogram. Then we can either choose $(1, 2A_{max} + 1, 1)$ or $(2A_{max} + 1, 1, 2A_{max} + 1)$ as the new residual. The amplitude A_{max} is deleted from the range-pair histogram.

The second step is to insert the remaining amplitudes A of the range-pair histogram into the rainflow matrix. To do this, we can assign a desired R-ratio R to each amplitude in the histogram. This requirement uniquely defines the lower and upper values S_{lower}, S_{upper} of the cycle by the conditions $S_{upper} - S_{lower} = 2A$ and $\frac{S_{lower}}{S_{upper}} = R$. While the first condition is given by the range-pair histogram, the second is an additional information which can be chosen arbitrarily (for example, a fixed ratio for all amplitudes or randomly according to a certain distribution).

Example 4.15 (Reconstruction From a Range-Pair Histogram)
As an example consider $RP_{cum} = (9, 6, 5, 3, 2)^T$ (9 cycles with range greater or equal 1, 6 with range greater or equal 2 ... 2 with range equal to 5). Since $A_{max} = 2.5$ (the largest range is 5), possible residuals are either $RES = (1, 6, 1)$ or $RES = (6, 1, 6)$. The remaining histogram after deletion of the residual cycle is $RP_{cum} = (8, 5, 4, 2, 1)^T$. The non-cumulative

histogram is $RP_{non-cum} = (3, 1, 2, 1, 1)^T$. That is we have one cycle of range 5, say, $RFM(1, 6) = 1$, one cycle of range 4, say, $RFM(2, 6) = 1$, two cycles of range 3, say, $RFM(2, 5) = 2$, one cycle of range 2, say, $RFM(3, 5) = 1$, and three cycles of range 1, say, $RFM(3, 4) = 3$. □

The *symmetric level crossing histogram* LC_{sym}: Again we reconstruct a symmetric rainflow matrix. First, we determine the starting point and the end point of the signal. Two cases need to be distinguished:

1. Both points are equal if all values of $LC_{sym}(k)$ are even. In that case, we can choose any point k_S with $LC_{sym}(k_S) > 0$. The residual then is either $(k_S, k_{min}, k_{max}, k_S)$ or $(k_S, k_{max}, k_{min}, k_S)$, where k_{min}, k_{max} denotes the smallest and largest non zero level respectively.
2. If LC_{sym} is not even, then there exists a pair of indices i_1, i_2 such that $LC_{sym}(k)$ is odd for $i_1 < k < i_2$. In that case, either $i_1 + 1$ or i_2 might be the starting point, the other one is the end point. Again the residual is chosen to include the largest range as in the first case.

Before we proceed, we have to remove the level crossings now contained in the residual from $LC_{sym}(k)$. After that, only even entries remain in $LC_{sym}(k)$.

We now choose an arbitrary probability distribution $p(i, j), i, j = 1, \ldots, N, i < j$ and draw pairs of indices (i, j) accordingly. If $LC_{sym}(k) > 0$ for $k = i, i + 1, \ldots, j - 1$, then all bin borders between i and j are crossed and we insert that cycle into the rainflow matrix by setting $RFM(i, j) = RFM(i, j) + 1$ and delete it from the level crossing vector: $L_{sym}(k) = L_{sym}(k) - 2$ for $k = i, i + 1, \ldots, j - 1$. This is repeated until the level crossing vector is empty.

The probability distribution can be adapted to fulfil certain desired requirements. For instance, if $p(i, j) \sim |i - j|^\alpha$, then the parameter α can be chosen to prefer either large cycles (leading to large damage) or small cycles (leading to small damage).

Example 4.16 (Reconstruction From Symmetric Level Crossings)

Consider the example $LC_{sym} = (4, 9, 13, 7, 4)^T$, where $N = 6$ and we have 5 levels to be crossed. The starting and end points are determined by the second rule using $i_1 = 1$, $i_2 = 5$, $k_{min} = 1$, and $k_{max} = 6$. Thus $RES = (2, 6, 1, 5)$ or $RES = (2, 1, 6, 5)$ are the possible residuals. After removing the level crossings contained in the residual we get $LC_{sym} = (2, 6, 10, 4, 2)^T$. The largest possible cycle remaining is $(1, 6)$. Thus if we set $RFM(1, 6) = 1$ and delete the corresponding crossings we get $LC_{sym} = (0, 4, 8, 2, 0)^T$. Next we set $RFM(2, 4) = 2$, delete both cycles, and get $LC_{sym} = (0, 0, 4, 2, 0)^T$. Then we set $RFM(3, 4) = 1$, delete the cycle and get $LC_{sym} = (0, 0, 2, 2, 0)^T$. Finally we set $RFM(3, 5) = 1$ and get an empty vector LC_{sym} after deletion of this cycle. A possible reconstruction is thus $RES = (2, 6, 1, 5)$ and $RFM(1, 6) = 1$, $RFM(2, 4) = 2$, $RFM(3, 4) = 1$, $RFM(3, 5) = 1$ (all other entries are zero). □

The examples illustrating the reconstruction of range-pair and level crossing histograms have been taken from previous work at TECMATH/LMS [147].

4.5.1.4 Multiaxial Rainflow Reconstruction

In the 1990s, some attempts were made to derive reconstruction algorithms for RP counting results (see Section 3.3 for the description of RP counting and Dressler *et al.* [79] for the multiaxial rainflow reconstruction). The difficulty here lies in the fact that the RP matrices are not independent. If there are n signals and we have N directions (RP combinations of the signals) with $N > n$, then only n out of N matrices are independent. The remaining ones depend on the others and the correlation of the signals. Thus, if we select the n matrices corresponding to the signals, reconstruct time signals from these (making sure that each signal has the same number of points), and perform an RP counting of that result, we get different matrices for the RP combinations which have not been taken into account during the reconstruction.

We can try to minimize the error by a suitable coordinate transformation. To explain this idea, consider two highly correlated loads $L_1(t)$ and $L_2(t)$. In this case, $\hat{L}_1(t) = L_1(t) + L_2(t)$ will tend to have higher amplitudes than $L_1(t)$ or $L_2(t)$, while $\hat{L}_2(t) = L_1(t) - L_2(t)$ will be smaller. If we apply a uniaxial rainflow reconstruction algorithm to the matrices of $\hat{L}_1(t)$ and $\hat{L}_2(t)$ and back-calculate L_1, L_2 from that, we get better results than if we applied a uniaxial reconstruction to the matrices of L_1 and L_2 directly.

In spite of the drawbacks of the multiaxial rainflow reconstruction, this method could be taken into account for the derivation of load signals acting on stiff components like a knuckle. If frequency properties become important, then the method of optimum track mixing (see Section 9.8) should be preferred.

4.5.2 Frequency-based Generation of Time Signals

The methods in this section are applicable if the frequency content of a signal is prescribed, for example, if a time series is required that fits a recorded PSD. The first two approaches – using the inverse discrete Fourier transform (IDFT) or linear time series models (ARMA) – are closely related to the algorithms for frequency-based editing discussed in Section 4.3.2. Other methods are available in case one is willing to use more complex mathematical models like those discussed in Chapter 6.

4.5.2.1 IDFT

The discrete Fourier transform (DFT) is an invertible function that maps a time series into the frequency domain (see Appendix C for details). Given the Fourier transform of a time series, it is possible to reconstruct the original signal exactly using the inverse transform IDFT. However, in practice, the complex-valued $\hat{x}_k$ are rarely retained. Usually, one calculates the real-valued PSD, which is proportional to $|\hat{x}_k|^2$ and contains only the amplitude information of the signal. And while it is no longer possible to reconstruct the original measurement from the PSD alone, one can generate signals with the same PSD using randomized phases, as shown below.

Given the PSD of a signal with N points sampled at frequency f_s, we know the absolute values $|\hat{x}_k|$ of the transformed series, $k \in \{0, \ldots, N-1\}$. As the time series we want to reconstruct is real-valued, we also know that $\hat{x}_k = \hat{x}_{N-k}^*$, $k \in \{1, \ldots, N-1\}$. Given arbitrary phases $\phi_k \in [0, 2\pi)$ for $k \in \{1, \ldots, \lceil \frac{N}{2} \rceil\}$ (the brackets denote rounding up), the Fourier transform

$$\tilde{x}_k := \begin{cases} |\hat{x}_0| & \text{if } k = 0 \\ |\hat{x}_k| e^{\frac{ik\phi_k}{N}} & \text{if } k \in \{1, \ldots, \lceil \frac{N}{2} \rceil\} \\ \tilde{x}_{N-k}^* & \text{else} \end{cases} \qquad (4.31)$$

satisfies both requirements. In fact, any set of ϕ_k yields $\tilde{x}_k$ resulting in the desired PSD. In the absence of further information, one can choose the phases as independent uniformly distributed random variables on $[0, 2\pi)$. Applying the IFFT algorithm to $\tilde{x}_k$ yields a time series with the given frequency content, although there is no guarantee that the amplitudes will be similar to those in the original signal; these depend on the exact phases.

Example 4.17 (Signal Reconstruction Using IDFT)
Consider the signal

$$x_1(t) = 2\sin(1.5 \times 2\pi t) + \sin(4 \times 2\pi t) \qquad (4.32)$$

sampled with $f_s = 10$ Hz in the interval from 0 s to 10 s. Figure 4.47 shows x_1 and two reconstructions based on the periodogram and IDFT. While the amplitudes are distinct, the periodograms for all three signals are exactly the same, as can be seen in Figure 4.48. Note that the original periodogram is distorted by leakage, and the spurious frequencies are replicated by the reconstruction.   □

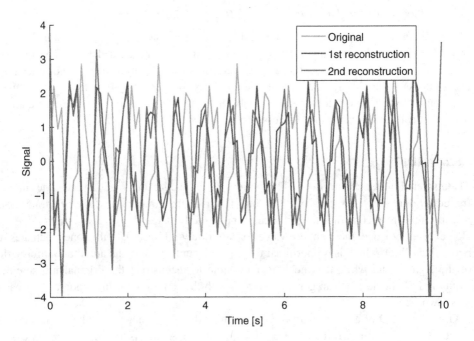

Figure 4.47   Signal reconstruction using IDFT

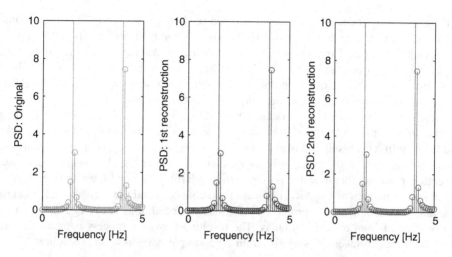

Figure 4.48   Periodogram before and after reconstruction

4.5.2.2   ARMA Processes

A different approach to generating a signal with a desired PSD uses linear time series models, so-called ARMA (autoregressive moving average) processes. The ARMA process of order (p, q) is a discrete time series defined by a difference equation:

$$x_k + \sum_{j=1}^{p} a_j x_{k-j} = \sum_{j=0}^{q} b_j \epsilon_{k-j}. \tag{4.33}$$

Here, the x_k are values of the process on an equidistant time grid, and the ϵ_k are independent, identically distributed random variables called *innovations*. Often, the innovations are assumed to be normally distributed with mean 0 and standard deviation $\sigma > 0$. In case $q = 0$, x_k is referred to as an AR(p) process, while $p = 0$ defines an MA(q).

Note that ARMA processes are closely related to the FIR and IIR filters introduced in Section 4.3.2. To be precise, an ARMA(p, q) is generated by filtering the sequence of innovations with an IIR filter of the same order, while an FIR filter yields a pure MA process. As for IIR filters, stability is a concern if the process contains an autoregressive component. In the context of time series, one requires *stationarity*, which means that the distribution of the x_k is invariant under a time shift (a precise definition is given in Chapter 6). For an ARMA process with independent normal innovations, stationarity is equivalent to BIBO stability of the IIR filter, or the condition that all roots of the polynomial $a(z)$ defined in Equation. (4.18) lie outside the complex unit circle.

To ensure that the coefficients b_k are uniquely determined, another requirement is that the roots of the polynomial

$$b(z) = \sum_{j=0}^{q} b_j z^j \tag{4.34}$$

are likewise located outside the complex unit circle. This condition is called *invertibility* of the MA process.

If the ϵ_t are independent normally distributed random variables, the PSD for the innovations is constant (white noise). As we know how the application of an IIR filter to a signal modifies its PSD (see Equation. (4.21)), we can use an ARMA process to approximate a target PSD by appropriate choice of the coefficients. Given a desired order for the process, this is a classic numerical optimization problem (e.g. least-squares fit) with additional constraints to ensure stationarity and uniqueness of the solution.

There exists a vast body of literature on ARMA processes and their properties. Comprehensive treatments can be found, for example, in the books by Priestley [192] or Brockwell and Davis [40]. One result that is of particular interest for the analysis of load signals can be found in Svensson [226]. There, it is shown that a second order AR process is sufficient to generate a random process with a prescribed irregularity factor and level crossing spectrum. The three parameters of the AR(2) process (two constants and the variance of the Gaussian innovations) are uniquely determined by the stability conditions and irregularity factor and can be derived from explicit formulas. The resulting Gaussian process is then transformed to get the required level crossing spectrum by a simple table-look-up procedure.

Remark 4.4

1. *An advantage of using ARMA processes for reconstruction instead of IDFT is that one can immediately generate signals of arbitrary length using the difference equation (4.33). To change the length of a signal derived by IDFT, one needs to interpolate the PSD to different frequencies. In the ARMA case, this is done implicitly by setting the order of the process which determines the behaviour of the PSD outside the set of Fourier frequencies.*
2. *Many of the algorithms for estimating the coefficients of an ARMA model actually work in the time domain, meaning that they require the original signal. In this context, the process representation can be used as a means of data compression, replacing a long signal by its ARMA coefficients. However, this is only possible off-line.*

   Note also that fitting an ARMA model in the time domain avoids any problems with leakage in the PSD.

4.5.2.3   Methods from Random Loads

While ARMA processes with normal innovations are already a very flexible class of time series models, they suffer from two drawbacks:

- They are linear stochastic models, which means that the dependence between observations is limited to correlation.
- The parameters depend on the time step Δt, and switching to a different grid is usually not straightforward.

The first point is actually not a disadvantage if we are interested in reconstructing a signal from its PSD. This is due to the fact that the PSD only contains information about correlation. In particular, an ARMA model is uniquely determined by its (rational function) PSD. Conversely, using a non-linear model for reconstruction requires additional assumptions about its non-linear behaviour before it can be adapted to the PSD via numerical optimization. And even then, a closed form solution for the PSD may not be available.

The second point can be addressed by using continuous time linear processes. An important class of models in this category are the stationary Gaussian processes introduced in Chapter 6 on random loads. These are completely characterized by their (continuous) PSD and can thus be fitted to a given target. In fact, such processes are closely related to ARMA models, as any discrete sample from a stationary Gaussian process is an MA process of infinite order and can be approximated by a finite order ARMA to any desired precision.

4.6 Summary

In this chapter, we considered methods to modify loads which either have been measured or calculated. Before describing the algorithms in detail, we first identified essentially three types of reasons for load editing, namely data correction, acceleration of rig tests or numerical simulation, and adjustment in view of specific tasks such as filtering out high frequent excitations or extrapolation to more severe events. As shown, a large variety of powerful methods for all these tasks exist. However, the application of the methods often is not straightforward in the sense, that obtaining the highest benefit with respect to a certain aspect (e.g. acceleration) might have an unwanted impact on others (e.g. the frequency content). Careful engineering judgment is thus required.

Therefore, we discussed which load properties are most important in which applications. For instance, when analysing a force acting on a stiff component, it is sufficient to consider the hysteresis cycles in the amplitude domain, while for a vibrating component such as a truck frame, the frequency domain of the load also needs to be taken into account. If several loads act on a component, the correlation of the signals needs to be considered in addition (see Section 4.1.1).

It is of great importance to keep in mind these relations during the editing process. Otherwise, too much effort may be spent on less important load features or, even worse, fatigue relevant load features might be neglected. Appropriate measures such as pseudo damage numbers, range pair spectra, frequency spectra, phase plots, or the RP method and their role in the process have been discussed too.

The methods in Section 4.2 are dedicated to cleaning the data prior to analysis. Due to the fact that there are many different sources of measurement errors, their impact on the data is manifold and can hardly be detected automatically although there are powerful software tools dedicated to fatigue oriented data analysis. In addition, the amount of data to be handled is considerable. Thus, the data checking and cleaning process might require a large effort. A few typical problems and some helpful methods for solving these have been presented.

The time domain editing methods described in Section 4.3 concentrate on fatigue oriented applications. General operations such as fade-in or out, cutting into pieces, interpolation, etc. have not been considered. The presentation has been roughly divided into, methods concentrating on the rainflow or amplitude domain and those dealing with the frequency domain.

One of the most important concepts is the hysteresis or rainflow filter. It is strongly motivated by local fatigue phenomena (stress-strain hysteresis cycles) and, not exclusively but most successfully, can be applied to loads acting on stiff components for accelerating tests or simulations. The reason for this is that the potential of acceleration (omission of signal segments with low damage potential), which is defined on the level of local stresses, can easily be back calculated to the level of external forces. The single as well as the multiple load case can be handled. Things are much more involved in the case of structures or systems with lower eigenfrequencies. In these cases, the dynamics of the structure needs to be taken into account. Although the back calculation of external forces from local stresses could also be done in such a setting, this approach is not supported by commercial tools and to our knowledge not or only rarely applied in vehicle industry. Instead, optimization methods such as those described in Chapter 9 can be used here.

Another important concept is based on the spectral or Fourier analysis. Frequency filter can be adjusted in order to restrain or remove a specific frequency band within the signal. Typical examples are the elimination of measurement noise or temperature drifts or the extraction of the rough road part from a signal using suitable low or high pass filters. These methods are not only useful in the context of fatigue loads but often applied in general signal processing problems, for instance, electrical applications. In our context, they are especially important when dealing with vibrating structures or systems.

An approach based on extreme value statistics for extrapolating a signal to more extreme events (e.g. deeper pot holes) has also been described. This method can be used, for instance, to first estimate the probability of higher amplitudes than those observed in the measurement and then to adapt the signal to be used on the test rig accordingly.

More comments and conclusions on time signal editing can be found in the rather detailed summary of Section 4.3.

Section 4.4 on load editing in the rainflow domain starts with two technical methods, namely re-scaling and superposition of rainflow matrices. These methods are rather straightforward and enable working with rainflow matrices without having to go back to the time domain.

More interesting are the methods for extrapolating rainflow matrices of measured signals to longer durations or more extreme events. In both cases, the underlying rainflow matrix is interpreted as the histogram of one realization of a stochastic process generating hysteresis cycles. The details of the methods are derived from that observation and the ideas commonly used in statistics for estimation of non-parametric distributions. As in the case of time signal editing, large amplitudes can be handled specifically using the theory of extreme value statistics. The need for these methods arises during the derivation of a customer load profile from a few measured service loads. Thus, for the discussion of the applicability and the benefit of these methods we refer also to Chapter 8.

The final section in this chapter is dedicated to the generation of time signals for rig testing or numerical evaluation based on some fatigue relevant information such as rainflow cycles or range pair spectra. Constructing a time signal from a given rainflow matrix is especially interesting if the matrix has been derived using, for instance, the extrapolation methods described above and a corresponding rig test is to be performed. This approach works very well for a single load and without controlling the frequency behaviour. If multi-input loads and vibrational aspects need to be covered, the methods described in Chapter 9 are better suited.

A second group of generation methods is based on spectral information, for example, a given PSD. There are a lot of results relating spectral properties to rainflow cycles under certain assumptions about the underlying stochastic process. If these assumptions are fulfilled (e.g. Gaussian-type processes), a time signal and thus also the hysteresis cycles can be reconstructed from a given PSD. In general, however, the PSD of a signal does not uniquely define the rainflow cycles and care must be taken in order not to get wrong amplitudes from a PSD reconstruction. Similar remarks apply to the other spectral reconstruction methods such as ARMA processes.

5

Response of Mechanical Systems

In this chapter we want to give a brief introduction to modelling techniques for mechanical systems, encompassing the entire load path from system loads to local stresses and strains. The description of the modelling techniques "Finite Element Method (FEM)" and "Multibody System Simulation (MBS)" focuses on their use for simulation and their potential in the field of load data analysis. Moreover, relations between simulation and load data analysis are pointed out. Finally, the issue of invariant system loads is addressed, that is, the question of getting realistic excitations before measurements on prototypes have been made.

For the reader's convenience, an introduction to some basics on FEM such as the kinematics of FE models, the derivation of the equations of equilibrium, a description of linear elastic material behaviour and the transition from continuous to discrete systems is included in Appendix D. Accordingly, Appendix E contains some basics about multibody simulation techniques.

5.1 General Description of Mechanical Systems

In the context of load data analysis for trucks, a mechanical system might be a subassembly or a component of the truck but it might also represent the full vehicle. It is rather natural to decompose a full vehicle or a large subsystem such as a cabin including its suspension into several parts, denoted here as bodies, which are connected to each other. Some of these connections can very well be represented as idealized joints, others are more complicated, for example, an air spring or the contact between a tyre and the ground. Having this in mind, we arrive at a type of model which is called a multibody system model (MBS model). In such a model, the bodies are assumed to be rigid (meaning that there is no inner deformation) and they are tied together by joints or force elements.

The rigidity assumption cannot be applied to all bodies under all possible loads. The tyre is a typical example, but also large structures such as the frame or the cabin cannot always be treated as rigid. Instead, we have to presume a certain flexibility of some of the bodies at least if the loads are high enough. This is where the finite element method (FEM) comes in, which is a discretization method for numerically handling continuum mechanical problems.

Guide to Load Analysis for Durability in Vehicle Engineering, First Edition. Edited by P. Johannesson and M. Speckert.
© 2014 Fraunhofer-Chalmers Research Centre for Industrial Mathematics.

A straightforward application of this theory leads to detailed models of single components. Often a linear approximation is chosen because of its simplicity and numerical performance, but if necessary, the method can also handle more sophisticated non-linear models. As we will see later, the finite element model of a single body can then be combined with the multibody model of the remaining system, leading to a combined MBS/FEM model.

Along this line, we arrive at MBS models, in which some of the bodies are represented by an FEM-based flexible model. This approach is typically taken to simulate a vehicle or a subassembly under specific loads in order to derive section forces acting on the bodies of the system. In these cases, the flexible bodies are usually represented by linear FE models and the non-linearities of the full system (if any) need to be captured as a part of the MBS model.

Alternatively, one could also use the finite element method as such to model not only a single body but even a full vehicle. The advantage of this approach is that there is only one method involved. However, the models typically become very large and especially in cases where non-linearities need to be taken into account, the performance of a simulation might be inadequate for durability applications.

The goal of the analysis of course plays an important role too. Calculating the response of a component under a static load is much easier than the computation under an irregular load of, say, 30 minutes sampled at 200 Hz which acts on the system and where the dynamic behaviour (mass, inertia tensor...) cannot be ignored.

Prior to the modelling of a mechanical system, it has to be clearly defined which results are needed, which effects (contact, flexibility, non-linearity of a spring...) have to be accounted for, and what type of model is both capable of accounting for the desired effects and fast enough for practical use.

Consequently, a large number of questions have to be answered and a specific formulation of the mechanical system can then be attempted, e.g.

- Do I have to take flexibility into account?
- How does the material respond under loading?
- Does temperature affect the materials response?
- Is one aware of structural/geometrical/material or other types of non-linearities affecting the system?

Answering these questions at an early stage helps to find the best-suited model set-up for the (numerical) analysis of the response of the component. Figure 5.1 schematically summarizes the general aspects to be considered in the description of mechanical systems.

Before describing such a model set-up in more detail, some more general introductory remarks are given in Section 5.1.1, for MBS, and in Section 5.1.2 for FEM. The most basic notions that will be employed in the descriptions in the forthcoming sections are briefly introduced in Appendices D and E.

5.1.1 Multibody Models

In this section we assume that the system under consideration (for instance a full vehicle) consists of many components which should be modelled as rigid bodies, tied together by joints or force elements. The decomposition of the system into different bodies including

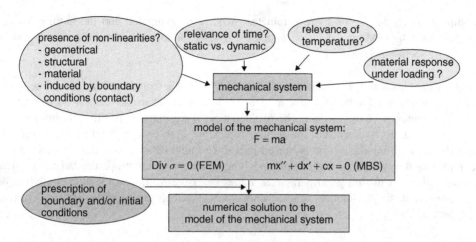

Figure 5.1 Schematic summary of the general aspects to be considered in the description of mechanical systems

their mass and inertia tensors is usually rather straightforward. At first glance, this seems to be the case also for the definition of the connections. But a closer look shows that this is not at all true.

Example 5.1 (Correct Force Flow: A Simple Model of a Door)

A door is connected to the door frame with two revolute joints. This model only works if the axes of both revolute joints are exactly the same. In that case both revolute joints imply the same restriction to the door body and one of the joints is redundant. A simulation of such a model might give good results for the motion of the door. However, the constraint forces in the joints (section forces of the door body) are not uniquely defined due to the redundancy.

In the redundant model only the sum of the vertical forces in both joints is uniquely defined by the force balance. The distribution of the force between upper and lower joint is arbitrary. In reality one would expect that the vertical section forces are approximately the same at the upper and the lower joint. A possible solution to this problem is to use suitable force elements instead of joints. In this case a bushing element without stiffness for the rotation about the vertical axis and appropriate stiffnesses for the remaining axes would be sufficient.   □

Example 5.2 (Adequate Model Complexity: An Air Spring)

As another example, consider an air spring between an axle and a frame. In reality there are many physical effects such as the flexibility of the rubber, the temperature-dependent properties of the air under pressure, the characteristics of valves and throttles, and many more. As a detailed physical model treating all these effects is usually not an option, semi-physical models with more or less parameters are often used. For the air spring, a model using non-linear (pressure dependent) characteristic curves for the force-displacement relation could be a solution.

No matter which model is finally chosen, there are typically some free parameters which need to be adapted to reality by fitting the model behaviour to some measured data. The more complex the model is, the more parameters need to be estimated and the more measured data is necessary. More details on another example, namely a hydraulic actuator, can be found in Section 5.2.2.   □

These examples show that building a MBS model is not a straightforward task. The mathematical toolboxes for building the models are very rich. There are almost no restrictions on the complexity of the models. The mathematical representation of the real physical connections needs to be such that the section forces are both uniquely defined and correctly distributed. In doing this, a compromise needs to be found between, on the one hand, highly complex models which are capable of modelling all the important properties but need a great amount of effort to find good parameter values (measurements and identification), and on the other hand, simpler models which cannot cover all effects but need less effort to estimate the parameters.

5.1.2   Finite Element Models

In this section we assume that the flexibility of the body (component) under consideration cannot be ignored. Thus, we need to deal with continuum mechanics and use the finite element method as a discretization method (see Appendix D).

Mechanical systems and thus also FE models can be characterized according to their dependence on time: a mechanical system is said to exhibit static behaviour if its response is independent of time t. If the system depends on time, it is called dynamic.

Further, FE models can be classified according to their material behaviour, that is, their response to loading. In idealized descriptions, material behaviour can range from purely elastic to ideally plastic or linearly viscous responses under applied external loads. However, combinations such as elasto-plastic behaviour or visco-elastic responses typically occur.

In the framework of durability and fatigue life assessment of components or parts of mechanical systems, it is sometimes sufficient to restrict attention to linear elastic material behaviour. Although this is not true for stress and strain calculations at local notches in general, it is often sufficient to calculate section forces acting on a component.

If temperature should be a relevant quantity in the analysis of components or parts due to, for example, thermal stresses, a coupled thermo-mechanical problem has to be solved which is typically non-linear due to the coupling of the balance of linear momentum (covering the mechanical part of the problem) and the balance of energy (covering the thermal part of the problem).

Yet, even in purely mechanical problems non-linearities can occur: Non-linearities can arise as structural-geometrical and material non-linearities, implying the occurrence of, for instance, large displacements. Non-linearities can also arise in the coupling of deformation and loading direction and as material laws relating stresses and strains in a non-linear fashion. Moreover, non-linearities can be induced by the specification of boundary conditions, for example, when contact between neighbouring surfaces of different components is accounted for.

Once there is sufficient information about the mechanical system under consideration, a (mathematical) model can be set up representing the mechanical system in a certain approximative way. Typically, this mathematical model involves the balance of linear momentum (Newton's second law, stating that the time rate of change of linear momentum equals the sum of all forces applied to it). Depending on what types of mechanical systems are considered, the balance of linear momentum appears in different mathematical representations. However, the content is always the same. Further, system boundary conditions must be specified for static systems, while initial and boundary conditions are required if dynamic systems are considered.

As the balance of linear momentum captures only properties that are common to all systems (independent of the material they consist of), a so-called material (or constitutive) law is needed to describe material-specific behaviour. Hooke's law, relating stress and strain in a linear fashion, is a well-known example of such a material law and reduces, in the one-dimensional case, to $\sigma = E\varepsilon$, where σ is the stress, ε is the deformation, and E is Young's modulus. The resulting mathematical model, representing a more or less idealized approximation of the mechanical system, can then typically be solved numerically by employing suitable (commercial) software tools.

5.2 Multibody Simulation (MBS) for Durability Applications or: from System Loads to Component Loads

In this section, we give a short introduction to the most important notions in multibody simulation. The emphasis is not on details or theoretical considerations, but on giving an idea of what can be done, what is easy, what is difficult, and what kind of results can be expected. Of course, the accuracy of these results is of great interest, but a general quantitative statement about the magnitude of errors cannot be made. This has to be analysed for each specific model of interest and is outside the scope of this guide.

5.2.1 An Illustrative Example

To illustrate the topics relevant in the context of load analysis, we first introduce an example of a front suspension of a truck including leaf springs, dampers, and bumpstops, see Figure 5.2. It is based on the model of a real truck. However, it has been simplified and modified for illustration purposes. Thus, the physical values of parameters and numerical results do not correspond to a specific truck. In the sequel, the parts of the model and its topology are described in detail.

The axle is modelled as a rigid body. There are connections to the stabilizer (two spherical joints), to the leaf springs (8 fixed joints for two leaf layers at front, rear, left, and right), to the dampers (two spherical joints), to the wheel hubs (two revolute joints for steering), and to the chassis (two bumper forces). The bumper forces are modelled as translational forces defined by a non-linear characteristic curve.

The wheel hubs are modelled as rigid bodies with revolute joint connections (y-axis) to the wheels, and revolute joint connections (z-axis) to the axle for steering. Universal and spherical joints respectively are used for the trail rod and steering rod connections.

The wheels are modelled as rigid bodies with revolute joints to the wheel hubs. They are used for exciting the whole system via enforced motions or forces.

Steering is enabled via the sequence chassis $\xleftrightarrow{\text{revolute}}$ steer arm $\xleftrightarrow{\text{spherical}}$ steer rod $\xleftrightarrow{\text{universal}}$ wheel hub left $\xleftrightarrow{\text{spherical}}$ trail rod $\xleftrightarrow{\text{universal}}$ wheel hub right.

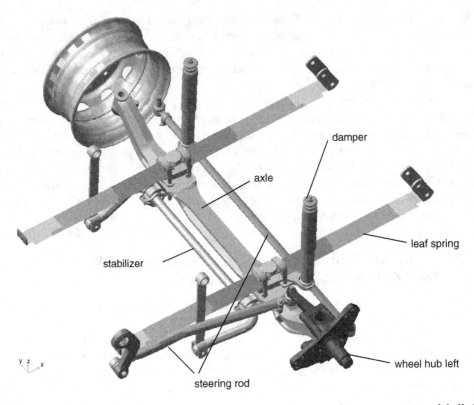

Figure 5.2 Sketch of the model of the front suspension. The most important parts are labelled. For a better overview, the wheel hub at the right side as well as the wheel at the left side are hidden. Also the chassis including the bumpers is hidden

The dampers are modelled as two rigid bodies. The lower part is connected to the axle via a spherical joint, the upper part is connected to the chassis via a universal joint. Between the lower and upper part, there is a cylindrical joint and a linear viscous damping force.

The stabilizer is modelled as two rigid bodies connected by a cylindrical joint and a rotational spring with a certain stiffness and damping modelling the stabilizer's flexibility. The connection to the chassis is given by the stabilizer arms, which are fixed to the chassis via universal joints and to the stabilizer via spherical joints.

The leaf springs are modelled as flexible bodies. Both springs to the left and to the right are split into a front and a rear part, which are mounted to the axle using fixed joints. There are two layers as can be seen in Figure 5.3. The leaf spring layers are attached to the axle using fixed joints. Both layers are divided into 3 flexible bodies (indicated by grey scales), which are connected to each other with fixed joints. At the end of the spring, both layers are connected to each other using a so-called in-plane joint, which permits a relative translation in the x- and y-direction and blocks the others. In the simulations below, 5–7 non-rigid body modes have been used for each flexible body. See Section 5.2.3 and the references therein for a discussion of flexible bodies in MBS.

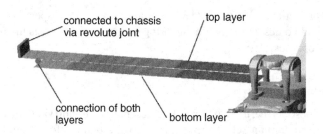

Figure 5.3   Sketch of the leaf spring model

5.2.2   *Some General Modelling Aspects*

The suspension model contains most of the typical entities used in multibody system simulation, namely rigid bodies, joints and constraints, force elements, and flexible bodies. Using this example, some aspects regarding the modelling of mechanical systems are discussed in some more detail.

5.2.2.1   Rigid Bodies

Rigid bodies are the simplest and most common elements representing parts or components of a mechanical system. Using this approach, one implicitly assumes that the components undergo no or at least only a very small deformation. For a component like the wheel hub, this assumption is well justified. If we are dealing with durability issues, the frequency range of interest is not too large, say, below 100 Hz. But if we are interested, for example, in acoustical properties, then this might no longer be valid and we could be forced to take into account the flexibility of the component.

It is not possible to calculate stresses or perform a fatigue life estimation on a rigid body, since stresses are a consequence of strains and strains are zero on a non-deformable body. Nevertheless, it makes perfect sense to model a 'stiff' component as rigid, even if a fatigue life assessment of the component is planned. The multibody simulation is used to calculate section forces. 'Rigid' body representation limits the MBS model size and shortens simulation time. Fatigue analysis is subsequently performed with an FE model of the component itself, created from detailed component geometry and material properties, with section forces from MBS used as load input.

In our suspension model all components with the exception of the leaf springs and the stabilizer are represented as rigid bodies. Whether this representation of the axle by a rigid body is justified depends on the type of the analysis we want to perform and the eigenvalues of the axle. If the eigenvalues are within the range of the excitations of the system, then we had better use a flexible model of the axle.

The property of a component being rigid or flexible with respect to the dynamical range of interest cannot be discussed without taking the connections to the remaining system into account. This means that the overall stiffness and damping of a subsystem (a component including bearings, etc.) could be achieved either with a flexible body and idealized joints, with a rigid body and a bushing type of connection (see below), or with a flexible body and bushing type connections. An example is the stabilizer in our suspension model, which is split into two rigid bodies and a rotational spring connection that gives the required stiffness and damping properties.

5.2.2.2 Built-in-flexibilities

Built-in-flexibilities can be used if a rigid representation is not valid. All multibody simulation tools offer a certain set of pre-defined flexible bodies (for example, linear beams) which need to be parameterized by the user. Again a stabilizer can serve as an example. As an alternative to the approach taken in our truck example, the stabilizer could be built from a sequence of beam elements, which are described by their length, cross-section, and material parameters. An example of a passenger car stabilizer is shown in Figure 5.4.

Even if linear beam elements are used, the entire component consisting of the sequence of beam elements behaves geometrically in a non-linear fashion because the deformation of the first beam leads to a new orientation of the frame of the next beam, and so on. Of course, the same is true if a sequence of flexible bodies is used, as it is the case for the leaf springs in our example model. Using only one flexible body for one part of the spring instead of three would lead to smaller deformations.

5.2.2.3 Constraints

Constraints including joints restrict the motion of a body with respect to other bodies or the ground (fixed inertia frame). These constraints are mathematically idealized in the sense that the restricted DOF like the translational in a spherical joint is exactly set to zero, whereas a motion of the free DOF is fully unrestricted and does not require the presence of a force or a torque (no work is spent during motion). In contrast to this, in a real physical connection, there are always some clearances and some friction. For this reason, all MBS packages offer the possibility of adding friction to joints. Coulomb friction is the simplest of these models.

5.2.2.4 Force Elements

Force elements are typically used to model springs, dampers, bumpstops, and many more connections that cannot be represented by an idealized joint. Usually they are defined by a relation between force and motion. In the simplest case of a linear spring this reads as $F = k \cdot \Delta s$, where Δs is the relative displacement, k is the spring constant, and F is the resulting force. The complexity of such elements ranges from such simple models up to

Figure 5.4   Sketch of a passenger car stabilizer

elaborate tyre models, which, from the viewpoint of multibody modelling, are nothing else than a sophisticated force element.

Another typical usage of force elements is to model bearings including rubber mounts. An example is the connection of the leaf spring of a trailer axle to the chassis. Often these force elements are called bushings. They are represented again by a formula of the type $F = K \cdot \Delta s$, where in this case the vector F contains 3 forces and 3 torques, Δs contains the translational and rotational displacements and K is a 6×6 -matrix of stiffnesses. K is non-diagonal if the stiffnesses depend on each other. Non-linear relations can be used as well. A similar expression can be used for damping.

5.2.2.5 Additional Modelling Elements

Additional modelling elements such as arbitrary state variables, algebraic or differential equations, or transfer functions can be employed to describe complex subsystems, for instance, a hydraulic cylinder. A rather simple hydraulic actuator model illustrates some of these concepts. The example model is taken from Dronka [85]. In Figure 5.5, the actuator and its most important components and state variables are sketched.

The valve dynamics is described by the second order linear differential equation

$$\ddot{y} = -\omega_0^2 y - 2D\omega_0 \dot{y} + k_p \omega_0^2 U, \tag{5.1}$$

where U is the electrical input to the valve, D, k_p, ω_0 are valve parameters, which have to be determined from experiments, and y is the resulting motion of the spool. The hydraulics can be described by

$$Q_A = \begin{cases} c_V y \sqrt{p_{sys} - p_A}, & y > 0 \\ c_V y \sqrt{p_A - p_0}, & y < 0 \end{cases}, \quad Q_B = \begin{cases} c_V y \sqrt{p_B - p_0}, & y > 0 \\ c_V y \sqrt{p_{sys} - p_B}, & y < 0 \end{cases} \tag{5.2}$$

$$\dot{p}_A = \frac{E}{V_A + Ax}(Q_A - A\dot{x}), \quad \dot{p}_B = \frac{E}{V_B - Ax}(-Q_B - A\dot{x}) \tag{5.3}$$

where E describes the compressibility of the oil, c_V is a valve constant describing the orifice, A is the effective area of the piston, V_A, V_B are constants, and x is the resulting

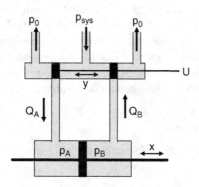

Figure 5.5   Sketch of a hydraulic actuator including valve, cylinder and piston

motion of the piston. The actuator force is then given by

$$F = p_A A - p_B B - F_R(x, \dot{x}, \ldots), \tag{5.4}$$

where F_R is a more or less complex expression for the friction forces. This actuator model contains the state variables y, Q_A, Q_B, p_A, p_B, the algebraic equations (5.2, 5.4), and the differential equations (5.1, 5.3). The model is "semi-physical" in the sense that some physical properties and quantities like pressure, flow, compressibility are addressed, but not all details like the geometry of the orifice, turbulent flow, etc., are modelled.

For the description of the valve dynamics, an equivalent transfer function could have been used alternatively. One could also try to model the whole actuator system from the electrical input U to the force F as a transfer function, but this simpler linear approach cannot cover the non-linear Bernoulli term in Equation (5.2) and only works for small displacements x and not too high pressures p_A, p_B.

5.2.3 Flexible Bodies in Multibody Simulation

5.2.3.1 Kinematics

A rigid body within a multibody model has 6 degrees of freedom at the most. It is connected to the remaining system by some joints or force elements. Each of these connections has a location (marker) on the rigid body, which is bodily fixed during motion. As it cannot deform, the difference vectors between the coupling markers are constant in the local frame of the body. The position of such a point x is given by

$$r(x, t) = r_0(t) + S(t)R(x), \tag{5.5}$$

where $R(x)$ is the position of the point x in the local frame of the body, $S(t)$ is the rotation matrix which transforms the local coordinates into the global reference frame, and $r_0(t)$ is the position of the local body frame expressed in the global reference frame. Since $r_0(t)$ contains the 3 translational DOF and $S(t)$ depends on 3 angles (for instance, Euler or Cardan angles), Equation (5.5) contains the 6 DOF of a rigid body. If we want to take flexibility or deformability into account, the position of a material point of the body (for instance, one of the markers) is given by

$$r(x, t) = r_0(t) + S(t)(R(x) + u(x, t)), \tag{5.6}$$

where $u(x, t)$ denotes the displacement of the point x with respect to the reference (undeformed) configuration due to the deformation. Getting $u(x, t)$ during a multibody simulation would require integrating the equations of motion of a continuum mechanical description (partial differential equations) of the body or the discretized equations based, for example, on finite element modelling. Since an FE model of a component often comprises between 10^4 to 10^6 DOF, this approach is not generally feasible.

5.2.3.2 Modal Representation

One therefore tries to describe the deformation of the component by a superposition of mode shapes in the form

$$u(x, t) = \sum_{k=1}^{n} \Phi_k(x) \cdot q_k(t), \tag{5.7}$$

where $\Phi_k(x)$ are the mode shapes, that is, the displacements of the material points x, n is the number of mode shapes, and $q_k(t)$ are the modal participation factors. In the following, the basic ideas are described briefly. Some more details are given in Section 5.3 and the references therein.

The approach of Equation (5.7) (called Ritz-Ansatz in the mathematical theory) separates the spatial variable x from the time-dependent variables $q_k(t)$. Of course, formula (5.7) cannot represent an arbitrary deformation of the component, but only such deformations lying in the linear span of the modes Φ_k. Since the true deformation of the component under a certain excitation of the system is not known in advance, it is not clear which modes and how many modes are required to get a reasonable approximation based on Equation (5.7).

Since a deformation requires a certain force acting on the component, it is rather natural to apply a force to one of the restricted DOF in the coupling points while keeping all other DOF in the coupling points fixed. This corresponds to a so-called unit load case in the FE analysis. A unit load is applied to one coupling DOF, and a static FE analysis is performed while keeping all the other coupling DOF fixed. This is done for all possible forces and torques acting on the component, that is all coupling DOF. If we superimpose the resulting mode shapes (static modes), based on Equation (5.7), we can describe the deformation for an arbitrary location of the coupling points. An inner vibration of the component keeping all coupling points fixed cannot be described by this approach so far. Thus, the so-called constraint normal modes, that is the vibration modes with fixed coupling points, are added to the set of mode shapes resulting in a description

$$u(x,t) = \sum_{k=1}^{n} \Phi_k^s(x) \cdot q_k^s(t) + \sum_{k=1}^{m} \Phi_k^c(x) \cdot q_k^c(t), \tag{5.8}$$

where Φ_k^s denotes the static modes and Φ_k^c denotes the constraint normal modes. Organizing the individual mode shape vectors Φ_k^s, Φ_k^c in a matrix and the modal coordinates q_k^s, q_k^c in a vector

$$q = \begin{pmatrix} q^s \\ q^c \end{pmatrix} \quad \Phi = (\Phi^s, \Phi^c) \tag{5.9}$$

we can write

$$u(x,t) = \Phi(x) \cdot q(t). \tag{5.10}$$

However, according to Equation (5.6), the rigid body degrees of freedom are explicitly handled by the multibody mechanism and must not be contained in the modal description according to Equation (5.8). In order to subtract the rigid body modes from the mode set in Equation (5.8), the modes are orthogonalized leading to a new set of modes with increasing "synthetical" frequency, among which the first 6 modes are an approximation of the rigid body modes.

The orthogonalization process works as follows. The mass and stiffness matrices M and K of the flexible body are transformed based on the mode set Φ. The reduced matrices are $\hat{M} = \Phi^T \cdot M \cdot \Phi$ and $\hat{K} = \Phi^T \cdot K \cdot \Phi$. Solving the eigenvalue problem $\hat{K} \cdot \hat{q} - \lambda \cdot \hat{M} \cdot \hat{q} = 0$ and arranging the corresponding eigenvectors in a transformation matrix T, we can compute a new basis of orthogonal modes and corresponding modal participation factors by

$$\hat{\Phi} = \Phi \cdot T \tag{5.11}$$

$$q = T \cdot \hat{q} \tag{5.12}$$

$$u = \Phi \cdot q = \Phi \cdot T \cdot \hat{q} = \hat{\Phi} \cdot \hat{q}. \tag{5.13}$$

The rigid body modes are eliminated from $\hat{\Phi}$ (the first 6 modes in that set in the order of increasing frequency) and the remaining set of modes can now be used as a representation of the flexibility of the component within the multibody mechanism.

This approach starting with the set (Φ^s, Φ^c) of static and constraint normal modes, orthogonalization of this set, and elimination of the rigid body modes is known as the Craig-Bampton approach and implemented in many FE and MBS packages.

5.2.3.3 Damping

So far, we have not taken damping into account. For the calculation of the static modes, damping is of no relevance. During the calculation of the constraint normal modes damping has deliberately been ignored. Although there are some methods for deriving a damping matrix D from material parameters, this approach is usually not taken. One reason is that the result of the FE-based damping calculation is often not satisfying. Pure material damping is small and does not take into account the effects of bearings, sealing, etc. Another reason is that the damping matrix, because it is not diagonal, leads to much higher numerical efforts.

As a consequence, the typical procedure is to define modal damping after the process of orthogonalization, that is to add a diagonal damping matrix $\hat{D}$ to the transformed mass and stiffness matrices $\hat{M}$ and $\hat{K}$. The amount of damping is usually defined in percentage relative to the critical modal damping. 2%, 5%, or 10% for higher frequency modes are typical values. Of course, the damping parameters could be determined experimentally, but this requires some effort and in most cases an experienced-based setting is used instead. Since this simple linear approach does not fully cover the physical effects (e.g. friction in joints), a rigorous justification of the common practice is missing. See, for example, Lion [144] for a discussion of this problem.

5.2.3.4 Selection of Modes

Another important topic, which has not been mentioned so far, is how many modes should be used. Regarding the static modes, this is easy. For each coupling DOF, we have to perform a static FE analysis. For a fixed joint we get 6 static modes, for a revolute joint we get 5, for a spherical joint 3, and so on. If a force element is attached to the flexible body, 6 modes according to 3 forces and 3 torques have to be used.

For the constraint normal modes the answer is not as straightforward. Here, the most obvious approach is to estimate the frequency content of the forces acting on the body, although these are usually unknown prior to a simulation. Since the modal representation of the body as described above is based on linear FE theory, the modes with a frequency higher than the highest excitation frequency are not excited. Thus, one often uses a crude estimate of the highest excitation frequency and selects all constraint normal modes up to twice that value. But one also has to take into account the elimination of the rigid body modes. If one uses too few modes, the (small) mode set $\Phi = (\Phi^s, \Phi^c)$ does not properly approximate the rigid body modes such that deleting the first six modes from the orthogonalized set $\hat{\Phi}$ 'removes more than just the rigid body modes'. Because of that, one often extends the frequency limit for the mode selection.

There are also practical aspects, namely the performance and stability of the multibody simulation including the flexible body. Using too many modes can reduce the performance significantly. The number of modes of course depends on the specific component, the system is a part of, and the excitation. Typical numbers (in addition to the number of static modes) range from 10 to 100.

The procedure we described to construct the mode set for a flexible body representation is called the 'Craig-Bampton method'. It is the most common method, but not the only way to do this. Another approach is to use so-called static attachment modes, which are the result of a unit load for one DOF keeping all the other coupling DOF free and using the inertia relief method to balance the applied unit force. In addition, the free normal modes instead of constraint normal modes are used. Sometimes also frequency response modes are added to the mode set in order to adapt the mode set to the applied load. A more detailed discussion of these methods and their properties is beyond the scope of this guide. We refer the interested reader to Craig and Bampton [60] and Craig and Kurdila [61] and the references therein.

5.2.4   Simulating the Suspension Model

5.2.4.1   Calculation of Section Forces

We now return to our truck suspension example. The task is to calculate section forces on the components for a subsequent fatigue life analysis. The underlying loading condition is given by measured wheel forces on a track of interest. The data we are using here as an illustration are based on some measurements on a public road with a different vehicle and have been modified before application to our model. Thus, as has been mentioned already in Section 5.2.1, the results below are typical of truck engineering but do not correspond to a specific vehicle.

The chassis of the model as described in Section 5.2.1 is 'fixed to ground' and the wheels are excited by the vertical wheel forces. The steering angle is set to zero. This corresponds to a 2 DOF suspension test rig. The section forces will be calculated under these simplifications, although this is not a 1-1-simulation of the real driving manoeuvre. In Figure 5.6, the left wheel hub and its joints to the suspension are shown. The type of joints and the resulting section forces and torques are indicated. In this case, we obtain 5 (revolute to wheel) + 5 (revolute to axle) + 4 (universal to steer rod) + 3 (spherical to trail rod) = 17 loads.

Doing the same thing for the axle, we end up with 72 section loads. This large number of loads is a result of the 8 fixed joints between the axle and parts of the leaf spring model. Of course, this type of constraint does not fully reflect the real connection between the leaf spring and the axle. The same is true for spherical joints between the stabilizer and the axle. In reality there are no or at least only small longitudinal and lateral forces at that point, whereas the joint we use here can, in principle, transfer both. This need not be a problem because the load path of the entire model in our case is such that these forces are small in the simulation as well. Thus, as has been mentioned already in the introductory section, 5.1.1, it is important to check that the model and its connections are capable of representing the correct load path between all parts of the machine.

Due to the fact that we only apply a vertical excitation, we mainly concentrate on vertical section forces F_z and torques M_x around the longitudinal axis in the following. As an example of section forces, the corresponding results at the left side of the axle are shown.

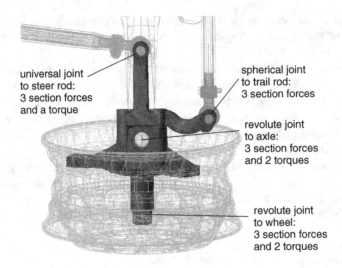

Figure 5.6   Sketch of the left wheel hub including the joints

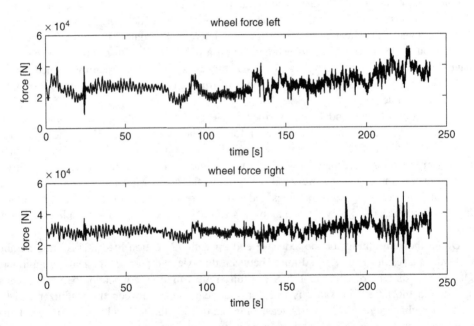

Figure 5.7   Vertical wheel forces

   In Figure 5.7 the excitations (load input) are shown. The range of forces is approximately between 10 kN and 50 kN. There are some pronounced peaks on the left side at about 25 seconds and some on the right side at about 190 seconds and 220 seconds. The resulting wheel motion is shown in Figure 5.8. The range of motion is between -150 mm and 120 mm. In particular, the effect of the force peak at 25 seconds can be seen clearly.

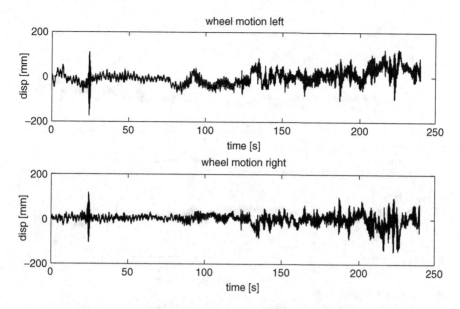

Figure 5.8   Vertical wheel motion

Figure 5.9 shows F_z and M_x at the wheel hub joint. Figure 5.10 shows the vertical section forces to the axle at the damper, the bumpstop, the stabilizer, and the leaf spring. Due to the simplified modelling of the connection of the leaf spring to the axle, only the sum of the vertical forces at the 4 fixed joints (both layers at front and rear side) is shown here. The forces and torques in the other directions are smaller and have not been plotted. As can be seen, the bumpstop is excited mainly in the last part of the track with the exception of one event at about 25 seconds.

In this manner, all loads acting on the axle can be derived from a simulation and applied to a suitable FE model of the axle. If the eigenfrequencies of the axle are well above the highest load frequency, a static superposition procedure will suffice. If not, a modal superposition procedure has to be used. See Section 5.3 for a discussion of static and modal superposition.

5.2.4.2   Transfer Function Analysis

At this point, we want to analyse the transfer behaviour between the vertical excitation force as input and the vertical wheel motion as output. In the following, calculations based both on a linearization of the mathematical model and the (non-linear) simulation results are done and compared with each other.

The model, as described in Section 5.2.1, has 124 degrees of freedom. This is calculated as follows: The front part of one of the leaf springs contains 6 flexible bodies with 36 non-rigid body modes. This results in $6 \cdot 6 + 36 = 72$ DOF. The connections between the flexible bodies and to axle and chassis comprise 6 fixed joints and one revolute joint between leaf and chassis, which together remove $6 \cdot 6 + 5 = 41$ DOF. The connection between lower and upper layer removes 2 DOF, such that there are in total $72 - 41 - 2 = 29$ DOF. For

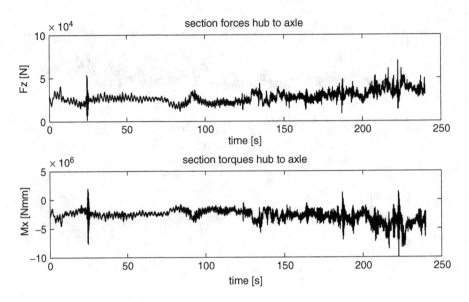

Figure 5.9   Section force F_z and torque M_x between hub and axle

the rear part of the leaf spring, an additional part between leaf spring and chassis, which is connected via a revolute joint (see Figure 5.2), introduces one more DOF, such that the leaf spring introduces in total $2 \cdot (29 + 30) = 118$ DOF. Since steering is blocked, the remaining part of the model has 6 DOF, such that we obtain 124 DOF for the entire model.

A linearization of the equations of motion of the model around an equilibrium state (here under 30 kN vertical wheel load) leads to a so-called state space representation in the form

$$\dot{x} = A \cdot x + B \cdot u \tag{5.14}$$

$$y = C \cdot x + D \cdot u, \tag{5.15}$$

where x denotes the vector $(q, \dot{q})^T$ of the DOF and the velocities, u is the vector of inputs, and y is the vector of outputs. In our example, we have chosen the vertical wheel forces as input u and the vertical wheel motion as output y. The matrices A, B, C, D describe the system for small motions around the equilibrium state.

The transfer functions (FRF) of such a system can be calculated easily. Since we have 2 inputs and 2 outputs, the FRF is 2×2 matrix of transfer functions. In Figure 5.11 the function between left force and left motion is shown.

The function looks very similar to a transfer function of a damped simple linear 1-DOF spring. In fact, in a small neighbourhood of the equilibrium state, the non-linearity of the system due to the leaf spring model is of no relevance and the non-linear bumpstops are not excited at all. The peak at about 3.3 Hz corresponds to the first eigenmode of the system (see Figure 5.12).

In addition, the transfer function can also be estimated based on simulation results using the methods described in Section 3.2. As is shown in Figure 5.8, the wheel motion is fairly large if the excitation from the modified public road measurement is used, and Figure 5.10

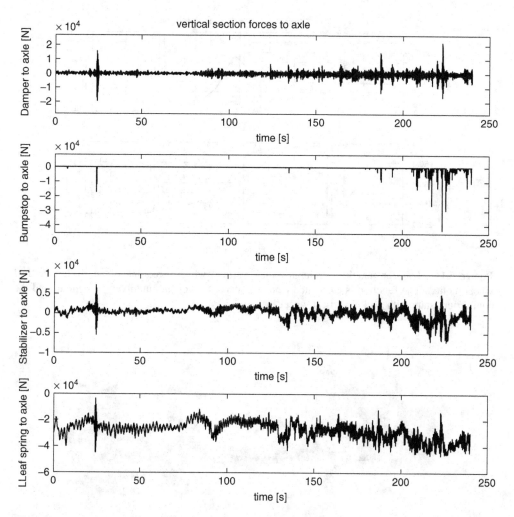

Figure 5.10 Vertical section forces to the axle at the damper (1), the bumpstop (2), the stabilizer (3), and the sum of vertical forces at the leaf spring joints (4)

shows that the bumpers are excited in that case. The system behaviour is thus no longer linear. Nevertheless, transfer functions can be calculated. In Figure 5.13 the corresponding estimation results are shown.

The overall shape resembles the theoretical result. The peak is at 3.3 Hz approximately. At higher frequencies the phase estimation deteriorates. Since the sampling rate used for the input and output time series is 100 Hz, we can only resolve the spectral content up to 50 Hz. To check the influence of the non-linear system behaviour, we re-estimate the transfer function using the section from 30 to 75 seconds only. The result is shown in Figure 5.14.

This estimation fits better to the theoretical result shown in Figure 5.11, indicating the influence of the non-linear effects at higher excitations.

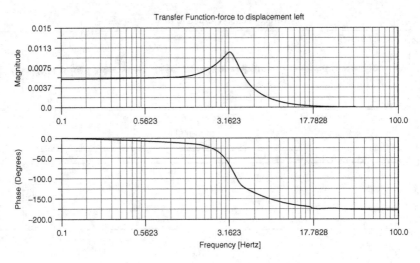

Figure 5.11 Transfer function for input = left vertical wheel force and output = left vertical wheel motion. The function is calculated from a linearization of the multibody system around the equilibrium state with 30 kN vertical wheel load

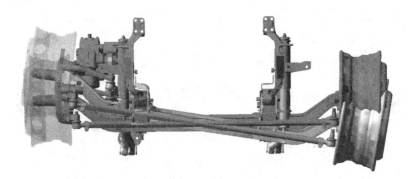

Figure 5.12 First eigenmode of the system at about 3.3 Hz

All estimations of the transfer functions in this subsection have been performed using a Hanning window, a (rather small) block length of 256 and an overlapping of half the block length. The frequency resolution is thus about 0.4 Hz.

5.3 Finite Element Models (FEM) for Durability Applications or: from Component Loads to Local Stress-strain Histories

This section describes the most widely used methods for the calculation of local stress tensor histories, namely the quasi-static and modal superposition. The continuum mechanical approach is described on the basis of the theory of linear elasticity and an appropriate discretization by the finite element method. The basic equation for the resulting linear FE

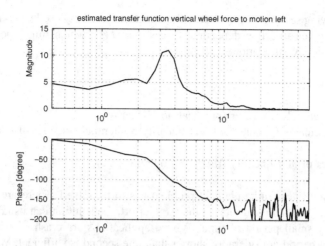

Figure 5.13 Transfer function for input = 'left vertical wheel force' and output = 'left vertical wheel motion'. The function is estimated from the modified public road excitation and its response. The horizontal frequency axis is in [Hz]

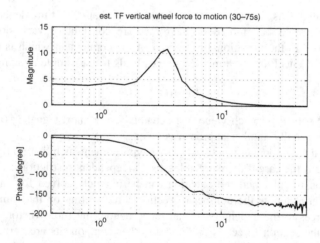

Figure 5.14 Transfer function for input = 'left vertical wheel force' and output = 'left vertical wheel motion'. The function is estimated from the modified public road excitation and its response using the section from 30 to 75 seconds only. The horizontal frequency axis is in [Hz]

models is given by (see Appendix D)

$$M \cdot \ddot{u} + D \cdot \dot{u} + K \cdot u = f, \tag{5.16}$$

where M, D, K denote the mass matrix, the damping matrix, and the stiffness matrix respectively, u is the vector of nodal displacements, and f is the vector of forces applied to the body. This equation is valid under the assumption of linear material behaviour ($\sigma = C \cdot \varepsilon$) as well as small strain ($\varepsilon_{ij} = \frac{1}{2}(\frac{\partial u_i}{\partial x_j} + \frac{\partial u_j}{\partial x_i})$, $i, j = 1, 2, 3$).

In the following sections, we will describe the most frequently used techniques for solving the equations given the time dependent forces $f = f(t)$, which might be the result of a preceding multibody simulation.

If the problem is such that the linear approach is not appropriate for any one of the following reasons:

- non-linear material behaviour (physical non-linearity),
- large displacements (geometrical non-linearity), potentially still small strains,
- contact,
- other sources of non-linearity (thermal properties, . . .),

then the full non-linear finite element problem has to be solved. There are a couple of FE solvers for these problems but due to the large effort, the solution can usually be computed only for a very small period of time. Typical applications are crash or misuse, where the simulation of a period considerably shorter than one second is sufficient. We do not further investigate this topic in the *Guide*.

5.3.1   *Linear Static Load Cases and Quasi-static Superposition*

The first technique is based on the observation that the time-dependent terms in Equation (5.16) might be ignored if the forces are constant or vary slowly. The latter property is related to the vibrational behaviour of the component which is described in the next subsection. Cancelling the time derivatives leads to the simple equation

$$K \cdot u(t) = f(t), \tag{5.17}$$

which could be solved for each value of t separately. For short signals $f(t)$, this might be appropriate, but usually another approach is taken.

Assume that the component is attached to the environment such that no rigid body motion is possible. This is the case if there are at least six independent restrictions. Then the stiffness matrix K is non-singular and the equation can be uniquely solved for an arbitrary right-hand side f. Let n denote the number of section forces acting on the component (without the reaction forces at the attachments) and let e_i denote the force vector such that the ith component of the section forces is one and all other components are zero. Then the time-dependent force vector $f(t)$ can be written in the form $f(t) = \sum_{i=1}^{n} f_i(t) \cdot e_i$, where $f_i(t)$ is a scalar time signal representing the i:th section force. If we now solve the equations

$$K \cdot u_i = e_i, i = 1, \ldots, n, \tag{5.18}$$

we can write the solution of Equation (5.17) in the form

$$u(x, t) = \sum_{i=1}^{n} f_i(t) \cdot u_i(x). \tag{5.19}$$

The loading represented in Equation (5.18) and the solution of the equation respectively are often called *unit load cases*. The representation of the solution in Equation (5.19) is called *the principle of quasi-static superposition*.

If a so-called direct solution technique based on the decomposition $K = L \cdot R$, where L, R are lower and upper triangular matrices, is chosen, then solving the set of n equations does not take much more time than solving a single static equation, as most of the effort is spent during the matrix decomposition, which needs to be performed only once.

In order to avoid unrealistic effects in the neighbourhood of the force transmission, so-called rigid body elements containing a master node, which is the point of application of the force, and a number of slave nodes are used to distribute the transmission of the force to the FE model. An example is shown in Figure 5.15.

If the component is not attached to the environment, for example, the cabin of a truck, then the matrix in Equation (5.18) is singular and the solution is defined up to an arbitrary rigid body motion. In that case, there is a special technique called *inertia relief*. The idea is to calculate an acceleration at all nodes of the component with the following properties:

- The resulting total force is in equilibrium with the applied unit force e_i. This leads to a unique solution of the unsupported unit load case.
- The nodal accelerations are calculated such that only a rigid body motion, no deformation (zero stresses or strains), is induced. The deformation calculated by the inertia relief unit load case is due to the applied unit load e_i only.

This procedure extends the solution of unit load cases and the principle of quasi-static superposition to free or only partially supported components. The inertia relief technique is available in all important FE solvers.

5.3.2   Linear Dynamic Problems and Modal Superposition

If the quasi-static approach is not appropriate due to excitations with a relatively high frequency, then we cannot drop the derivative terms in Equation (5.16) because the component is capable of a vibrational motion. Consider the equation

$$M \cdot \ddot{u} + K \cdot u = 0, \tag{5.20}$$

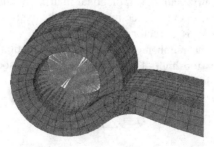

Figure 5.15   Detail of an FE model of a trailer arm. A rigid body element has been used to distribute the application of the force to the FE model. If a revolute joint is attached to the master node of the rigid body element, there are 3 forces and 2 torques which can be transmitted to the arm. Accordingly, 5 unit load cases have to be generated in that case.

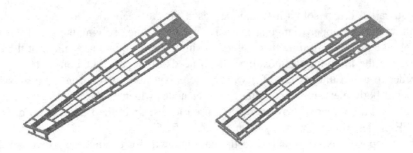

Figure 5.16 Two eigenmodes of a trailer frame (left: torsion, right: bending)

where the excitation force is zero and the damping term of Equation (5.16) has been neglected. To solve this equation, we write the solution $u(x, t)$ in the form $u(x, t) = \phi(x) \cdot e^{i\omega t}$. Inserting this into the equation results in the eigenvalue problem

$$-\omega^2 M \cdot \phi + K \cdot \phi = 0. \tag{5.21}$$

The solutions of this equation are called the *normal modes*, see Figure 5.16 for an example. For each normal mode there is an associated frequency f (corresponding to the eigenvalue ω^2 of the mode) given by $2\pi f = \omega$. If the component is unconstrained, then the first six normal modes represent rigid body motions without a deformation. The corresponding frequencies are zero. In that case we speak of free or unconstraint normal modes, otherwise we have constraint normal modes.

The normal modes are usually ordered by increasing frequency $f_1 \leq f_2 \leq f_3 \cdots$ and arranged in the modal matrix $\Phi = [\phi_1\ \phi_2 \ldots]$.

A very important fact is that the modes are orthogonal with respect to the mass and stiffness matrix. From Equation (5.21) it is clear that the modes are defined only up to an arbitrary factor. Therefore they can be normalized by mass and we obtain the orthogonality relations

$$\Phi^T \cdot M \cdot \Phi = I \tag{5.22}$$

$$\Phi^T \cdot K \cdot \Phi = \Lambda, \tag{5.23}$$

where I is the identity matrix and Λ is a diagonal matrix containing the eigenvalues ω_i^2.

An arbitrary deformation $u(x)$ of the component can be expressed as a linear combination of the modes, that is, there is a vector q such that $u = \Phi \cdot q$. In practice, this does not help very much because the number of modes in that equation (which is the number of DOF in the FE model) is too high. Thus, only a subset of modes is used leading to the (approximate) representation

$$u(x, t) = \sum_{i=1}^{m} \phi_i(x) \cdot q_i(t) \tag{5.24}$$

for the time-dependent deformation of the component.

So far, we have not taken the loads into account. But if we insert the representation of u into the equation

$$M \cdot \ddot{u} + K \cdot u = f \tag{5.25}$$

and multiply the equation with Φ^T from the left, we get

$$\ddot{q} + \Lambda \cdot q = \Phi^T \cdot f = \hat{f}. \tag{5.26}$$

This is a system of uncoupled linear differential equations which can be solved very efficiently for an arbitrarily given set of load time histories $f = f(t)$. Usually, only a certain number of these equations, say the first m, are solved. The time dependent coefficients $q_i(t), i = 1, \ldots m$, are called *modal participation factors*. Then the solution u of Equation (5.25) is approximated by formula (5.24), where the modal participation factors $q_i(t)$ are calculated by Equation (5.26).

This approach to solving the linear transient Equation (5.24) is called *the modal transient solution technique* and is offered in all important FE packages. An enhanced version of the modal transient analysis is to use static as well as normal modes in the representation of Equation (5.24). This has been discussed briefly in Section 5.2.2 (the Craig-Bampton approach).

A comparison of the lowest non-zero eigenfrequency f_{low} with the highest load frequency f_{load}, observed in the excitations of the component, decides whether this technique or a quasi-static approach as mentioned above is appropriate. A rule of thumb is to use the quasi-static approach if $f_{load} \leq 0.5 f_{low}$ and the modal transient approach otherwise.

So far, we have not taken damping into account. There are essentially two difficulties which arise in that case: While the mass and stiffness matrices are constructed automatically from any FE pre-processor using the density, Youngs modulus, and Poisson's ratio of the material, there is no generally accepted, recipe for doing the same for the damping matrix D (see also Section 5.2.3 for some more explanations). Even if we had the damping matrix, then, due to the mathematical structure of Equation (5.16), there is in general no set of modes leading to the decoupling of the equations for the modal participation factors. Thus, we cannot solve the equations of motion as efficiently as described above without damping. A way to circumvent both difficulties is to perform the normal modes analysis without damping as described above and add a diagonal modal damping matrix $\hat{D}$ after the decoupling in the form

$$\ddot{q} + \hat{D} \cdot \dot{q} + \Lambda \cdot q = \hat{f}. \tag{5.27}$$

This simple approach still leads to uncoupled equations which can be solved efficiently. The modal damping parameters in $\hat{D}$ are chosen empirically (see also the remarks in the corresponding subsection of Section 5.2.3).

Frequently, the so-called *frequency response approach* is used when solving the general equation of motion Equation (5.16). Here, it is assumed that the excitation forces can be expressed as a sum of harmonic functions

$$f(t) = \sum_{k=1}^{m} f_k \cdot e^{i\omega_k t}. \tag{5.28}$$

Inserting the representation

$$u(x, t) = \sum_{k=1}^{m} u_k(x) \cdot e^{i\omega_k t} \tag{5.29}$$

into the equations of motion leads to the set of n equations

$$(-\omega_k^2 \cdot M + i\omega_k \cdot D + K) \cdot u_k = f_k, \tag{5.30}$$

which can be solved in the same way as the static equations defined in Equation (5.17).

5.3.3   From the Displacement Solution to Local Stresses and Strains

In the two subsections above, we have briefly described how to (approximately) solve the equations of motion given in formula (5.16). There are two approaches, namely the quasi-static and the modal superposition approach. In both cases we end up with a representation of the time-dependent displacements $u(x, t)$ in the form

$$u(x, t) = \sum_{i=1}^{m} \psi_i(x) \cdot p_i(t),$$ (5.31)

where the modes ψ_i are the solutions of either static unit load cases or a modal analysis, and the time-dependent functions $p_i(t)$ are either the section forces acting on the component or the modal participation factors calculated on basis of the section forces. The representation Equation (5.31) is valid also in the frequency response case, as Equation (5.29) is of the same form as Equation (5.31).

In all cases, we can also obtain the strains and stresses $\varepsilon_i(x), \sigma_i(x)$ corresponding to the modes $\psi_i(x)$ by applying the superposition formula to these quantities too. Thus, we get the stress and strain-time histories for all nodes and elements of the component in the form

$$\sigma(x, t) = \sum_{i=1}^{n} p_i(t) \cdot \sigma_i(x), \quad \varepsilon(x, t) = \sum_{i=1}^{m} p_i(t) \cdot \varepsilon_i(x).$$ (5.32)

Due to the specific techniques of the finite element solvers, the stresses are computed at the so-called Gauss integration points of an element of the mesh. Usually these stresses can be extrapolated to the nodes of the element and reported as 'element nodal' results. As a node typically belongs to several neighbouring elements, there are several 'element nodal' results for one node. Depending on the types of elements used in the FE modelling, the stresses at one node are different for different adjacent elements. For that reason, the stresses are often averaged to get a smooth stress distribution over the component. However, this averaging procedure decreases the stress peaks and is somewhat non-conservative with respect to a subsequent fatigue life calculation. It is therefore recommended that the 'element nodal' results are used instead. Sometimes, also element centre stresses are reported by FE solvers.

5.3.4   Summary of Local Stress-strain History Calculation

The following overview summarizes the basic steps needed to calculate the local stress time histories:

1. Derive the section forces acting on the component by either
   (a) measuring,
   (b) multibody simulation (see Section 5.2), or
   (c) empirical considerations.
2. If the section forces can be represented by a sum of harmonics (see Equation (5.28)), then proceed with the frequency response approach (solve Equation (5.30) and write the solution in the form (5.29)).
3. Otherwise, perform a modal analysis to get the eigenfrequencies of the component.

4. Analyse the frequency content of the section forces and check the relation $f_{load} \leq 0.5 f_{low}$, where f_{load} denotes the highest relevant load frequency and f_{low} denotes the lowest non-zero eigenfrequency of the component. The factor 0.5 in this relation is an empirical constant.

5. Depending on the result of the frequency comparison, proceed either with the

   (a) quasi-static superposition approach:

      i. Create and solve the m unit load cases for all section forces acting on the component (see Equation (5.18)). Use the inertia relief method if the component is not sufficiently supported.

      ii. Read the stresses $\sigma_i(x)$ corresponding to the unit load cases from the FE solver results and calculate the local stress histories $\sigma(x, t) = f_1(t) \cdot \sigma_1(x) + \ldots + f_m(t) \cdot \sigma_m(x)$. This part of the process is typically performed in a dedicated fatigue life evaluation solver, which needs the static modes and the section forces as input.

   (b) or the modal superposition approach:

      i. Calculate the normal modes up to a maximum frequency of interest, which might be defined by a simple rule like $f_{max} \geq 2 f_{load}$.

      ii. Define the modal damping parameters for the selected modes and solve the decoupled system (5.27). This step is usually performed in the FE solver. The result is the set of modal participation factors.

      iii. Read the stresses $\sigma_i(x)$ corresponding to the modes from the FE solver results and calculate the local stress histories $\sigma(x, t) = q_1(t) \cdot \sigma_1(x) + \ldots + q_m(t) \cdot \sigma_m(x)$. This part of the process is typically performed in a dedicated fatigue life evaluation solver, which needs the modal participation factors and the modes as input.

5.4 Invariant System Loads

This section addresses the problem of exciting simulation models in the early development stages when no measurements from prototypes are available yet. If the simulation model contains a proper interface to the 'outer world' (like a tyre model interacting with a rigid road profile), the vehicle-independent description of this 'outer-world-interface' (i.e. the road profile) serves well as an invariant excitation for the simulation of different vehicles (see approach 2 below). However, tyre models are still a challenge and digital road profiles are only available for selected test tracks.

Moreover, modelling the driver as another important interface between truck and 'outer world' is difficult also. In addition to dedicated driver models, driving simulators may be used to handle such questions. A recent approach to estimate system loads using a 'human-in-the-loop-concept' is described in Kleer et al. [136]. However, representation techniques of the driving behaviour and its impact on loads are not further addressed in this chapter.

In many cases, only vehicle-specific measured data like wheel forces, accelerations, or local strains from previous vehicles are available. Such data, of course, must not be used as input for other (new) vehicle models without modification. There are essentially the following ways to deal with this problem.

1. **Rough estimation of scaling factors:** If the system to be analysed is not too different from a previous variant with available measured data, it is tempting to re-use the measured

data with slight adaptations. One example is a weight scaling of wheel forces. It is clear that this simple procedure does not necessarily take all important effects into account. For example, possible changes in the vibrational behaviour are neglected.

There is no general answer to the question how large this error is since this heavily depends on the system. In order to estimate the magnitude of this error, one could apply the procedure to two former variants A and B, where measurements exist for both. In this case, the re-scaled loads from system A can be transferred to system B and compared to the measurements for system B. This kind of load estimation based on data of a predecessor system is a crude version of the more sophisticated approach 3 described below.

2. **Digital road:** If we have a full vehicle MBS model including tyres, we can use road profiles and some assumptions about the driver's behaviour to excite the model and derive load data for durability purposes (section forces, etc.).

   This approach needs several prerequisites, namely a full vehicle model, a driver model, a tyre model, and the road profile. As tyre modelling for durability applications is not yet solved in a fully satisfying way (accurate longitudinal or lateral forces, parameter identification, . . . , see below), one sometimes needs alternative solutions for the estimation of durability loads.

3. **Back calculation of invariant substitute loads:** This approach is based on an estimation of some kind of invariant loading for the new model based on the data of the old model. An example of such an invariant loading is an 'effective vertical road profile', which can be calculated using a model of the old vehicle including a more or less simple tyre model. This effective road profile can be adapted in such a way that some important target quantities (for example, wheel forces) are close to the measured ones. The profile adapted in this way and the simple tyre model are used for the excitation of the new model. Using this approach, the requirements on the quality of the tyre model are reduced, because the effective road profile can account for defects of the tyre model.

In the following subsections, these questions are addressed in more detail.

5.4.1   Digital Road and Tyre Models

If the emphasis is not on the tyre or wheel itself, it is sufficient to model the tyre as a subsystem linking the wheel hub to the road. This subsystem can be seen as a complicated force element similar to, but more involved than, for example, a bushing. To this end the contact between tyre and road needs to be modelled as well as the complicated non-linear stiffness and damping properties of the tyre.

The simplest models use a single contact point to the ground and assume a functional relation between slip and displacement at the contact point and the forces at the wheel hub (magic formula). Others describe the belt of the tyre as a rigid or flexible ring which is connected to the hub via a number of spring/damper elements. Usually several contact points to ground are used in these models.

One approach to cover also lateral forces is to use multiple rings connected to each other (again with spring/damper elements), another is to use beam elements in lateral direction (see bottom left part of Figure 5.17).

The most complex type of model mentioned here is an FE model which captures the detailed geometry of the tyre including the different components (belt, carcass, rubber

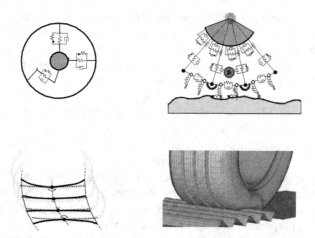

Figure 5.17 Examples of tyre models. Rigid ring (top left), flexible ring (top right), beam elements to capture lateral excitation (bottom left), and an FE model (bottom right)

Table 5.1 Typical range of application for tyre models

Type of model	Typical applications
Empirical models such as Magic formula, Pacejka	Handling
MBS-type models such as Rigid ring or Flexible ring	Ride and comfort NVH and durability
Finite element models	Cleat test
	Curb strike
	Misuse

material, steel layer, . . .). Some of the models including the typical range of applications are sketched in Table 5.1 and in Figure 5.17.

The complexity of the models increases from top to bottom. See, for example, Pacejka [182] and Schmeitz [210] for details on tyre modelling. All models need a set of parameters which have to be identified based on more or less expensive physical experiments. Special test rigs based on drums are used, for instance, where the tyre even rolls over cleats while pressed against the drum. The larger the tyre is, the more difficult it becomes to perform the required test.

One drawback of the empirical and MBS-type models is the poor transferability of parameters. Since the models are tailored to describe the force transfer behaviour of the physical tyre rather than the geometric and material details, parameters identified for a specific tyre can hardly be used for another tyre which is structurally similar but differs, for example, with respect to the geometry. Thus the work that was done during the required experiments and the numerical procedure of parameter identification needs to be done again for another tyre.

For the FE-type models, one needs the parameters of the material used in the tyre. Once these are identified (which is generally rather difficult), they can, in principle, be transferred to other tyres that differ only with respect to geometric properties.

For most of the important MBS-type models, there are interfaces to commercial MBS codes. The performance of these models is good enough for the simulation of several minutes (for instance, test track events) or even an hour. This is not true for the FE-type models. There are no common interfaces to MBS codes and the performance is such that the simulation time is limited to a few seconds at the most.

One way to overcome this problem is to apply the modal representation technique of flexible bodies in an MBS environment (see Section 5.2.3). But this is a linear theory and non-linear extensions are needed for the tyre representation. There are mathematical concepts that can be used to do this (see, for example, Herkt [112]), but more research is needed, especially with respect to contact mechanics.

For durability applications, the requirements on the tyre models are stringent, since rough road applications have to be covered as well as curves and specific events such as emergency braking, potholes etc. It is widely accepted that there is currently no MBS-type model covering all these purposes in a fully satisfying manner.

Nevertheless, more and more test tracks have been geometrically measured and digitized meanwhile in order to use them for the multibody simulation in early development phases. Applications in the passenger car industry are reported for example in Lion and Eichler [145]. More information on MBS-type models as well as recent developments and applications can be found, for instance, in Oertel [174], Gipser [103], Gallrein and Bäcker [96], Bäcker *et al.* [9], Bäcker *et al.* [10], or Gallrein and Bäcker [97].

5.4.2 Back Calculation of Invariant Substitute Loads

Even if tyre models do further improve with respect to accuracy and performance, there are important applications where the third approach (the back calculation procedure) seems to be more appropriate. This is obvious when the simulation of a drive is required but the road profile is not digitized (public road or offroad terrain). Another example will be described below in Section 5.4.3.

5.4.2.1 Formulation of the Problem

A mathematical description of the problem can quite easily be formulated in the following abstract setting. Let $x = x(t)$ denote the (time-dependent) vector of variables, describing the state of the system of interest. For a MBS vehicle model, this is the vector of displacements and velocities of all bodies of the model. Other quantities like the pressure or flow in a hydraulic component may also be part of the vector x of state variables. Let $u = u(t)$ denote the (time-dependent) vector of input variables exciting the system. For a full vehicle model including tyres, this is the driver input (steering wheel angle, acceleration, and braking signal).

In this general setting, the road profile is a complex non-linear contact constraint. To simplify this for illustration purposes, we consider a straight forward drive at constant speed over a rough road and concentrate on the vertical excitation only. We can then ignore the driver input and use $u(t) = \zeta(t \cdot v)$ as the input signals, where v is the velocity, $s = t \cdot v$ is the position on the road, and $\zeta(s)$ is the vertical road profile.

The general mathematical description of the system is then given in the form

$$F(x, \dot{x}, u, t) = 0. \tag{5.33}$$

This system of equations, differential algebraic equations (DAE), is completed with an appropriate set of initial conditions, which are omitted here for the sake of simplicity. This rather general framework covers the usual case of MBS systems described in Appendix E.

We are often not only interested in the state variables themselves (displacements and velocities), but also in some other quantities y which can be derived from x in the form

$$y = g(x, \dot{x}, u, t). \tag{5.34}$$

Important examples are the section forces at the joints or force elements. The problem of forward simulation is then to solve Equation (5.33) for x given the input u and in a post-processing step calculate y using Equation (5.34). We abbreviate that procedure by simply writing

$$y = S(u), \tag{5.35}$$

where the operator S includes the initial conditions, solves Equation (5.33), and calculates the response based on Equation (5.34).

The problem of back calculation is to find u based on a predefined output $y = y_{ref}$, that is inverting S to derive an input that produces a desired system response. An important and well-known example of such an inverse problem arises at the test rig. Consider a suspension test rig which is supposed to reproduce measured wheel forces. In that case, the vector y contains the measured wheel forces and the input u is the electric current (drive file) going into the valves of the hydraulic actuators.

The task to find u is solved in that case using the method of iterative learning control (ILC). To be able to apply the method for calculating the drive file, it must not use the mathematical description of the system given by Equation (5.33), but merely the input/output quantities u, y, since only these are at hand at the rig. The following subsection is a brief introduction to the main ideas.

5.4.2.2   Iterative Learning Control

The problem is to invert the (generally very complex and non-linear) mapping $y = S(u)$ using only observations of the output y given the input u. The following Newton-type procedure based on a linearization of the system is widely used.

First, a so-called setpoint $y_0 = S(u_0)$ is determined. In the suspension test rig, this might be the constant valve current u_0 which is needed to produce the constant wheel load y_0, measured at the vehicle at rest. Next, a noise signal u_{noise} is added to the setpoint and the response

$$y_{noise} = S(u_0 + u_{noise}) - y_0 \tag{5.36}$$

of the system is calculated. From that response and the corresponding input, the linearized system around the setpoint $H = H(u_0)$ is identified using the methods described in Section 3.2.4.

To find the input u corresponding to a desired output y_{ref}, an iterative procedure is applied which calculates a new input according to

$$u_{i+1} = u_i + \Delta u_i, \tag{5.37}$$

$$H \cdot \Delta u_i = y_{ref} - y_i. \tag{5.38}$$

Thus, the update is calculated such that it fills the gap between the current output y_i and the reference output y_{ref} for the linearized system H. Since the numbers p, q of input and output signals often differs (typically $q > p$, which is the more stable over-determined case), Equation (5.38) is solved in a 'least squares sense' (the residual is minimized).

This procedure need not converge. If not or not quickly enough, it might help to re-identify the system based on a certain iterate u_i instead of the initial setpoint u_0 and proceed with $H = H(u_i)$. This reflects the fact that the linearization is able to describe the system only in a certain limited neighbourhood of the linearization point.

There are many additional refinements of the procedure. In most cases, the update Δu_i is not fully applied for safety reasons. Instead we have $u_{i+1} = u_i + \gamma_u \cdot \Delta u_i$, with $\gamma_u < 1$. In addition, the factor γ_u often depends on the component of the input signal (e.g. smaller for the horizontal actuators, somewhat larger for the vertical ones). Another technique to improve the convergence behaviour is to weight the components of the error while solving $H \cdot \Delta u_i = y_{ref} - y_i$ (an error in the vertical wheel force might be more serious than an error in the lateral wheel force). We will not go into details of the method of iterative learning control but refer to the detailed overview given in de Cuyper [67].

5.4.2.3 Alternative Methods for the Inverse Problem

The method of iterative learning control described above is a 'black box method' in the sense that only the input/output behaviour of the system is used. This is essential if the method is applied to the drive file iteration for a test rig. But if the method is used for the back calculation of the input to a numerical model, it could be very helpful to use the mathematical description of the system at hand instead of only relying on the observations $y = S(u)$. In Burger et al. [47] and Burger [46], some of these methods are described in view of invariant vehicle loads and briefly mentioned here.

1. *Trajectory prescribed path control/Method of Control Constraints*: The method combines Equation (5.34) containing the desired reference output and Equation (5.33) and tries to solve that system for u and x given y. The structure of the resulting system of equations is such that considerable numerical difficulties arise during solving and a careful analysis is needed.
2. *Indirect Optimal Control Approach*: The method introduces a cost functional of the form $J(x, u) = \int_0^T (y_{ref}(t) - y(t))^2 dt$ and derives the necessary conditions for u to minimize J. The resulting equations are combined with Equation (5.33) and solved with an appropriate solver. For more details, see Bryson and Ho [45], Callies and Rentrop [49], Oberle and Grimm [171], or Betts [26].
3. *Direct Optimal Control Approach*: Again a cost functional of the form described above is introduced. The unknown input u is discretized using suitable time grids and the integral of the cost functional is approximated by a finite sum using only the values of

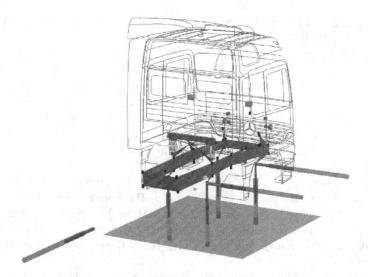

Figure 5.18   MBS model of the cab/frame/rig-system in SIMPACK

the unknown variables u at the grid. The state variables x are calculated by a direct simulation of the system using Equation (5.33). Then the discretized cost functional is minimized as a function of the discretized vector u. This results in a large non-linear constrained optimization problem (see Gerdts [100, 101]). An alternative is to discretize the state variables x as well and use the system equations Equation (5.33) as constraints, leading to an even larger optimization problem. More details may be found in von Stryk [236] or Betts [26].

The application of such techniques to the invariant loading problem in automotive engineering is rather new and will probably develop in the next few years. To our knowledge, there are no published references including interesting real-life applications in this area. A simulation study can be found in Burger *et al.* [48]. The situation is different for the older iterative learning control method. Here a couple of applications are reported in, for instance, Bäcker *et al.* [11], Bäcker *et al.* [8], Mauch *et al.* [158], or the example mentioned below.

5.4.3   An Example

The iterative learning control (ILC) method has been applied to the derivation of an excitation for a truck cabin with a frame on a rig (see Weigel *et al.* [243]). The goal was to define a test for a new cabin, for which no prototype and no measurements exist yet. Figure 5.18 shows a sketch of the rig to be used for the test. There are four vertical cylinders and two lateral ones. In the physical rig, the longitudinal cylinder is replaced by a rod.

The starting point was a test track measurement for the truck with the former cabin. Although the frame displacements during the test track drive are of course not fully invariant under modifications of the cabin, it has been decided to identify them for the existing truck variant and use these for the early test of the new prototype. Alternatively one could also have used a full vehicle model and the road profile itself as invariant loading, but in this specific case, the simpler approach has been considered good enough for the definition of the test in that early development phase.

Thus, the already existing truck cabin, the frame, and the test rig have been modelled as an MBS system in the software tool SIMPACK. Here, the longitudinal rod in the physical rig has been replaced by a cylinder to check how much can be gained by using 7 instead of 6 actuators. The frame has been modelled as a flexible body. This has also been checked for the cabin, but it was sufficient to use a rigid body in that case. The hydraulics of the real rig have not been modelled. Instead, the input quantities u of that MBS model are the displacements of the seven test rig cylinders, which determine the motion of the frame at the coupling points.

As outputs we have four spring displacements at the suspension between cabin and frame and several accelerations at the frame and the cabin. There are reference values available for all output quantities gained by measurement. The SIMPACK model has been used as a pure black box model, which produces the corresponding outputs by simulation. No analysis of the mathematical equations has been made. The ILC procedure has been applied via MATLAB routines.

Since a numerical model is always only an approximation of the real vehicle, the virtual drive file iteration sometimes exhibits more difficulties than the real iteration on the physical rig. To improve the convergence behaviour in our case, the identification of the linearized model has been split into a low-frequent part and a high-frequent part. In addition, a setpoint u_0 based on the rigid frame motion has been used which is as close as possible to the flexible frame motion.

Based on this, the cylinder displacements that yield the best approximation of the measured output quantities have been identified and transferred to the new cabin. More details and results of the calculations can be found in Weigel et al. [243].

5.5 Summary

In this chapter we gave a brief introduction to modelling techniques for mechanical systems, considering the entire load path from system loads to local stresses and strains. One reason for doing this is that load data analysis can be very beneficially applied during numerical simulation of systems or components. Engineers concerned with simulation may gain more insight into their simulation results by looking not only at the time signals as they arise from the simulation, but also at derived quantities such as rainflow matrices or range pair spectra. Another reason is that engineers concerned, for instance, with the construction of customer load profiles or test rig signals, often have to complement measured signals by calculated signals from the computation department. Knowing some principles of the underlying simulation techniques, their strengths and weaknesses, helps in the assessment of the signals and drawing the right conclusions from the corresponding analysis results.

The description of the modelling techniques started with multibody system simulation. This technique is often applied in order to derive section forces for certain components. The corresponding vehicle model may either be excited by signals measured on a prototype vehicle or by a driver model guiding the vehicle model including tyres over a digitized road profile. In the former case, we are restricted to the model of the measurement vehicle and use the simulation to derive quantities which have not been measured due to limitations of the measuring equipment. In the latter case, we can even predict loads for vehicles without having a physical prototype. A well-known specific problem arises if a full vehicle model is excited by measured wheel forces. In this case, the vehicle model starts to drift off after a small period of simulation time due to inaccuracies in the model or the measured forces. In Speckert *et al.* [221], this problem and possible solutions have been recently addressed from a mathematical point of view.

The modelling techniques and the types of elements used to build up a model have been illustrated using an example of a truck suspension. It has been shown that this is not straightforward. There are many different models for the same vehicle or suspension. Which model complexity is best suited depends on the required results and the amount of knowledge about the involved elements (e.g. parameters of an air spring). For instance, getting the section forces right for durability applications requires a more careful modelling of rubber or hydro-bushings than a handling simulation. Without going into detail, special attention has been paid to flexible bodies within multibody system simulation.

Once we have section forces of a component, local stress or strain signals can be calculated using Finite Element Analysis. Since the computational effort for a direct dynamic simulation becomes very large in case of many degrees of freedom, some specific techniques for linear FE models have been presented briefly, namely the quasi-static and modal superposition. Both approaches are well developed, rather straightforward and supported by many commercial FE packages. The question how to build up a good FE model (which elements to choose, how to model welds or bolts...) is beyond the scope of this guide. The FE section concluded with a rough guideline which steps have to be taken to calculate local stresses from section forces.

In Section 5.4 a specific question has been addressed, namely how to calculate excitation signals for vehicles without having corresponding measurements. The straightforward approach is to use a digital driver model together with a digitized road profile, and a vehicle model including tyres. Since this approach requires the digitized road profile, which does not always exist, as well as the tyre models which are not always good enough, some interesting alternative approaches have been described. The idea is to identify the road profile (virtual profile) from a set of measurement signals of a reference vehicle, an MBS model of the reference vehicle, and a simplified tyre model. Different mathematical algorithms may be used and are briefly described. One approach is similar to the iterative learning control mechanism, which is well known in the context of drive file iteration for rig testing. Recent approaches try to explore the underlying MBS model to derive algorithms which are more robust and better adapted to possible non-linearities. Anyway, the result is the pair consisting of the simplified tyre model and the virtual profile, which can be used to excite a variant of the vehicle model for which no measurements are available. Such techniques are rather involved and not yet applied as a matter of routine in the vehicle industry.

The presentation of the simulation techniques given above is extremely short and touches only a small part of all important issues. However, understanding the basic ideas of numerical models can be an important further step towards a meaningful application of the various load data analysis methods described in Chapters 3 and 4. This has been the main motivation for including a chapter like that in this guide to load analysis.

6

Models for Random Loads

6.1 Introduction

The theory of random processes is often used to model or describe unpredictable variability of functions, and has found applications in estimation of risks for fatigue failures of structures, in quality assessments when tests of parts or whole products have to be performed, or at early design stages when one wishes to analyse the reliability and safety of systems or components subjected to realistic loads. In our case we will use random processes to model load signals, such as the ones in Figure 6.1. Typical problems are what the expected damage is, how long a load measurement should be to achieve sufficient precision in damage predictions, and the estimation of the probability of failure or a safety index.

In Chapter 3, ways to analyse measured load signals were discussed in great detail. The rainflow counting method was introduced in order to extract the essential load information for the fatigue damage accumulation process. By using fatigue damage accumulation together with the Basquin equation, the damage D and the pseudo damage d were introduced, see formulas (3.23–3.25), to measure the severity of a recorded load signal for a specific component. The damage is used to describe customer loads and to estimate the risk of fatigue failures in Chapter 7.

In order to assess more easily how "dangerous" the recorded load signal is, life damage and equivalent range were also introduced. The life damage D_{life} is defined as the damage inflicted on a component by the measured load applied repeatedly for the target design life of the vehicle, i.e. by means of blocked loading. Next, the equivalent load range R_{eq} is a range of the sinusoidal load that would give the computed life damage D_{life} in $N_0 = 10^6$ cycles, see Section 3.1.12 for details and Example 3.18, where computations of R_{eq} are exemplified. Note that in this chapter we use the equivalent range R_{eq}, defined as twice the equivalent amplitude, $R_{\text{eq}} = 2A_{\text{eq}}$.

Somewhat simplistically one could define the endurance limit of the material as a range ΔS_e such that, say, 10^6 cycles with range ΔS_e are needed before fatigue failure occurs. Then a direct comparison of R_{eq} to ΔS_e gives a crude measure of the safety level. One could say that the measured load is "safe" if $R_{\text{eq}} < \Delta S_e$. If the sequential effects can be neglected, then the sequence of "safe" loads has $R_{\text{eq}} < \Delta S_e$, i.e. it is safe too.

Guide to Load Analysis for Durability in Vehicle Engineering, First Edition. Edited by P. Johannesson and M. Speckert.
© 2014 Fraunhofer-Chalmers Research Centre for Industrial Mathematics.

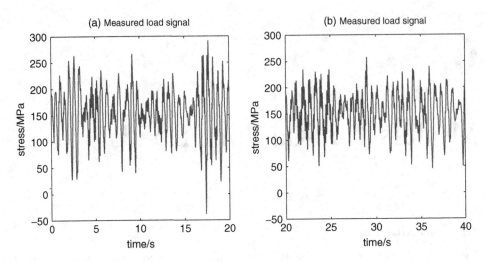

Figure 6.1   Part of a measured stress signal where two consecutive 20-second periods are shown

However, one of the most important features of environmental loads is that one does not know the values of load signals in advance, and that repeated field measurements (under almost identical conditions) will always give somewhat different records, see plots in Figure 6.1. The two plots presents measured stress signals under stationary conditions for consecutive 20-second periods. Clearly the two measured signals look somewhat different, but we do not expect that their equivalent ranges R_{eq} are essentially different. This is because we think that the loading conditions were **stationary**. This question will be further discussed in the following example where the whole measured stress signal is analysed.

Another important problem is how much can be changed in the field experiment and still the resulted estimated R_{eq} would be about the same. More precisely, suppose that one wishes to estimate the load severity during a "task", for example, driving 10 km on a forest road. Will the equivalent ranges R_{eq} be the same if one repeats the experiment and measures stresses again the same day, another season, and with another driver, in another country, etc? If there is no essential difference between the estimated equivalent ranges, which means that principally only one measurement of stress signals was needed to measure in order to find R_{eq} for the task, then we say that the "stress process" is **ergodic**.

In this chapter we will primarily consider damage accumulation processes for stationary and ergodic loads. Non-ergodic loads can be seen as mixtures of ergodic loads which all have to be treated separately. The important issue how to split the loading conditions into classes such that ergodicity can be assumed is discussed in Chapter 8, see Sections 8.5 and 8.6. Some special types of non-stationary loads will also be considered, namely loads that can be split into stationary parts, called **locally stationary** loads, and stationary loads with an added variable mean or scaled by a deterministic function.

Example 6.1 (Measured Stress)
In this example we will analyse measured stresses for 3 minutes (sampling frequency 250 Hz), see Figure 6.2a. Longer measurements of the stress signal are not available as the loading conditions have changed. However, we would like to check if the signal can be

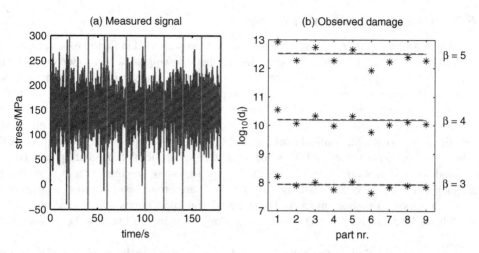

Figure 6.2 (a) Measured stresses during 180 seconds. (b) The stars represent computed pseudo damage for 20-second-long stress measurements for $\beta = 3, 4, 5$, the solid lines are the average of the 9 damage values, the dashed line the pseudo damage computed for whole 180-second-long record divided by 9

seen as stationary. In order to investigate this issue, we split the stress signal into 9 parts, each 20 seconds long. (The first two parts are presented in Figure 6.1.) Then nine pseudo damages $d_i(\beta)$, $i = 1, \ldots, 9$, for three values of the Wöhler exponent $\beta = 3, 4$ and 5, are calculated. The logarithms of the damage, $\ln(d_i(\beta))$, are presented as stars in Figure 6.2b. The solid lines are the logarithms of the average damage $\overline{d}(\beta) = (1/9) \sum_{i=1}^{9} d_i(\beta)$. We can see that the damage values d_i are close to the average values and conclude that the measured stress signal seems to be stationary. The variability around the lines is attributed to randomness of the stress signal.   □

In the following sections, we will discuss general properties of random processes and means to estimate the expected pseudo damage, $\mathbf{E}[d]$. Then specific methods will be discussed for some narrower classes of random loads; Gaussian, Markov and some typical non-stationary loads. The selected classes of stochastic processes provide a reasonable balance between mathematical simplicity and the ability to model actual real-life load histories. We also devote a section to the problem of estimating the coefficient of variation for the pseudo damage d, $\mathbf{R}[d] = \sqrt{\mathrm{Var}[d]}/\mathbf{E}[d]$, needed in the evaluation of safety indices. Besides we will answer the question: "How long do we need to measure?"

Two main examples will be used to illustrate the methods discussed. The first example, which has already been introduced above, deals with a typical measurement of a stress at a hotspot. The stress signal is measured with 250 Hz sampling frequency for 3 minutes, when the loading conditions can be considered stationary. In the second example, artificial road surfaces are generated based on a model for the random process. The road surfaces are the input to a simple mechanical model representing a quarter vehicle, and its response is evaluated. Sections marked by "*" contain more mathematically technical material.

6.2   Basics on Random Processes

Even after a very careful analysis of the measured stress presented in Figure 6.1a, it would be hard to predict the exact path of the stress during the following 20 seconds (see the curve in Figure 6.1b). This property is verbalized by saying that the stress varies in an unpredictable manner. However, the plots have similar average properties, for example similar mean stresses, variances, and even the PSD functions, defined in Equation (3.38), seems to be close to each other. From the stress measurements presented in Figure 6.2, the nine PSD functions are estimated, using the methods discussed in Section 3.2.2, (see Figure 6.3), where also the mean values and the standard deviations of the nine parts are presented.

Random processes are mathematical models for functions varying in an "unpredictable manner" but having common average properties. These properties can sometimes be used as parameters of a random process. There are several applications of the random process models for loads in fatigue safety assessment. Some of them are listed in the following:

- **Generating the signals for fatigue tests based on random load models**. The use of randomly generated sequences of turning points may reduce bias in the estimation of fatigue life, if the process model is good. In such a case one uses a larger number of realistic patterns of the turning points than is the case when the measured signal is blocked.
- **Computing the equivalent range, i.e. the expected pseudo damage**. The method may reduce the uncertainty in the predicted fatigue life. However, the possibility of modelling errors should not be overlooked.
- **Producing a variety of realistic loads at the early design stage**. Here, Gaussian loads, modelling the frequency content and correlation of the load signals, can be especially useful.

In this section, we will denote processes by capital letters $X(t)$, $Y(t)$, $Z(t)$. For example, $X(t)$, $0 \leq t \leq T$ will denote a random function defined on the interval $[0, T]$. Loads are

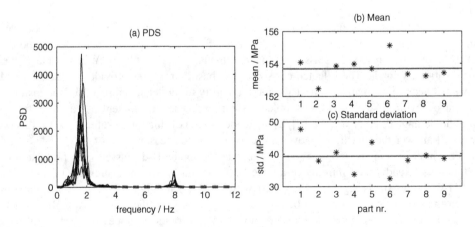

Figure 6.3   (a) The thin lines show the nine PSDs estimated for the stresses discussed in Example 6.1. The dashed line is the average PSD. (b) Mean values. (c) Standard deviations

usually smooth functions that have derivatives denoted by $X'(t)$, $X''(t)$, and so on. (The derivatives are also random processes.) The value of a process at some time instance, for example, for $t = 100$, $X(100)$ is a random variable. Sometimes it is convenient to define a random process as a sequence of random variables indexed by t.

Variability of process values at any time t, i.e. of a random variable $X(t)$, is described by its cumulative distribution function (CDF). If the process is stationary, all random variables $X(t)$ have the same CDF. In practice, such a distribution is not known in advance and has to be postulated or estimated on the basis of observed (historical) data.

If the stress signal is modelled by a random process, then important quantities such as pseudo damage d or the equivalent range R_{eq} are random variables too, and one would like to know their distributions. This is a complex statistical problem and more often than not, only the expected damage $\mathbf{E}[d]$ and the coefficient of variation $\mathbf{R}[d]$ are used to describe the variability of the pseudo damage d.

6.2.1 Some Average Properties of Random Processes*

Mean, variance, covariance: The variability of random processes at a fixed time t, is often described by average values, i.e. by the expectation

$$m(t) = \mathbf{E}[X(t)] \tag{6.1}$$

and by the variance

$$\sigma^2(t) = \text{Var}[X(t)] = \mathbf{E}[(X(t))^2] - m(t)^2. \tag{6.2}$$

The dependence between process values at two time points t and s, is measured by the covariance function

$$r(t, s) = \mathbf{E}[X(t)X(s)] - m(t)m(s). \tag{6.3}$$

If the process is **stationary** (roughly: the mechanisms generating the load does not change with time), the mean and the variance are constant $m(t) = m$, $\sigma^2(t) = \sigma^2$, while the covariance function, also called autocovariance function, depends only on the time difference $\tau = s - t$, $r(t, s) = r(\tau)$, see also Equation (3.39). As mentioned in Section 3.2.1, for stationary processes the power spectrum density (PSD), $S(\omega)$, is a Fourier transform of $r(\tau)$.

Spectral moments: In the following, we will use the so-called spectral moments, which are defined as follows

$$\lambda_i = \int_0^\infty \omega^i S(\omega) \, d\omega. \tag{6.4}$$

Note that $\lambda_0 = \sigma^2 = r(0)$, $\lambda_2 = \text{Var}[X'(0)] = -r''(0)$ while $\lambda_4 = \text{Var}[X''(0)] = r^{iv}(0)$.

An important class of spectra, often used in applications, are concentrated around a peak frequency, f_p. Such spectra are called narrow band spectra. The spread around the peak frequency is often measured by the following bandwidth parameters

$$\alpha_1 = \frac{\lambda_1}{\sqrt{\lambda_0 \lambda_2}}, \qquad \alpha_2 = \frac{\lambda_2}{\sqrt{\lambda_0 \lambda_4}}. \tag{6.5}$$

The parameter α_1 is sometimes called groupness, and is used in studies of the envelope process, while α_2 is the irregularity factor. For Gaussian loads, α_2 is equal to the fraction of intensities of mean level upcrossings and the intensity of local maxima. Note that both bandwidth parameters are bounded by 1. For the single cosine process $\alpha_1 = \alpha_2 = 1$.

Crossing intensity: An important characteristic of load variability is the expected number of level upcrossings $\mathbf{E}[N^+(u)]$. For a stationary process $X(t)$, $0 \leq t \leq T$, the expected number of upcrossings can be computed if one knows the intensity of upcrossings $\mu^+(u)$, i.e. the rate u is crossed in upward direction, viz.

$$\mathbf{E}[N^+(u)] = T \cdot \mu^+(u). \tag{6.6}$$

The intensity $\mu^+(u)$ can be computed if the joint probability density function (PDF) of the process and its derivative $(X(0), X'(0))$ are known. The intensity is given by Rice's formula, Rice [195, 196], viz.

$$\mu^+(u) = \int_0^\infty z f_{X(0),X'(0)}(u, z) \, dz, \tag{6.7}$$

where $f_{X(0),X'(0)}(u, z)$ is the joint PDF of $(X(0), X'(0))$.

Gaussian processes: An important class of random processes are Gaussian processes, also called normal processes. A useful property of Gaussian loads is that the response to a linear filter is a Gaussian load too.

A Gaussian process $X(t)$ has normal CDF, with mean and variance that may depend on time t. The PDF of $X(t)$ is given by

$$f_{X(t)}(x) = \frac{1}{\sqrt{2\pi}} \frac{1}{\sigma(t)} e^{-\frac{(x-m(t))^2}{2\sigma(t)^2}}. \tag{6.8}$$

Gaussian processes have many convenient properties and there is a variety of tools to compute probabilities of interest. A stationary Gaussian process is defined by the mean m and the PSD $S(\omega)$. Consequently, if the PSD function $S(\omega)$ of the load $X(t)$ is known, then a response $Y(t)$ of linear filter having transfer function $H(\omega)$, is also a stationary Gaussian process that has PSD $|H(\omega)|^2 S(\omega)$.

For a stationary Gaussian process $X(t)$, the level upcrossing intensity is given by means of its spectral moments

$$\mu^+(u) = \frac{1}{2\pi} \sqrt{\frac{\lambda_2}{\lambda_0}} e^{-\frac{(u-m)^2}{2\lambda_0}}. \tag{6.9}$$

As will be shown later on, the intensity of mean level upcrossings, $\mu^+(m)$, is an important parameter when computing the expected damage for Gaussian loads. The so-called apparent frequency f_z is defined as the mean upcrossing intensity

$$f_z = \mu^+(m) = \frac{1}{2\pi} \sqrt{\frac{\lambda_2}{\lambda_0}}. \tag{6.10}$$

and is often interpreted as the average frequency of cycles.

Skewness, kurtosis: Although Gaussian processes are often used in practice, the measured load may often be non-Gaussian, which may be manifested by the fact that $X(t)$ has a non-Gaussian distribution. The departure of $X(t)$ from Gaussianity is often measured by two parameters, skewness and kurtosis excess, which are defined in the following.

The k:th central moment is defined by

$$\mu_k = \mathbf{E}[(X - \mathbf{E}[X])^k]. \tag{6.11}$$

Then it can be seen that $\mu_2 = \sigma^2$, i.e. the second central moment is equal to the variance. The **skewness**, denoted by γ_1, which is a measure of the asymmetry of $X(t)$, is defined by

$$\gamma_1 = \frac{\mu_3}{\sigma^3}. \tag{6.12}$$

For a symmetrical $X(t)$, the skewness is zero. Thus, a Gaussian process has zero skewness. Finally, the **kurtosis**, which is a measure of the "peakedness" of the PDF, is defined as μ_4/σ^4, while the *excess kurtosis* is given by

$$\gamma_2 = \frac{\mu_4}{\sigma^4} - 3. \tag{6.13}$$

Note that excess kurtosis is equal to zero for a Gaussian PDF, and the fact that higher kurtosis means that more of the variance is due to infrequent extreme deviations from the mean. The influence of skewness and kurtosis on the expected nominal damage has been studied in several papers; some recent publications are Wang and Sun [241], Gao and Moan [98], Åberg *et al.* [1].

Obviously the expected pseudo damage $\mathbf{E}[d]$, which is often used in fatigue performance and safety evaluations, is one of the most important characteristics of the random loads. However, computing $\mathbf{E}[d]$ is a much harder problem than the other average properties discussed in this section. There are two main methods to estimate $\mathbf{E}[d]$. The first is a statistical method based on a measured signal, and the second uses models of loads as random processes. More precisely, a particular type of random processes is chosen as a description of the variability of loads, and then the "theoretical" $\mathbf{E}[d]$ for the chosen model is computed.

6.3 Statistical Approach to Estimate Load Severity

For stationary and ergodic loads, only one sufficiently long measured load is needed to estimate the expected pseudo damage $\mathbf{E}[d]$ (accumulated during measuring period T, say). Then it is possible to estimate the "true" equivalent range $\tilde{R}_{\mathrm{eq}}$ by means of

$$\tilde{R}_{\mathrm{eq}} = \left(\frac{K \cdot \mathbf{E}[d]}{N_0} \right)^{1/\beta}, \tag{6.14}$$

where K is the extrapolation factor, and N_0 usually 10^6, see Section 3.1.12 for details. In this section we will discuss some methods to estimate the expected pseudo damage $\mathbf{E}[d]$. This subject, called extrapolation methods, has already been discussed in Section 4.4.3. We will here limit ourselves to more general comments and illustrate them by examples.

6.3.1 The Extrapolation Method

In practice, the so-called extrapolation method is often used, which means that $\mathbf{E}[d]$ is estimated by the observed damage d, then

$$R_{eq} = \left(\frac{K \cdot d}{N_0} \right)^{1/\beta}. \tag{6.15}$$

When using this estimation method, an important question is: "How long should we measure so that $R_{eq} \approx \tilde{R}_{eq}$?" We will only give a partial answer to this question by examining the relative error, i.e.

$$R_{eq}/\tilde{R}_{eq} = \left(\frac{d}{\mathbf{E}[d]} \right)^{1/\beta}. \tag{6.16}$$

Employing Gauss' approximation formulas for mean and variance, we have that

$$\mathbf{E}[R_{eq}/\tilde{R}_{eq}] \approx 1 \quad \text{while} \quad \text{Var}[R_{eq}/\tilde{R}_{eq}] \approx \frac{1}{\beta^2}(\mathbf{R}[d])^2, \tag{6.17}$$

where the coefficient of variation squared is $(\mathbf{R}[d])^2 = \text{Var}[d]/(\mathbf{E}[d])^2$.

Consequently, the coefficient of variation for the equivalent range is approximated by $\mathbf{R}[R_{eq}] \approx \mathbf{R}[d]/\beta$. As will be explained later, the coefficient of variation decreases as the inverse of the square root of the length of the measurement. So roughly speaking, if one wishes to decrease the relative error by half, one should measure four times longer. Later on, in Section 6.7, we will present methods to estimate $\mathbf{R}[d]$ for ergodic loads when only one measurement of the load is available. If the accuracy of the estimate R_{eq}, defined in Equation (6.15), is too low (coefficient of variation to high), one can try to extend the measuring period, or use other estimation methods of $\mathbf{E}[d]$.

Example 6.2 (Measured Stress, Cont)
For the measured stress, as will be shown later on in Example 6.13, the coefficient of variation for the damage ($\beta = 3$) is estimated to $\mathbf{R}[d] = 0.14$. From Equation (6.17), a 95% confidence interval for the relative error $R_{eq}/\tilde{R}_{eq}$ can be calculated according to

$$\left[1 - 2\frac{\mathbf{R}[d]}{\beta}; 1 + 2\frac{\mathbf{R}[d]}{\beta} \right] = \left[1 - 2 \cdot \frac{0.14}{3}; 1 + 2 \cdot \frac{0.14}{3} \right] = [0.90; 1.10].$$

In other words, the equivalent range estimated on the basis of the measured 3-minute stress can have errors up to 10%.   □

6.3.2 Fitting Range-pairs Distribution

One possible approach, which is sometimes taken in the literature, is to fit a CDF or PDF to the observed ranges. Sometimes the Weibull law is used; for Gaussian loads Dirlik proposed a particularly well-suited CDF (Dirlik [73]), while in Nagode and Fajdiga [169] a multimodal density was proposed. Having estimated a particular distribution with PDF $f_S(s)$, then the average damage caused by a cycle is given by

$$\mathbf{E}[S^\beta] = \int_0^\infty s^\beta \, f_S(s) \, ds = \beta \int_0^\infty s^{\beta-1} \, \mathbf{P}(S > s) \, ds. \tag{6.18}$$

The expected pseudo damage is then given by

$$\mathbf{E}[d] = T\nu\,\mathbf{E}[S^\beta],\tag{6.19}$$

where ν is the intensity of rainflow cycles, often having unit Hertz. However, the observed ranges often have a complex CDF and it can be difficult to find a suitably wide class to CDF to make a good fit.

Example 6.3 (Measured Stress, Cont)

In the measured stress signal, presented in Figure 6.2a, 1205 rainflow ranges were found, and thus the intensity of cycles is $\nu = 1205/(1800 \quad \text{s}) = 0.67 \quad$ Hz. For the ranges, we have checked most of the standard models for CDF and the Weibull distribution gave the best fit, see Figure 6.4, see also the right plot where the Weibull PDF is compared with normalized histogram. The PDF and the histogram resemble each other. The fitted line estimates probabilities

$$\mathbf{P}(S > s) = e^{-\left(\frac{s}{27.9}\right)^{0.69}}.$$

For a Weibull distributed S, viz.

$$\mathbf{P}(S > s) = e^{-\left(\frac{s}{a}\right)^c},\tag{6.20}$$

the expected damage is

$$\mathbf{E}[S^\beta] = a^\beta\,\Gamma(1 + \beta/c).\tag{6.21}$$

Hence for the fitted CDF

$$\mathbf{E}[d] = T\ \nu\ \mathbf{E}[S^\beta] = 1800 \cdot 0.67 \cdot 27.9^\beta\,\Gamma(1 + \beta/0.69).$$

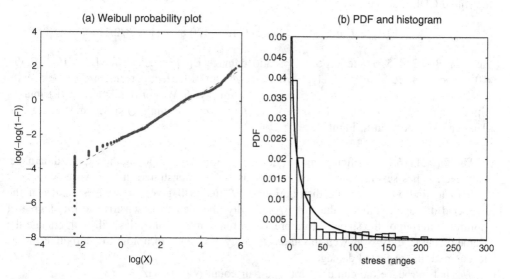

Figure 6.4   (a) The observed ranges in the stress, from Figure 6.2, plotted on Weibull probability paper. (b) Comparison of fitted Weibull PDF with the normalized histogram

For $\beta = 3, 4, 5, 6$ the expected pseudo damage $\mathbf{E}[d] = 1.08 \cdot 10^9, 3.61 \cdot 10^{11}, 1.7 \cdot 10^{14}$ and $1.05 \cdot 10^{17}$, respectively, while the observed damage d was $7.67 \cdot 10^8, 1.43 \cdot 10^{11}, 3.0 \cdot 10^{13}$ and $7.13 \cdot 10^{15}$. Obviously, the Weibull model overestimates the probabilities of getting high ranges and should not be used for the data. $\square$

The above example is given as a warning example illustrating a typical situation when visual inspection gives the impression that the PDF (or CDF) fits reasonably well to the data, but the expected damage computed by means of Equation 6.19 is strongly biased. This is caused by the fact that values of high moments $\mathbf{E}[S^\beta]$ depend heavily on the upper tail of the distribution. Consequently, a chosen model (CDF) should describe accurately the variability of large cycles, while a good fit in the range of moderate and small cycles is less important. In practice, the small cycles are removed by means of rainflow filtering, which helps to solve the above-mentioned problem (see the following example for illustration).

Example 6.4 (Measured Stress, Cont)
In this example all the cycles with a range below 10% of the maximum observed range (354) are removed. The intensity of cycles decreased from 0.67 to 0.184 Hz, i.e. we reduced the number of cycles to almost $1/4$. It is important to notice that the accumulated damage remains almost unchanged after the removal of the small cycles; $7.64 \cdot 10^8, 1.43 \cdot 10^{11}, 3.0 \cdot 10^{13}$ and $7.1 \cdot 10^{15}$, respectively.

Next a Weibull CDF is fitted to the ranges. This gave parameters which are very close to the Rayleigh CDF ($c \approx 2$, see Lord Rayleigh [194]), viz.

$$\mathbf{P}(S > s) = e^{-\left(\frac{s}{120.8}\right)^{2.1}}.$$

In Figure 6.5a, the normalized histogram of the rainflow filtered ranges are compared with the fitted CDF. For the new model

$$\mathbf{E}[d] = T \ v \ \mathbf{E}[S^\beta] = 1800 \cdot 0.184 \cdot 120.8^\beta \ \Gamma(1 + \beta/2.1)$$

and for $\beta = 3, 4, 5, 6$, the expected pseudo damage $\mathbf{E}[d] = 7.39 \cdot 10^8, 1.29 \cdot 10^{11}, 2.49 \cdot 10^{13}$, and $5.18 \cdot 10^{15}$, respectively. Now we have almost perfect agreement for $\beta < 5$, with only 17% underestimation for $\beta = 5$, while 27% for $\beta = 6$. We conclude that the distribution fits the data better, but the probability of getting cycles with high ranges is underestimated. This can also be seen in Figure 6.5b. $\square$

The formula (6.19) is often proposed in the literature for computing the expected damage. However, it has several drawbacks. Firstly, as was demonstrated in the two examples, it is hard to find an appropriate distribution for S that will give unbiased estimates of the expected damage for high values of β. Secondly, the method is sensitive to the degree of rainflow filtering of the signal. More precisely, removing small cycles will both change the shape of the CDF of S as well the intensity of cycles v. We turn now to alternative means to compute the expected damage.

We first introduce the cumulative range-pair count $N^{rp}(h)$, viz.

$$N^{rp}(h) = \text{"number of rainflow cycles having range } S_i \text{ exceeding } h\text{"}. \qquad (6.22)$$

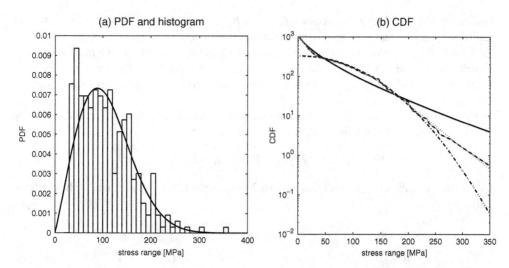

Figure 6.5   (a) Comparison of the Weibull PDF, $c = 2.1$ (close to Rayleigh having $c = 2$) and the normalized histogram of ranges (rainflow filtered with threshold 35.4). (b) The observed range-pair count, dashed line, the fitted using the Weibull model, solid line, the fitted Rayleigh model, dashed dotted line, and the approximation in Equation (6.27), red dots

Now the pseudo damage can be computed as

$$d = \sum S_i^\beta = \beta \int h^{\beta-1} N^{rp}(h) \, dh, \qquad (6.23)$$

and hence

$$\mathbf{E}[d] = \beta \int h^{\beta-1} \mathbf{E}[N^{rp}(h)] \, dh. \qquad (6.24)$$

Plotting the observed cumulative range-pair count together with the expected $\mathbf{E}[N^{rp}(h)] = T \, \nu \, \mathbf{P}(S > h)$ helps avoid the errors mentioned above.

6.3.3   Semi-parametric Approach

The semi-parametric approach to computing $\mathbf{E}[d]$ employs Equation (6.24) and the fact that, often, $\mathbf{E}[N^{rp}(h)] \approx N^{rp}(h)$ for small and moderate values of h. (By this we mean that the coefficient of variation of $N^{rp}(h)$, i.e. $\mathrm{Var}[N^{rp}(h)/\mathbf{E}[N^{rp}(h)]]$, is small and, thus, for h, $0 \le h \le h_0$, the relative error is close to zero.)

For $h > h_0$, the error in using $\mathbf{E}[N^{rp}(h)] \approx N^{rp}(h)$ in Equation (6.24) is not negligible and may cause large errors in the estimated expected damage. In order to decrease this error, one may use a parametric model for $\mathbf{E}[N^{rp}(h)]$, $h > h_0$. Here the variance of the estimate of the expected damage will be smaller but the method may introduce a bias due to possible model errors.

In order to approximate the expected cumulative range-pair count $\mathbf{E}[N^{rp}(h)]$ for the levels $h > h_0$, we adopt the peak over threshold (POT) method, see also Appendix B.3, often used to estimate the tails, small/large quantiles, of a CDF. In the method asymptotic results are used to model variability of excesses $R = S - h_0$. More precisely, let S_i be the ranges of rainflow cycles extracted from a measured load. For all $S_i > h_0$, one defines the excesses r_i as $S_i - h_0$. Then the variability of r_i is modelled by means of the GPD (Generalized Pareto Distribution) CDF. Often the special case of the exponential CDF fits the data very well. (A related method was proposed in Johannesson and Thomas [123], see Equation (4.30) in Chapter 4.)

Having fitted a suitable CDF, one can compute probabilities $\mathbf{P}(R > r)$, $r \geq 0$. Then the expected cumulative range-pair count is approximated by

$$\mathbf{E}[N^{rp}(h)] \approx N^{rp}(h_0) \cdot \mathbf{P}(R > h - h_0), \quad h > h_0. \tag{6.25}$$

Combining Equation (6.24) with Equation (6.25) gives the following estimate of the expected damage

$$\mathbf{E}[d] \approx \sum_{S_i \leq h_0} S_i^\beta + N^{rp}(h_0) \int_0^\infty (r + h_0)^\beta f_R(r)\, dr$$

$$= \sum_{S_i \leq h_0} S_i^\beta + N^{rp}(h_0) \, \mathbf{E}[(R + h_0)^\beta], \tag{6.26}$$

where $f_R(r)$ is the PDF of R.

Example 6.5 (Measured Stress, Cont)
For the stress under study, choose the threshold h_0 such that 2.5% of the ranges are higher than h_0. Here, $h_0 = 182$ and $N^{rp}(h_0) = 0.025 \cdot N$, where N is the total number of cycles. There were 29 such ranges and the exponential CDF with mean 39.9 was fitted to the data. The approximation using Equation (6.25)

$$\mathbf{E}[N^{rp}(h)] \approx F(h) = \begin{cases} N^{rp}(h), & h \leq 182, \\ 29 \cdot e^{(h-182)/39.9}, & h > 182, \end{cases} \tag{6.27}$$

is presented as red dots in Figure 6.5b. We can see that the fit is now excellent. Suppose that $\beta = 3$, then $\sum_{S_i \leq h_0} S_i^\beta = 4.15 \cdot 10^8$, while

$$N^{rp}(h_0) \int_0^\infty (r + h_0)^\beta f_R(r)\, dr = 29 \cdot \int_0^\infty (r + 182)^\beta \frac{1}{39.9} \exp(-r/39.9)\, dr = 3.79 \cdot 10^8.$$

Consequently, $\mathbf{E}[d]$ is estimated by $7.94 \cdot 10^8$, while the observed damage is $7.67 \cdot 10^8$, i.e. the extrapolation gave a damage increase by only 3%. However, for $\beta = 6$, the increase in the estimate of the expected damage is 22% higher than the observed damage. It can be asked how reliable such an extrapolation is. To answer such a question is difficult because we wish to extrapolate the size of rainflow cycle range above the observed levels. Moreover, the physical system should also be considered in order to see if there are any limitations of the signal, for example, a spring compression that is limited by bumpstops. In that case, these limitations should be applied to the extrapolation. However, what one can check is how well the proposed model agrees with the observations. This will be done in the following.

Table 6.1 Column 1 – the Wöhler exponent β; Column 2 – the observed pseudo damage d; Column 3 – estimate of d by means of Equation (6.28); Column 4 – the estimate of $\mathbf{E}[d]$ by means of Equations (6.26–6.27)

β	d	d using Equation (6.28)	$\mathbf{E}[d]$ using Equations (6.26–6.27)
3	$7.67 \cdot 10^8$	$7.65 \cdot 10^8$	$7.94 \cdot 10^8$
4	$1.43 \cdot 10^{11}$	$1.41 \cdot 10^{11}$	$1.53 \cdot 10^{11}$
5	$3.02 \cdot 10^{13}$	$2.94 \cdot 10^{13}$	$3.42 \cdot 10^{13}$
6	$7.13 \cdot 10^{15}$	$6.69 \cdot 10^{15}$	$8.72 \cdot 10^{15}$

The largest observed range is 354 MPa. If we truncate $F(h)$, defined in Equation (6.27), at that level, the damage becomes

$$\int_0^{354} \beta \, h^{\beta-1} \, F(h) \, dh. \tag{6.28}$$

For $\beta = 3$, the difference between the observed damage is

$$\left| \sum_i^{\beta} S - \int_0^{354} \beta \, h^{\beta-1} \, F(h) \, dh \right| = 2.26 \cdot 10^6, \tag{6.29}$$

which means that the error is less then 0.3%. A similar accuracy is achieved for other β values, see Table 6.1. We conclude that the model explains the observed data perfectly. $\square$

6.4   The Monte Carlo Method

If a random model for the load $X(t)$ is chosen, one can using appropriate software simulate a function of time which we call a realization of the random load. Then the pseudo damage d estimated from the simulated load signal is a realization of the random variable d. Repeating this procedure N times, one obtains the damage values $d_1, \ldots, d_N$, and then $\mathbf{E}[d]$, $\mathrm{Var}[d]$ or even CDF of d can be estimated. Such an approach is called a *Monte Carlo method*.

This method is very general and can be combined with simulations of responses in a vehicle under random excitations. For example, one can model road surface variability by means of random processes, and then simulate the stresses in a vehicle driven on the road by means of dedicated software. In the second main example of this chapter we will present such a Monte Carlo analysis of equivalent range estimation. The example follows results presented in Bogsjö [30].

Example 6.6 (Road With Pot Holes)
The road surface is modelled as a Gaussian stationary process with ISO 8608 standard spectrum, ISO [117]

$$S_{ISO}(\kappa) = 10^q \left(\frac{\kappa}{0.1} \right)^{-w}, \quad 0.01 \leq \kappa \leq 10, \tag{6.30}$$

and zero otherwise. Here κ is the spatial frequency in m^{-1}. Let $Z_0(x)$ be a simulation of the Gaussian process with the spectrum S_{ISO}, (see Figure 6.6), upper curve. Often $Z_0(x)$ is called a standard model. However, as is shown in Bogsjö [30], it is not a very accurate road model for durability evaluations. In order to have a more realistic description of road surfaces, one adds irregularities, $Z_i(x)$, see Figure 6.6, which are non-stationary Gaussian processes, placed at random along the road with constant frequency. The resulting road surface, $Z(x)$, x has a unit metre, is the sum

$$Z(x) = Z_0(x) + \sum_i Z_i(x). \tag{6.31}$$

Vehicle fatigue damage is assessed by studying a quarter-vehicle model travelling at constant speed on road profiles, see Figure 6.7b for a schematic description of the quarter-vehicle. This very simple vehicle model cannot be expected to predict loads on a physical vehicle exactly, but it will highlight the most important road characteristics as far as fatigue damage

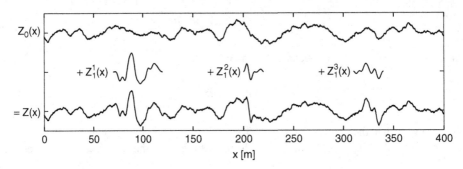

Figure 6.6   A synthetic road profile $Z(x) = Z_0(x) + \sum_i Z_1^i(x)$; $Z_0(x)$ the standard Gaussian road profile, $Z_1^i(x)$ the extra rough parts added to $Z_0(x)$

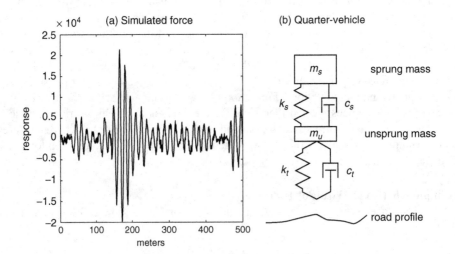

Figure 6.7   (a) The simulated force acting on the sprung mass of the quarter-vehicle. (b) Schematic description of the quarter-vehicle

accumulation is concerned; it might be viewed as a "fatigue load filter". In this example the model comprises masses, linear springs and linear dampers. The parameters are set so that the dynamics of the model resembles a heavy vehicle.

In the model, let $Y(x)$ denote the force acting on the sprung mass of the quarter-vehicle model, when the vehicle is at position x on the road profile modelled by $Z(x)$, defined in Equation (6.31). Then rainflow cycle counting is used to evaluate the pseudo damage. Figure 6.7a shows a 500-metre part of a 44 km long simulation of $Y(x)$. The process $Y(x)$ is stationary but non-Gaussian, as can be seen in Figure 6.8a. In the figure we can see that, except for extreme responses, most of the values follow a Gaussian CDF.

In Figure 6.8b, logarithms of 40 simulated pseudo damage values d are presented. The distances driven at a speed of 70 km/h are 45 km. The estimates of $\mathbf{E}[d]$ are just the average of the 40 values and are equal to $3.5 \cdot 10^{38}$. Finally we also check the distribution of d by plotting the logarithms of damage on normal probability paper. We can see that the damage follows a log-normal CDF reasonably well, viz. $\ln(d)$ is approximately N(88.5, 0.46) distributed.   □

The above example illustrates the applicability of Monte Carlo methods to evaluate the effect of road quality on the pseudo damage. The method is useful at the early design stages. Despite the undisputable advantages of the Monte Carlo approach, there is still a need for simpler, semi-explicit formulas that give expected pseudo damage as a function of suitably chosen characteristics of a random process. This is because environmental loads are often non-stationary, i.e. statistical properties of $X(t)$ changes in time. One can often describe such a load as a sequence of stationary random loads evolving from one stationary load to the next one in a random manner. If parameters of all the stationary subloads were

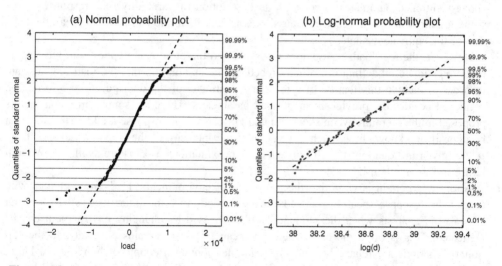

Figure 6.8   (a) Force acting on the sprung mass of the quarter-vehicle, sampling period 50 m, plotted on normal probability paper. (b) The logarithms of the 40 independently simulated pseudo damages, $\beta = 8$, of the force acting on the sprung mass during 45 km drive with speed 75 km/h. The red circle in the plot represents the logarithm of the pseudo damage accumulated in the load for which the normal plot is given to the left

known, the Monte Carlo approach could be used to compute estimated damage. However, the parameters of these stationary subloads are often unknown in advance and should be considered random sequences themselves. When this is the case, fast computational methods to screen the possible scenarios to get the order of magnitude of accumulated pseudo damage can be very useful.

6.5 Expected Damage for Gaussian Loads

At early design stages when the reliability of components has to be estimated, one often models the external forces by means of stationary Gaussian processes while stresses can be computed by means of linear filters of the external loads. Then responses are Gaussian too and their means and PSD are easily computable from load spectra and the transfer function of the systems, see Pitoiset and Preumont [186], Pitoiset *et al.* [188, 189] and Gupta and Rychlik [108] for some applications. This property makes modelling Gaussian processes a very attractive tool for modelling randomly varying loads and a large part of the literature on random fatigue is devoted to finding the expected pseudo damage $\mathbf{E}[d]$ for Gaussian load.

Having specified the PSD of a Gaussian load, one can employ the Monte Carlo method to find the CDF of the pseudo damage. Consequently, one can ask the question why there are so many formulas proposed to approximate $\mathbf{E}[d]$. One possible explanation is that the formulas employ only some characteristics of the spectrum. Therefore, if the characteristics can be easily estimated, for example by means of experience, then the formulas can be practically useful. The parameters used are the variance of the process, i.e. λ_0, the mean upcrossing frequency f_z, the irregularity factor α_2 measuring how many cycles in average there are per one mean level upcrossing, and the groupness parameter α_1. Beside α_1, the above-mentioned parameters form a crude description of variability of a response. Now, if the assumption that the load is Gaussian is added, then one can have an idea about the size of the equivalent range.

However, the real loads are seldom Gaussian and therefore it is important to have some tools to estimate modelling errors introduced by the assumption that the load is Gaussian. In order to include this feature of a load in an estimate of the damage, more information is needed besides the previously mentioned parameters. There are some correction factors proposed in the literature when skewness and kurtosis of the response are known. More can be done if one knows how much the true crossing intensity of the response deviates from $\mu^+(u)$ given in Equation (6.9). Then several approximations of $\mathbf{E}[d]$ are available. We will return to this problem later on.

Another important issue is that the properties of environmental loads often change with time. (The response spectrum depends on road conditions and other factors.) It is easier to make an analysis when the sequential effects caused by changing environment can be neglected. This is often the case when the mean response remains constant and the stationarity periods are relatively long, for example, contain a couple of hundreds of larger cycles. In such a situation, the expected pseudo damage $\mathbf{E}[d]$ can be approximated by the total amount of expected pseudo damage d_i accumulated during periods when loads are stationary, viz.

$$\mathbf{E}[d] \approx \sum \mathbf{E}[d_i]. \tag{6.32}$$

Now, if one adds the assumption that the responses can be modelled as Gaussian processes, then $\mathbf{E}[d_i] \approx g(f_z(i), \lambda_0(i), \ldots)$, and the analysis of the expected damage reduces to modelling the time variability of some spectral parameters. We will give an example of such analysis later on in this section.

6.5.1 Stationary Gaussian Loads

A constant amplitude load $X(t)$ is defined by two parameters: the frequency of peaks (or a period) and the range, i.e. the vertical distance between maximum and minimum. For a random load, one would like to have a similar concept to measure its "average" frequency and range. It will be shown that R_s, equal to four times the standard deviation, viz.

$$R_s = 4\sqrt{\lambda_0}. \tag{6.33}$$

will be a convenient way to define the range. This range will be called the *significant range*. Further, one often uses the mean upcrossing frequency f_z, see Equation (6.10), as the "average" frequency of cycles.

Example 6.7 (Gaussian Load With One Frequency)
Consider a Gaussian process where the spectrum contains only one harmonics

$$X(t) = \sigma R \cos(2\pi t/T_p + \phi), \tag{6.34}$$

where R is a standard Rayleigh distributed variable, $\mathbf{P}(R > r) = \exp(-r^2/2)$, and ϕ a uniformly distributed phase, independent of R. For the process $\lambda_0 = \sigma^2$, $f_z = 1/T_p$ and $R_s = 4\sigma$. Then, for all $\beta \geq 1$,

$$\mathbf{E}[d(\beta)] = T \, \nu \, R_s^\beta, \qquad \nu = 2^{-\beta/2}\Gamma(1 + \beta/2) \, f_z. \tag{6.35}$$

$\square$

6.5.1.1 Narrow Band Approximation

The cosine process in Equation (6.34) is extremely narrow-banded as it contains only one frequency. However, Equation (6.35) is well defined for any stationary Gaussian process and because it is exact for the simple cosine process, it is called the *narrow band* approximation. The approximation postulates that the random range is distributed as $2\sqrt{\lambda_0} R$, where again R is Rayleigh distributed (Lord Rayleigh [194]) and that there are $T \cdot f_z$ cycles during time T. Under these assumptions

$$d^{nb}(\beta) = T f_z \mathbf{E}[(2\sqrt{\lambda_0}R)^\beta] = T \, \nu \, R_s^\beta, \qquad \nu = 2^{-\beta/2}\Gamma(1 + \beta/2) \, f_z. \tag{6.36}$$

It may look strange that the above-defined *rate of cycles* ν depends on β. However, this is the price for having R_s independent of β. In addition, for the range of Wöhler exponent $2 \leq \beta \leq 4$, $\nu \approx 0.5$ (actually $0.47 \leq \nu \leq 0.5$).
The narrow band approximation of the equivalent range R_{eq}, see Equation (6.14), is

$$R_{eq}^{nb} = R_s \cdot \left(\frac{K \cdot \nu}{N_0} \right)^{1/\beta}. \tag{6.37}$$

The narrow band approximation is frequently used in practice whenever the loads can be assumed to be approximated by Gaussian processes. One reason for the popularity is that responses quite often have narrow band spectra. Another reason is that it is simple, and that it actually is a conservative estimate of the expected damage. More precisely, an important property of the narrow band approximation is that d^{nb} is an upper bound for $\mathbf{E}[d(\beta)]$, i.e.

$$\mathbf{E}[d(\beta)] \leq d^{nb}(\beta), \qquad \mathbf{E}[d(1)] = d^{nb}(1), \tag{6.38}$$

see Rychlik [202] for proof. Note that it can still happen that $d(\beta) > d^{nb}(\beta)$, particularly if T is short.

Example 6.8 (Gaussian Load: Expected Damage)
In this example we use a Gaussian process with unimodal spectrum, named Pierson-Moskowitz, which is often used in ocean engineering. The spectrum is normalized to have the significant range $R_s = 1$ and is shown in Figure 6.9. First, the Monte Carlo method was used to find $\mathbf{E}[d]$, based on 1000 realizations of the load each 1800 seconds long. The results are presented in Table 6.2, column 2. In column 3 of Table 6.2 the narrow band approximation of expected pseudo damage, Equation (6.36), is given. For this particular spectrum the approximation is quite satisfactory.

We conclude that, for typical values of β, d^{nb} is only slightly conservative, and the narrow band equivalent range R_{eq}^{nb}, Equation (6.37), thus seems to be an accurate approximation of R_{eq}, where T is only 30 minutes.   □

6.5.1.2 Correction Factor for Broad Band Spectra*

For the narrow band processes, the bound in Equation (6.36) is very close to the expected value and the conservatism of the bound increases with spectrum wideness. However, even for processes with broader spectra, the approximation is often sufficiently accurate. For some more complex systems, for example, when the dynamics of a system is described by means

Table 6.2 Narrow band approximation: Column 1 – the Wöhler exponent β; Column 2 – 'true' value of expected damage computed using the Monte Carlo method; Column 3 – narrow band approximation of expected damage derived by means of Equation (6.36)

β	$\mathbf{E}[d]$	d^{nb}
1	140.8	140.8
2	108.7	112.3
3	100.7	105.6
4	105.1	112.3
5	120.4	132.0
6	149.0	168.5
7	196.8	230.9
8	274.5	336.9

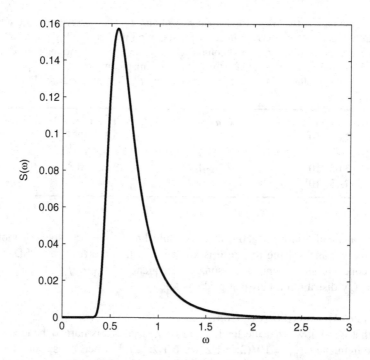

Figure 6.9 PSD used in Example 6.8

of lumped-mass model, then the response spectrum may have several peaks centred on some resonance frequencies. The spectrum may have energy distributed both in low and high frequency ranges making the response look very noisy. For such responses, the conservatism of the narrow band approximation may become too severe. In the literature, there are many correction factors proposed to reduce the conservatism of d^{nb}. Principally, those methods propose to use lower frequencies of the significant range R_s. (In Bengtsson and Rychlik [19], many of the proposed approximations are presented and their accuracy tested.) Here we will present only one of them.

The following approximation was proposed in Tovo [234], viz.

$$d^{BT}(\beta) = (p + (1 - p)\alpha_2^{k-1}) \cdot d^{nb}(\beta), \qquad (6.39)$$

with $p = \min((\alpha_1 - \alpha_2)/(1 - \alpha_1), 1)$. In Benasciutti [15] the following improvement for the constant p is given

$$p = (\alpha_1 - \alpha_2)\frac{1.112e^{2.11\alpha_2}(1 - \alpha_1 - \alpha_2 + \alpha_1\alpha_2) + (\alpha_1 - \alpha_2)}{(\alpha_2 - 1)^2}. \qquad (6.40)$$

As reported in Bengtsson and Rychlik [19] this approximation works well in many cases. The drawback is that it requires more detailed information about the spectrum $S(\omega)$.

We finish this subsection with an example which is partially based on the measured stresses analysed in Example 6.1. We will assume that only very limited information of the

Table 6.3 Column 1 – the Wöhler exponent β; Column 2 – the observed pseudo damage d; Values in Columns 3 – 5 are derived under the assumption that the load is Gaussian. Column 3 – the expected nominal damage $\mathbf{E}[d]$ computed using the Monte Carlo method; Column 4 – the narrow band approximation (bound) for the expected damage employing the approximative frequency $f_z = 0.2$, Column 5 – the approximation d^{BT} of $\mathbf{E}[d]$ by means of Equations (6.39 and 6.40)

β	d observed	$\mathbf{E}[d]$	d^{nb}, $f_z = 0.2$	d^{BT}
3	$7.67 \cdot 10^8$	$7.25 \cdot 10^8$	$6.54 \cdot 10^8$	$6.91 \cdot 10^8$
4	$1.43 \cdot 10^{11}$	$1.23 \cdot 10^{11}$	$1.09 \cdot 10^{11}$	$1.14 \cdot 10^{11}$
5	$3.02 \cdot 10^{13}$	$2.28 \cdot 10^{13}$	$2.02 \cdot 10^{13}$	$2.10 \cdot 10^{13}$
6	$7.13 \cdot 10^{15}$	$4.59 \cdot 10^{15}$	$4.04 \cdot 10^{15}$	$4.22 \cdot 10^{15}$

response is available, more exactly, that one only knows the first spectral moment of the PSD, i.e. the variance of the response is known, and that the response PSD is a bi-modal spectrum concentrated around the resonance frequencies $f_{p1} < f_{p2}$. Finally one assumes that the energy distribution between peaks is known.

Example 6.9 (Measured Stress, Cont)
Suppose that one knows that the load acting on a structure is narrow banded concentrated around the frequency $f_{p1} = 10/(2\pi)$ Hz ≈ 1.6 Hz, say. Further, the system has a resonance frequency about $f_{p2} = 50/(2\pi)$ Hz ≈ 8 Hz. Consequently, the response spectrum will have two peaks with energy distributed between the two components of the spectrum as 0.94:0.06. The variance of the response, i.e. $\lambda_0 = 1.54 \cdot 10^3$. Note that this is a crude description of the spectrum estimated for the measured response presented in Figure 6.3, dashed line. We wish to have a crude estimate of the expected damage for exposure period $T = 3$ minutes. Since the crossing spectrum of the response is not known, we assume that the response is Gaussian.

We propose to use the narrow band approximation d^{nb}, defined in Equation (6.36), and therefore we need to estimate the frequency f_z. Here we propose to weight the two peak frequencies and approximate

$$f_z = 0.94 f_{p1} + 0.06 f_{p2} = 2 \quad \text{Hz}.$$

The resulting d^{nb} is presented in Table 6.3, Column 4. These results will be compared with the approximation d^{BT} of $\mathbf{E}[d]$ defined by Equations (6.39–6.40).

In order to compute d^{BT}, much more information about the spectrum is needed. For the spectrum from Figure 6.3, dashed line, we have computed the following spectral moments $\lambda_0 = 1.54 \cdot 10^3$, $\lambda_1 = 1.98 \cdot 10^3$, $\lambda_2 = 4.16 \cdot 10^3$ and $\lambda_4 = 4.180 \cdot 10^3$. Next the mean upcrossing frequency f_z and the bandwidth parameters are evaluated

$$f_z = 2.6 \text{ Hz}, \quad \alpha_1 = 0.78, \quad \alpha_2 = 0.25.$$

The values of d^{BT} are given in Column 5 of Table 6.3 and we can see that the two approximations are very close. In order to evaluate the quality of the approximation, we employed the Monte Carlo method to find the expected pseudo damage during 3 minutes assuming that the response is Gaussian with the PSD from Figure 6.3. The values obtained are given in Column 3 in Table 6.3. We conclude that the d^{nb} approximation with $f_z = 2$ Hz underestimates the expected damage by less than 10%. This is a very good agreement. In order to investigate the possible modelling errors by assuming that the loads are Gaussian, we have given in Column 2 the observed damage, i.e. calculated from the measured stress. We can see that for high values of β the errors are quite large. This is in line with other results presented in Examples 6.3 and 6.4. $\qquad$ □

6.5.2 Non-stationary Gaussian Loads with Constant Mean*

An example of simulated forces acting on the sprung mass generated by a road surface Equation (6.31), discussed in Section 6.4, is an example of a non-stationary Gaussian load. Here we assume that the location and length of the added irregularities are known. Note that the mean force is constant. This type of non-stationary loads was studied in Bogsjö and Rychlik [32] and it was demonstrated that for a non-stationary Gaussian load $X(t)$, one can accurately approximate the damage by using time-variable significant ranges

$$R_s(t) = 4\sqrt{\text{Var}[X(t)]}, \quad \text{and local period} \quad f_z(t) = \frac{1}{2\pi}\sqrt{\frac{\text{Var}[X'(t)]}{\text{Var}[X(t)]}}, \quad (6.41)$$

then

$$\mathbf{E}[d(\beta)] \le \int_0^T v(t)\, R_s(t)^\beta\, dt, \qquad v(t) = 2^{-\beta/2}\Gamma(1 + \beta/2)\, f_z(t). \quad (6.42)$$

(For $2 \le \beta \le 4$, we have $v(t) \approx 0.5 f_z(t)$.)

In the last formula one assumes that the values of significant ranges and cycle rates $R_s(t)$, $v(t)$ are known at all time instances and thus, in Equation (6.42), we just average the variability of $R_s(t)$, $f_z(t)$.

However, the most important special case is when values of $R_s(t)$, $f_z(t)$ change slowly in time so that the load can "locally" be considered to be stationary. Its coefficient of variation could then be approximated by means of Equation (6.69). Often the periods of stationarity are so long that the coefficient of variations of the damage during stationarity periods is close to zero. Then we may suppose that

$$d \approx \mathbf{E}[d] \approx \int_0^T v(t)\, R_s(t)^\beta\, dt. \quad (6.43)$$

In other words, the variability of accumulated damage is primarily caused by the variability in encountered significant ranges and cycle rates. In practice, values of $R_s(t)$, $v(t)$ are uncertain and sometimes even unknown. These can depend on some factors like road surface quality, vehicles velocity, driver behaviour and cargo. In order to incorporate these uncertainties into reliability evaluations, it can be convenient to model $R_s(t)$ and $f_z(t)$ as random processes. In order to compute the expected nominal damage, one needs to find the long-term distribution of significant ranges and cycles rates. Suppose that the distribution has the PDF $f(s, v)$ then

$$\mathbf{E}[d(\beta)] = T \cdot \int\int v\, s^\beta f(s, v)\, ds\, dv. \quad (6.44)$$

In the last formula we tacitly assumed that the distribution of (R_s, v) describes the variability of load severity encountered by a specific component. However, in general, (R_s, v) can also take into account variation between components, variation that may depend on many factors such as drivers, market, type of operation, etc. Estimating the long-term distribution, and consequently the expected damage, is a hard but feasible problem. However, we want to point out again that the methods discussed here are only applicable for locally stationary Gaussian loads with constant mean.

Example 6.10 (Locally Stationary Gaussian Load, Constant Mean)
This is a constructed example which is intended to illustrate the concepts discussed in this section. Assume that the response is given by means of linear filtering of an external force and that the transfer function is narrowly concentrated around the peak frequency f_r. Consequently, we may approximate $f_z(t) \approx f_r$ as long as the transfer function remains unchanged and the external force have "flat" spectrum. Further, assume that the force can be modelled as a locally stationary Gaussian process with constant mean. Then with $v = 2^{-\beta/2}\Gamma(1 + \beta/2) f_r$

$$\mathbf{E}[d(\beta)] = T \cdot v \int s^\beta f(s) \, ds = T \cdot v \cdot \mathbf{E}[R_s^\beta], \tag{6.45}$$

where $f(s)$ is the PDF describing long term variability of $R_s(t)$. Suppose that R_s is Weibull distributed, viz.

$$P(R_s > s) = e^{-\left(\frac{s}{a}\right)^c}, \tag{6.46}$$

then, using Equation (6.21),

$$\mathbf{E}[R_s^\beta] = a^\beta \, \Gamma(1 + \beta/c). \tag{6.47}$$

Consequently Equation (6.45) can be evaluated,

$$\mathbf{E}[d] = T \, f_r \cdot (a/\sqrt{2})^\beta \Gamma(1 + \beta/2) \, \Gamma(1 + \beta/c). \tag{6.48}$$

$\square$

6.6 Non-Gaussian Loads: the Role of Upcrossing Intensity

Simple measures of non-Gaussianity of the load are skewness and kurtosis defined in Equations (6.12) and (6.13), respectively. If skewness and/or excess of kurtosis are not equal zero, then one may consider abandoning a pure Gaussian model. One possibility is to use transformed Gaussian processes, see Rychlik et al. [205], Rychlik and Gupta [204]. Here we give some other means to treat non-Gaussian loads.

6.6.1 Bendat's Narrow Band Approximation

The *narrow band* approximation defined in Equation (6.36) was first introduced in Bendat [16] for random loads having unimodal symmetrical upcrossing intensity $\mu^+(x)$, which means $\mu^+(x)$ satisfies the following conditions: $\mu^+(x)$ has only one local maximum and

$\mu^+(-x) = \mu^+(x)$. Obviously, for symmetrical $\mu^+(x)$, the upcrossing intensity attains its maximum for $x = 0$.

Bendat proposed computing the expected damage by means of the following approximation; for $h > 0$

$$\mathbf{P}(S > h) \approx \mu^+(h/2)/\mu^+(0). \tag{6.49}$$

Further, Bendat chose the intensity of cycles ν to be equal to the zero upcrossing intensity, viz. $\nu = \mu^+(0)$. Thus, using Equations (6.18, 6.19), Bendat's approximation d^{Be}, writes

$$d^{Be}(\beta) = T \cdot \beta \int_0^{+\infty} h^{\beta-1} \mu^+(h/2) \, dh. \tag{6.50}$$

Obviously, the upcrossing intensity of a stationary Gaussian load is symmetrical, see Equation (6.9), and therefore, by some simple calculus, we obtain that

$$d^{Be} = T \cdot \beta \int_0^{+\infty} h^{\beta-1} \mu^+(h/2) \, dh = d^{nb}. \tag{6.51}$$

In the following subsection we will generalize Bendat's approach to asymmetrical and even multimodal crossing spectra.

6.6.2 Generalization of Bendat's Approach*

First we will introduce a generalization of the upcrossing count $N^+(u)$ of a level u to upcrossings of an interval $[u, v]$. More precisely, for any constants $u \le v$ define the *crossing of interval count* by

$$N^+(u, v) = \text{"number of times the load passes from below } u \text{ to above } v\text{"}. \tag{6.52}$$

There is an important connection to rainflow cycles, in that $N^+(u, v)$ is equal to the number rainflow cycles with a maximum above v and a minimum below u. Using the interval crossing count the pseudo damage $d = \sum_{i=1}^{N} S_i^\beta$ can be computed in the following alternative way

$$d = \beta(\beta - 1) \int_{-\infty}^{+\infty} \int_{-\infty}^{v} (v - u)^{\beta-2} N^+(u, v) \, du \, dv, \quad \beta > 1, \tag{6.53}$$

see Rychlik [201] for proof.

Example 6.11 (Interval Crossings)
Here we will demonstrate that Equation (6.53) works. Suppose that $X(t) = 1 + \cos(2\pi t)$, $t \in [0, 5]$. It is easy to see that $N^+(u, v) = 0$ if $v \ge 2$ or $u \le 0$ and 5 otherwise and consequently, by Equation (6.53), with $\beta = 3$

$$d = 3 \cdot 2 \int_0^2 \int_0^v (v - u) \cdot 5 \, du \, dv = 30 \int_0^2 \left[uv - \frac{u^2}{2} \right]_0^v dv$$

$$= 30 \int_0^2 \frac{v^2}{2} \, dv = 15/3 \cdot 8 = 40. \tag{6.54}$$

This signal has 5 cycles with tops $v_i = 2$ and bottoms $u_i = 0$, hence the pseudo damage $d = 5 \cdot 2^3 = 40$, which is the same as derived in Equation (6.54). $\qquad\square$

It is easy to see that, if a load crosses an interval $[u, v]$, it upcrosses all levels z, $u \leq z \leq v$ at least once. Hence

$$N^+(u, v) \leq \min_{u \leq z \leq v} N^+(z) = N^{bnd}(u, v). \tag{6.55}$$

As before, $N^+(z)$ is the number of upcrossing of the level z by the load. The last inequality combined with Equation (6.53) gives the following important bound

$$d \leq \beta(\beta - 1) \int \int (v - u)^{\beta - 2} N^{bnd}(u, v) \, du \, dv = d^{lc}. \tag{6.56}$$

Finally, the upper bound for the expected pseudo damage $\mathbf{E}[d]$ can be computed by using Equation (6.56) with the following $\mu^{bnd}(u, v)$ function

$$\mu^{bnd}(u, v) = \min_{u \leq z \leq v} \mathbf{E}[N^+(z)] \geq \mathbf{E}[N^+(u, v)] \tag{6.57}$$

then

$$\mathbf{E}[d] \leq \beta(\beta - 1) \int \int (v - u)^{\beta - 2} \mu^{bnd}(u, v) \, du \, dv = d^+. \tag{6.58}$$

More details about rainflow cycles and interval crossings are found in Rychlik [201, 203] and Scheutzow [208].

For a stationary load, if the joint PDF of $X(0)$ and $X'(0)$ is known (can be evaluated for the used model), then the expected number of upcrossing can be calculated by means of Rice's formula in Equation (6.7) and

$$\mathbf{E}[N^+(u)] = T \cdot \mu^+(u). \tag{6.59}$$

For symmetrical loads, when $\mu^+(-u) = \mu^+(u)$, if $\mu^+(u)$ is unimodal, then one can show (only calculus is needed) that $d^+ = d^{Be} \geq \mathbf{E}[d]$.

In Section 6.5.2 we mentioned that simulated forces acting on the sprung mass, discussed in Section 6.4, are non-stationary Gaussian loads, under the assumption that the location and length of the added irregularities are known. Further, one could use local equivalent loads to get accurate approximations of the expected damage, see Bogsjö and Rychlik [31, 32], Bengtsson *et al.* [17]. However, the assumption that the positions and lengths of irregularities are known is impractical since all calculations need to be redone for new positions of the irregularities. Suppose that this information is not available and that the irregularities are placed at random, then the force becomes a stationary process, albeit non-Gaussian. In the following example, we compute the values of d^{lc} for one realization of forces during a 45-km-long drive.

Example 6.12 (Road With Pot Holes, Cont)
In Figure 6.10, we show the observed crossing count in one simulation of the sprung force, see the solid line. We compare it with the expected crossing count under the assumption that the force is a stationary Gaussian process. The difference is very large and we conclude that the significant range $R_s = 4 \sqrt{\text{Var}[X(0)]}$ (and d^{nb}) is too small in this case.

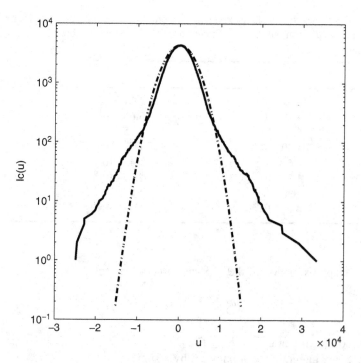

Figure 6.10 Comparison between observed crossing count $N^+(u)$, solid line, and the $\mathbf{E}[N^+(u)]$, see Equation (6.9), i.e. under assumption that stress is a stationary Gaussian process, dashed dotted

Instead we will compare d^{lc} with the observed damage in order to check if the bound is too conservative or not.

From Table 6.4, we can see that the narrow band approximation could be used for small values of β (below 3). For higher values it becomes unacceptably unconservative. The upper bound d^{lc} overestimates the observed damage by at most 32%, which is an acceptable accuracy. However, we need to remember that d^{lc} depends on the particular simulation, because the 45 km length is far too short for ergodicity to work for high values of β. □

6.6.2.1   Non-stationary Symmetrical Loads*

Often, properties of environmental loads vary with time, for example, the type of road changes from time to time. The upper bound d^+ for the expected damage is defined for any type of load, stationary or not, whenever the expected number of upcrossing $\mathbf{E}[N^+(u)]$ can be evaluated.

A special case of locally stationary loads, which means that driving conditions change slowly, is of interest. In such a case $\mathbf{E}[N^+(u)] = \int_0^T \mu_t^+(u)$, where $\mu_t^+(u)$ is the intensity of upcrossings of level u during stationarity conditions. If all stationary loads are symmetrical and unimodal with the top at zero, i.e. locally one can compute Bendat's significant range $R_s(t)$ and the apparent frequency of cycles $\nu(t)$, viz.

$$R_s(t) = \left(\beta \int s^{\beta-1} \mu_t^+(s)/\mu_t^+(0) \, ds \right)^{1/\beta}, \qquad \nu(t) = \mu_t^+(0) \qquad (6.60)$$

Table 6.4 Column 1 – the Wöhler exponent β; Column 2 – observed pseudo damage d; Column 3 – bound d^{lc} computed using Equation (6.56); Column 4 – bound d^{nb} computed using Equation (6.36)

β	d	d^{lc}	d^{nb}
1	$3.25 \cdot 10^7$	$3.24 \cdot 10^7$	$3.65 \cdot 10^7$
2	$3.16 \cdot 10^{11}$	$3.63 \cdot 10^{11}$	$3.93 \cdot 10^{11}$
3	$5.47 \cdot 10^{15}$	$6.15 \cdot 10^{15}$	$4.97 \cdot 10^{15}$
4	$1.39 \cdot 10^{20}$	$1.57 \cdot 10^{20}$	$0.71 \cdot 10^{20}$
5	$4.67 \cdot 10^{24}$	$5.43 \cdot 10^{24}$	$1.13 \cdot 10^{24}$
6	$1.87 \cdot 10^{29}$	$2.28 \cdot 10^{29}$	$0.19 \cdot 10^{29}$
7	$8.4 \cdot 10^{33}$	$1.07 \cdot 10^{34}$	$0.04 \cdot 10^{34}$
8	$4.07 \cdot 10^{38}$	$5.38 \cdot 10^{38}$	$0.07 \cdot 10^{38}$

then Bendat's approximation of the expected pseudo damage is

$$d^{Be} = \int_0^T v(t) \, R_s(t)^\beta \, dt. \tag{6.61}$$

Now it can be shown that $d^{Be} = d^+ \geq \mathbf{E}[d]$.

In practice, the integral is approximated by a sum

$$d^{Be} = \sum_{i=1}^N N_i \cdot R_s^\beta(i), \qquad N_i = \Delta t_i \, v(i). \tag{6.62}$$

where Δt_i, $v(i)$ and $R_s(i)$ are the duration of the i:th stationarity period, its apparent frequency and Bendat's significant range, respectively. As before in the case of locally stationary Gaussian loads, even here the values of $v(i)$ and $R_s(i)$ are not known, and can thus be considered random. Then an important issue is to find the long run CDF of significant ranges and apparent frequencies that are encountered during period T then

$$d^{Be} = T \int_0^\infty \int_0^\infty v \, s^\beta f(v, s) \, ds \, dv. \tag{6.63}$$

6.6.3 Laplace Processes

The Gaussian model is frequently used for modelling environmental loads, e.g. sea elevation, wind loads and road profiles. However, the Gaussian model is often only valid for short sections of the load. For example, for roads profiles, short sections of roads, say, 100 m, is well modelled by a Gaussian process, whereas longer sections of roads, say, 10 km, typically contain shorter sections with high irregularity, and the variability between sections is higher than can be explained by the stationary Gaussian model. This phenomenon can be captured by a Laplace process, which can be seen as a Gaussian process with randomly varying variance. Thus, the Gaussian process is a special case of the Laplace process. The Laplace process may be described by its PSD, skewness and kurtosis. A stationary Laplace process

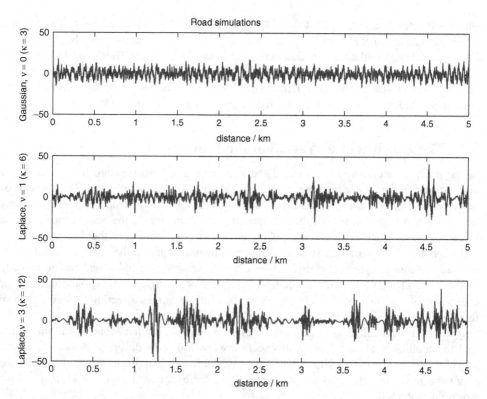

Figure 6.11   Simulations of road profiles; Laplace processes with different kurtosis, $\kappa = 3, 6, 12$; $\kappa = 3(\nu + 1)$. (Note that Laplace with $\nu = 3$ gives a stationary Gaussian process)

can be constructed as a Laplace Moving Average (LMA), where the variance of the process is continuously varying according to a so-called Laplace motion. A non-stationary Laplace process can be constructed by randomly varying the local variance for each 100-metre section according to a gamma distribution. By using the standardized ISO road spectrum ISO [117], this Laplace model can be described by only two parameters, namely its mean roughness, C and the Laplace shape parameter, ν. The parameters can be estimated from a sequence of IRI (International Roughness Index), which is often available, and can then be used for reconstruction of road profiles. Simulated profiles for different Laplace shape parameters are shown in Figure 6.11. More details on Laplace road models are found in Bogsjö *et al.* [33], Johannesson and Rychlik [122].

A practically important theoretical result on the expected damage for the Laplace-ISO model is presented in Johannesson and Rychlik [122]. The expected damage, with damage exponent β, for a Laplace road with ISO spectrum and parameters (C, ν) can be approximated by an explicit algebraic expression

$$\mathbf{E}[D_\nu(\beta, C, \nu)] \approx 0.07093 e^{13.92\,\beta} \left(\frac{C}{C_0}\right)^{\beta/2} \left(\frac{\nu}{\nu_0}\right)^{\beta/2-1} \nu^{\beta/2} \frac{\Gamma(\beta/2 + 1/\nu)}{\Gamma(1/\nu)}, \qquad (6.64)$$

where the reference values are $C_0 = 14.4$ m^3 and $v_0 = 10$ m/s. The first factor depends on the parameters of the quarter-vehicle response, and the coefficients have been estimated from a very long simulation of a Gaussian road. The coefficients 0.07093 and 13.92 will be different for another filter. The second factor is a correction for the average roughness C. The third factor is a correction for the vehicle speed v. The last factor is a correction for the Laplace model, depending only on the Laplace shape parameter v.

6.7   The Coefficient of Variation for Damage

In the previous sections we have used the equivalent range to measure the severity of a load. More often than not the equivalent range, defined by Equation (6.14), is estimated by Equation (6.15), which is just a transformation of the observed pseudo damage d. As we have shown in Equation (6.17), the relative error in estimation of the equivalent range can be measured by $\mathbf{R}[d]/\beta$, where $\mathbf{R}[d]$ is the coefficient of variation of the damage.

The coefficient of variation $\mathbf{R}[d]$ is also approximately equal to the variance of $\ln(d)$ and used in evaluation of the safety index (see Chapter 7, where the index will be introduced). In the same chapter we will also compare the size of statistical errors, measured by the coefficient of variation of d, with other sources of uncertainties affecting the prediction of fatigue life.

If the load is modelled as a random process, one can employ the Monte Carlo method to find the coefficient of variation. In this section we will discuss different means to estimate $\mathbf{R}[d]$ when only one measurement of stress is available or when one has limited information about the load properties, which is discussed in, for example, Bengtsson *et al.* [18] and Bogsjö and Rychlik [32].

6.7.1   Splitting the Measured Signal into Parts

In Example 6.1 the stationarity of a measured signal was investigated by splitting the measured stress into N parts (subloads). The pseudo damage values d_i, $i = 1, \ldots, N$, were computed for the subloads and their variability investigated in order to test whether the assumption of stationarity of the measured stress signal is valid.

Here we will use the computed pseudo damage values d_i to estimate the coefficient of variation. If the assumption of stationarity of the load is reasonably supported by variability of d_i, then standard methods, like Equation (6.67) below, can be employed to estimate $\mathbf{R}[d_i]$. Obviously $\mathbf{R}[d] \neq \mathbf{R}[d_i]$. However, for an ergodic load, if the length of a subload T/N is sufficiently long, then the damage values d_i become approximately uncorrelated. Further, the influence of the rainflow residual becomes negligible, i.e. $d \approx \sum_{i=1}^{N} d_i$. In such a case $\mathrm{Var}[d] \approx N \cdot \mathrm{Var}[d_i]$ and $\mathbf{E}[d] \approx N \cdot \mathbf{E}[d_i]$ giving

$$\mathbf{R}[d] = \frac{\sqrt{\mathrm{Var}[d]}}{\mathbf{E}[d]} \approx \frac{\sqrt{N\,\mathrm{Var}[d_i]}}{N\mathbf{E}[d_i]} = \frac{1}{\sqrt{N}}\mathbf{R}[d_i]. \tag{6.65}$$

The values of $\mathbf{E}[d_i]$ and $\mathrm{Var}[d_i]$ can be estimated by standard methods, i.e.

$$\overline{d} = \frac{1}{N}\sum_{i=1}^{N} d_i, \qquad s^2 = \frac{1}{N-1}\sum_{i=1}^{N}(d_i - \overline{d})^2, \tag{6.66}$$

giving

$$\mathbf{R}[d] \approx \frac{1}{\sqrt{N}} \cdot \frac{s}{\bar{d}}. \tag{6.67}$$

By splitting the stress signal into smaller records, we are able to both estimate the equivalent range and to give a measure for the uncertainty of the derived estimate. This is a natural approach to estimate the statistical uncertainties in an estimate of R_{eq} when a measured signal is very long. The only difficulty is the choice of the number of subloads. The number of divisions, i.e. N, is a compromise between the uncertainty in the value of the estimated coefficient of variation and the possible bias. Here the bias can be caused by the long dependence structure in the load, i.e. the damage values d_i computed for the shorter records can actually be correlated, and thus violate the main assumption behind Equation (6.67).

Example 6.13 (Measured Stress, Cont)
In this example, we have split the load into $N = 9$ subloads, all 20 seconds long. First we check whether the division introduces some bias due to the influence of the residual. In Figure 6.2b the dashed lines represent the level $d(\beta)/9$ and we can see that it is practically identical with the average $\bar{d}(\beta)$, which means that the non additivity of the rainflow method (represented by the residual) is negligible. (The bias will increase with the parameter β.)

Assuming that damage values d_i are independent, then, using Equation (6.67), the estimates of the coefficient of variation $\mathbf{R}[d(\beta)]$ are 0.14, 0.19, 0.25, for $\beta = 3, 4, 5$, respectively. Those values are not negligibly small. ☐

Example 6.14 (Road with Pot Holes, Cont)
Since the sprung force is computed by means of a computer program (simulation of a random process model), the Monte Carlo method can be used to evaluate $\mathbf{R}[d]$. Then a study of the accuracy of the estimator Equation (6.67) can also be done.

In order to use Equation (6.67), one first needs to choose the division of the load into subloads. Here we used $N = 45$, i.e. we have split the simulated force (shown in Figure 6.7a) into 1-km-long parts. Employing Equation (6.67), we obtained the following estimates of $\mathbf{R}[d(\beta)]$ 0.23, 0.42 and 0.56, for $\beta = 4, 6, 8$, respectively. The Monte Carlo method showed that the 'true' values of the coefficients were 0.23, 0.48 and 0.9, respectively.

The estimates of $\mathbf{R}[d(\beta)]$, for $\beta = 4, 6$, are very close to the true values. However, this could be only accidentally good agreement because for the other 1000 simulated forces the coefficients, computed by means of Equation (6.67), varied in the interval [0.12,0.37], for $\beta = 4$, and in [0.2,0.7] for $\beta = 6$. For $\beta = 8$, the difference between the true value 0.9 and the estimated 0.56 is very large. ☐

As we have seen in the above example the splitting method may not work very well for high values of β unless the measured signal is very long. In the following example we will show that similar effects are observed even in broad-band Gaussian load.

6.7.2 Short Signals

For short measured stress records, it is not recommended to use Equation (6.67) to estimate the coefficient of variation. A related approach, presented next, can be used instead.

Since the pseudo damage $d = \sum_{i=1}^{N} S_i^{\beta}$, where S_i is the range of the i:th rainflow cycle, then $\text{Var}[d] = \sum_i \sum_j \text{Cov}[S_i^{\beta}, S_j^{\beta}]$. Now for stationary and ergodic loads, the covariances $\text{Cov}[S_i^{\beta}, S_j^{\beta}] = r(\tau)$, $\tau = j - i$, where the autocorrelation function $r(\tau)$ can be estimated by standard statistical methods. The rainflow ranges are ordered by times when maxima of the rainflow cycles occur. In Bogsjö and Rychlik [32], there is a comparison between the two approaches, i.e. splitting a measured signal into parts and then computing the variance of $d = \sum d_i$, and the one discussed in this section, i.e. computing variance of $d = \sum_{i=1}^{N} S_i^{\beta}$. For long signals the methods give similar estimates. Only for very short signals, the second method is more reliable.

A more explicit but crude estimate of the coefficient of variation is obtained by assuming that the ranges are independent, then

$$\mathbf{R}[d(\beta)] \approx \sqrt{\frac{\mathbf{E}[d(2\beta)]}{(\mathbf{E}[d(\beta)])^2} - \frac{1}{N}} \approx \sqrt{\frac{\mathbf{E}[d(2\beta)]}{(\mathbf{E}[d(\beta)])^2}}. \tag{6.68}$$

Nevertheless the expectations $\mathbf{E}[d]$ in Equation (6.68) are, in general, unknown and need to be approximated. For example $\mathbf{E}[d]$ can be evaluated using methods discussed in the previous section.

6.7.3 Gaussian Loads

A simple approximation of the coefficient of variation for ergodic Gaussian loads is obtained by assuming that the ranges are independent. Then, after some algebra, we get

$$\mathbf{R}[d(\beta)] \approx \frac{1}{\sqrt{v \cdot T}} \sqrt{\frac{\Gamma(1+\beta)}{\Gamma(1+\beta/2)^2}} \tag{6.69}$$

where v is given in Equation (6.36). We can see that the coefficient of variation decreases with the loading time T. The formula works well except in cases when the load is a very narrow band process. Then Equation (6.69) tends to underestimate the coefficient of variation.

Example 6.15 (Gaussian Load: Coefficient of Variation of Damage)
The approximation in Equation (6.69) is based on assumed independence of the damage increments. This property can be violated for very narrow band processes. In this continuation of Example 6.15 we use a Gaussian process with unimodal spectrum, named Pierson-Moskowitz and often used in ocean engineering, see Figure 6.8. Note that Gaussian loads having PSD are ergodic. First, the Monte Carlo method was used to find $\mathbf{E}[d]$ and the coefficient of variation. It is based on 1000 realizations of the load each 1800 seconds long and therefore we have a relatively small statistical error. Consequently, we call the derived values $\mathbf{R}^{true}$. The results are presented in Table 6.5 columns 2 and 3, respectively. In columns 4 and 5 of this table the narrow band approximation of expected pseudo damage and coefficient of variation Equation (6.69) is given. For this particular spectrum, the approximation is satisfactory.

We conclude that, for typical values of β, d^{nb} is only slightly conservative, the coefficient of variation is small, and consequently the narrow band equivalent range R_{eq}^{nb}, Equation (6.37) seems to be an accurate approximation of R_{eq}, where T is only 30 minutes.   □

Table 6.5 Coefficient of variation for the Gaussian load: Column 1 – the Wöhler exponent β; Columns 2 and 3 – 'true' values of expected damage and coefficient of variation computed using the Monte Carlo method, respectively; Columns 4 and 5 – narrow band approximations of values in columns 2 and 3, respectively, derived by means of Equations (6.36, 6.69)

β	$\mathbf{E}[d]$	true $\mathbf{R}$	d^{nb}	$\mathbf{R}$, using Equation (6.69)
1	140.8	0.03	140.8	0.04
2	108.7	0.06	112.3	0.09
3	100.7	0.10	105.6	0.15
4	105.1	0.14	112.3	0.21
5	120.4	0.19	132.0	0.27
6	149.0	0.25	168.5	0.34
7	196.8	0.32	230.9	0.40
8	274.5	0.41	336.9	0.45

6.7.4   Compound Poisson Processes: Roads with Pot Holes

Suppose that pot holes occur with a constant rate $\lambda > 0$, and that the pot holes cause most of the damage, see Bogsjö [30]. Denote by d_i the damage caused by the i:th pot hole. The values of d_i depend on many factors, the shape of the pot hole, driving pattern, etc., and thus cannot be predicted in advance. This uncertainty in the value of d_i is modelled by means of a random variable with some CDF. Here we will need to estimate $\mathbf{E}[d_i]$ and $\mathbf{E}[d_i^2]$.

Suppose that there are $N(T)$ pot holes during time period T, and that the mechanism generating damage is stationary. Then d_i, $i = 1, 2, \ldots$, have the same CDF. Further, we assume that the values of d_i are independent. Consequently, if $N(T)$, the number of pot holes encountered during period T, is Poisson distributed with mean λT, while $d_0 = 0$, $d_i > 0$, are independent and identically distributed, then the pseudo damage due to the pot holes

$$d(T) = \sum_{i=0}^{N(T)} d_i, \qquad (6.70)$$

forms a so-called *compound Poisson process*. Here λ is the intensity of pot holes. (Note that each d_i is a sum of a number of ranges (raised to power β) that can be associated with the i:th pot hole.)

For the compound Poisson process the expected pseudo damage is

$$\mathbf{E}[d(T)] = \lambda T \; \mathbf{E}[d_1], \qquad (6.71)$$

while the coefficient of variation is equal to

$$\mathbf{R}[d(T)] = \frac{1}{\sqrt{\lambda T}} \frac{\sqrt{\mathbf{E}[d_1^2]}}{\mathbf{E}[d_1]}. \qquad (6.72)$$

Note that the coefficient of variation decreases as the square root of the expected number of holes, λT.

Obviously, the mean and the variance of the damage, together with the intensity of pot holes, have to be estimated. This can be done by first identifying the locations in the load where substantial damage accumulates, and then by some cluster identification algorithm finding the ranges that can be associated with a particular "pot hole". Finally, one can compute the associated damage increases d_i, and fit a CDF to the data, see the examples below.

Computation of Equation (6.72) becomes particularly simple when the CDF is log-normal. Then $\ln(d_i)$ are normally distributed, e.g. $N(m, \sigma^2)$, and

$$\mathbf{R}[d] = \frac{1}{\sqrt{\lambda T}} \frac{\sqrt{\mathbf{E}[d_1^2]}}{\mathbf{E}[d_1]} = \frac{1}{\sqrt{\lambda T}} e^{\sigma^2/2}. \tag{6.73}$$

Equation (6.72) can also be used for estimation of the coefficient of variation of $d(\beta)$, when β is large, for many types of stationary responses. This is because, often, the occurrences of large cycles form a Poisson point process and those dominate the pseudo damage for large values of β.

Example 6.16 (Road with Pot Holes, Cont)

Let's consider damage $d(\beta)$, $\beta = 8$, computed for the simulated force acting on the sprung. In Figure 6.12a we can see that there are 9 sections that induced most of the damage. This resulted in 9 values of the variables d_i. The values of $d_i \cdot 10^{-6}$ are equal to

$$1.683, \; 1.494, \; 0.186, \; 0.095, \; 0.071, \; 0.064, \; 0.043, \; 0.041, \; 0.036.$$

Now employing Equation (6.72), we get that

$$\mathbf{R}[d] \approx \frac{1}{\sqrt{9}} \frac{\sqrt{0.5693}}{0.4125} = 0.61. \tag{6.74}$$

Clearly the estimate based on only 9 observations is very uncertain and still is considerably smaller than the "true" value of $\mathbf{R}[d]$ obtained using the Monte Carlo approach.

One way to reduce the uncertainty is to introduce prior assumptions about the CDF for d_1. More precisely, first a suitable model for CDF has to be chosen, and then the 9 observations are used to choose parameters specifying the CDF. In Figure 6.12b, we show the logarithms of d_i plotted as dots on a normal probability paper. The fit is not very good but we still assume that $\ln(d_1)$ is $N(-2.02, 2.23)$. Now using Equation (6.73), we obtain

$$\mathbf{R}[d] \approx \frac{1}{\sqrt{9}} e^{2.23/2} = 1.02, \tag{6.75}$$

which is closer to the "true" value of the coefficient (0.9) obtained by means of the Monte Carlo method in Example 6.14.   □

Example 6.17 (Measured Stress, Cont)

For the measured three-minute stress signal we found 8 large damage increases (here again $\beta = 8$). The values of $d_i \cdot 10^{-20}$ are

$$2.597, \; 0.790, \; 0.439, \; 0.267, \; 0.186, \; 0.126, \; 0.105, \; 0.046.$$

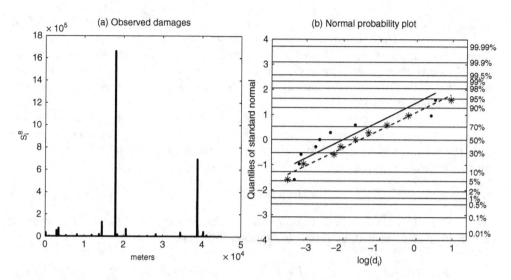

Figure 6.12 (a) Observed range pairs S_i^8 for the simulated force. (b) Logarithms of damage increases d_i plotted on normal probability plot; dots are $\ln(d_i)$ for the simulated force and solid line the fitted model; stars are $\ln(d_i)$ for the measured stress and dashed line is the fitted model

As in the previous example, we fit a CDF to the observed damage. The logarithms of d_i are plotted on a normal probability paper, see stars in Figure 6.12b. The fitted model is represented by the dashed line. Since $\ln(d_1)$ is estimated as $N(-1.31, 1.606)$

$$\mathbf{R}[d] \approx \frac{1}{\sqrt{8}}\, e^{1.606/2} = 0.79. \tag{6.76}$$

Here we do not have a "true" value for the coefficient of variation to check the accuracy of the estimate. Under the additional assumption that the measured stress is a Gaussian process, with the spectrum given in Figure 6.3, the Monte Carlo method gave a "true" value of $\mathbf{R}[d] = 0.37$. However, as we have shown before, the measured stress is non-Gaussian. It is a slightly skewed process and we observed that, for large values of β, the assumption of Gaussianity leads to serious underestimation of the expected pseudo damage. We suspect that the same happens with the coefficient of variation. $\square$

6.8 Markov Loads

A random load satisfying the following property: if the value $X(t_0) = x_0$ is known, then the variability of $X(t)$, for $t > t_0$, does not depend on what happens before t_0, is called a Markov process. The class of Markov processes is very broad and flexible and Markov processes have found applications in diverse scientific problems such as speech recognition, biology, reliability, finance and many other areas. Typical examples are AR(1)-processes, Markov chains, and diffusions.

Only Markov chains (MC) will be considered here. Markov chains are sequences of random variables $X_0, X_1, X_2, \ldots$. In our application, the variables X_k are often the sampled values of the stress $X(t)$ at times $t_0 = 0 < t_1 < t_2 < \ldots$, i.e. $X_0 = X(t_0)$, $X_1 = X(t_1)$, $X_2 = X(t_2)$, ... Note that t_k may not be equidistant and thus the frequency content in $X(t)$ is lost. Further, X_k is discretized to J levels $x_1 > \ldots > x_J$. Maybe the most important special case is when t_k are locations of local extremes in $X(t)$. Then X_k forms a sequence of turning points (TP). If the sequence satisfies the Markov property, it is called a Markov chain of turning points (MCTP).

There are mainly two properties of MCTP that make them an important class of models in fatigue studies. The first one is that simulation algorithms of MCTP are very simple and fast. The second property is that, for the models, there are explicit formulas for the expected number of interval upcrossing

$$\mu^+(u, v) = \mathbf{E}[N^+(u, v)], \tag{6.77}$$

where $N^+(u, v)$ were defined in Equation (6.52). Consequently, also the expected damage $\mathbf{E}[d(\beta)]$ can be computed, by means of Equation (6.53), viz.

$$\mathbf{E}[d(\beta)] = \beta(\beta - 1) \int_{-\infty}^{+\infty} \int_{-\infty}^{z} (v - u)^{\beta - 2} \mu^+(y, z) \, du \, dv, \quad \beta > 1. \tag{6.78}$$

Obviously, in order to compute the damage, one still needs to numerically compute the double integral. However, this can be approximated by a summation in the same way the pseudo damage is computed from the rainflow matrix in Chapter 3.

Example 6.18 (Simple Markov Chain)
Maybe the simplest example of an MC is a "white noise" process, i.e. a sequence of independent identically distributed variables X_k. Then it can be shown that

$$\mu^+(u, v) = \frac{\mathbf{P}(X_0 < u) \cdot \mathbf{P}(X_0 > v)}{\mathbf{P}(X_0 < u) + \mathbf{P}(X_0 > v)}, \quad u \leq v. \tag{6.79}$$

□

The load presented in Example 6.18 is rather only of theoretical interest because real loads are seldom as irregular as white noise processes. Some dependence between the consecutive values of X_k is usually observed. However, explicit results are always useful as tests for accuracy of algorithms, approximations, etc. Further, a minor modification of the white noise model, presented in the next two examples, will lead to far more useful models for durability evaluation. More precisely, one still assumes that the values of X_k are independent but no more equally distributed. In the following remark, we explain why the proposed models can be useful. (In Karlsson [132], these types of loads were used.)

Remark 6.1 *Consider any load function $X(t)$, $0 \leq t \leq T$, and let level u be the most often crossed by $X(t)$, say $u = 0$ for simplicity. Then define a sequence of turning points X_k by extracting the highest and lowest points between the following upcrossings of the level u. (In ocean engineering this sequence is called the sequence of crests and troughs of apparent sea waves.) It can be shown that the rainflow damage computed for the sequence of crests and troughs X_k is always smaller than the rainflow damage found in $X(t)$. In the following* *examples, two simple random models for the sequences of crests and troughs will be given. For very noisy load signals, the damage found in $X(t)$ and X_k can be very close.*

Example 6.19 (MCTP Based on Rayleigh Variables)

Consider a stationary Gaussian load $X(t)$ having mean zero and variance λ_0. Suppose that the frequency f_z is also known. We turn now to the construction of a very simple MCTP X_k that could be used in tests as an approximation for $X(t)$.

Consider a sequence of independent standard Rayleigh variables R_k, $k = 0, 1, \ldots$, i.e. $\mathbf{P}(R_k > r) = \exp(-r^2/2)$, and let

$$X_0 = -\sqrt{\lambda_0} R_0, \quad X_1 = \sqrt{\lambda_0} R_1, \quad X_2 = -\sqrt{\lambda_0} R_2, \quad X_3 = \sqrt{\lambda_0} R_3, \ldots \tag{6.80}$$

It it easy to see that X_k with even indices are local minima while X_k with odd indices are local maxima. Further, the local maxima and minima are independent. Let us denote a local minimum by m and a maximum by M, then

$$P(M \le u) = 1 - \exp(-u^2/2\lambda_0), \quad u \ge 0, \qquad P(m \le u) = \exp(-u^2/2\lambda_0), \quad u \le 0. \tag{6.81}$$

Some calculations, see Rychlik [200] and Karlsson [132], are needed to demonstrate that

$$\mu^+(u, v) = \frac{1}{2} \begin{cases} \mathbf{P}(M > v), & 0 \le u < v, \\ \mathbf{P}(m < u), & u < v \le 0, \\ \frac{\mathbf{P}(m < u)\mathbf{P}(M > v)}{1 - \mathbf{P}(m \ge u)\mathbf{P}(M \le v)}, & u < 0 < v. \end{cases} \tag{6.82}$$

Note that the level upcrossing intensity $\mu^+(u) = \mu^+(u, u)$ of X_k is proportional to the level upcrossing intensity of the Gaussian $X(t)$. The proportionality constant is $1/(2f_z)$. Consequently, the load $X(t)$ acting during interval T corresponds to $X_0, X_1, \ldots, X_N$ for $N \approx 2 f_z T$. Next we compare the expected damage for the two models for the process $X(t)$ with the spectrum defined in Example 6.8.

In Table 6.2 the expected damage for the Gaussian process was given. If we now combine formulas (6.82) and (6.78), the values of $\mathbf{E}[d(\beta)]$ can be computed for the sequence X_k. In Table 6.6 the expected damage for $X_0, X_1, \ldots, X_N$ is presented where $N \approx 2 f_z T$. We can see that X_k causes somewhat less damage than $X(t)$ does (about 15% underestimation for $\beta = 8$). A part of the load is shown in Figure 6.13a. □

In the following example, we will generalize the definition of the sequence of X_k, Equation (6.80), to the case of non-Gaussian loads $X(t)$ having unimodal $\mu^+(u)$.

Table 6.6 Row 1 – the Wöhler exponent β; Row 2 – 'true' value of expected damage computed using the Monte Carlo method; Row 3 – $\mathbf{E}[d(\beta)]$ for the sequence X_k defined in Equation (6.80)

β	3	4	5	6	7	8
Gaussian model	101	105	120	149	197	274
Markov approximation	95	96	107	131	172	239

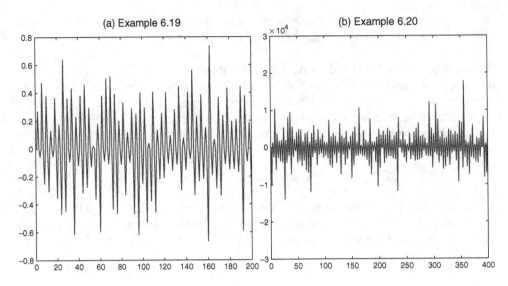

Figure 6.13 Plots of parts of loads from Examples 6.19 and 6.20, (left) and (right), respectively

Table 6.7 Row 1 – the Wöhler exponent β; Row 2 – observed pseudo damage $d(\beta)$; Row 3 – the expected damage $\mathbf{E}[d(\beta)]$ for the sequence of independent minima and maxima with distributions in Equation (6.83)

β	3	4	5	6	7	8
$d(\beta)$	$5.5 \cdot 10^{15}$	$1.4 \cdot 10^{20}$	$4.7 \cdot 10^{24}$	$1.9 \cdot 10^{29}$	$8.4 \cdot 10^{33}$	$4.1 \cdot 10^{38}$
$\mathbf{E}[d(\beta)]$	$5.2 \cdot 10^{15}$	$1.2 \cdot 10^{20}$	$3.7 \cdot 10^{24}$	$1.4 \cdot 10^{29}$	$6.0 \cdot 10^{33}$	$2.8 \cdot 10^{38}$

Example 6.20 (MCTP Based on Level Crossings)
Suppose that the level upcrossing intensity $\mu^+(u)$ (or $N^+(u)$) is known. Also assume that $\mu^+(u)$ is unimodal with maximum at zero. We will employ Bendat's approach, see Section 6.6.1, and let m, M in Equation (6.82) have the following distributions

$$\mathbf{P}(M \leq v) = 1 - \mu^+(v)/\mu^+(0), \ v \geq 0, \text{ and zero otherwise,} \tag{6.83}$$

$$\mathbf{P}(m \leq v) = \mu^+(v)/\mu^+(0), \ v \leq 0, \text{ and zero otherwise.} \tag{6.84}$$

Now, by replacing in Equation (6.80) $R_0, R_2, R_4,..$ by independent random numbers having the m-distribution while $R_1, R_3, R_5,..$ by independent random numbers having the M-distribution, we have constructed a sequence of turning points having level upcrossing intensity $\mu^+(u)$ and the interval upcrossing intensity $\mu^+(u, v)$ is given by Equation (6.82). Again if we combine formulas (6.82) and (6.78) the values of $\mathbf{E}[d(\beta)]$ can be computed.

In Table 6.7, we give the values of $\mathbf{E}[d(\beta)]$ in the case when $\mu^+(u)$ is taken to be the observed $N^+(u)$ from Figure 6.10 (solid line). The results are compared with the second column in Table 6.4. Again, the expected damage of X_k underestimates the damage for the force studied in Example 6.12. The underestimation increases with β reaching level about 40% for $\beta = 8$.   □

In the last two examples, we constructed sequences of independent local extremes. The sequences gave less damage than the corresponding values of the continuous load models. An important reason for the observed underestimation is the assumption of independence of crests and troughs. In real loads a high crest is often followed by a deep trough which will increase damage for high values of β. This effect can be seen when we compare the sequences of turning points shown Figure 6.13b and Figure 6.14 right top plot. The loads in both figures have the same crossing intensity and almost the same irregularity factor, but in the first figure the turning points are independent, while in the second it can be seen that high cycles often come in groups.

The simplest way to introduce the dependence is to let the distribution of the height of a local extremum depend only on the height of the previous one. This way of introducing dependence into the sequence of turning points means that the sequence is a Markov chain.

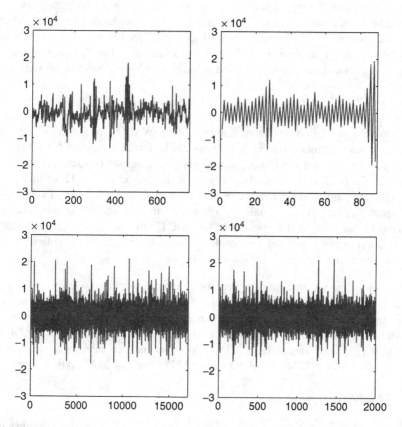

Figure 6.14 Illustration to Example 6.21. Top – simulations turning points corresponding to a 2 km drive for MCTP defined by the corresponding Markov matrices presented in Figure 6.15 top. Bottom – the same as top but for a 45 km drive

For continuous valued MC there are very few explicit results, but the Monte Carlo method is straightforward to use. However, for discrete (sampled) values of loads the computations of rainflow matrices and expected damage values require only some basic matrix operation. The difference between the damage values computed for continuous and discretized values of load is negligible if the discretization levels are chosen dense enough. In the following, we will therefore deal only with the discretized X_k. More precisely, we assume that X_k can take one of J possible values $x_1 > x_2 > \cdots > x_J$, say. As already mentioned, for the discrete valued MC there are many useful explicit formulas to compute rainflow matrices when the MC chain is specified.

In the following example we will consider the simulated force for a truck model driving on 45 km road, see Example 6.6.

Example 6.21 (Road with Pot Holes: MCTP Approximation)
In order to check whether a Markov chain of turning points (MCTP) could be used to model variability of the force described in Example 6.6, we will estimate the Markov matrix from the signal. We choose grid x_k with $J = 201$ levels, starting with $x_1 = 4 \cdot 10^4$ and with $x_{201} = -4 \cdot 10^4$. The contour lines of the Markov matrix F^{mM} are shown in Figure 6.15 top left plot. We can see that the signal is quite noisy and, consequently, the matrix contains many transitions between nearby extremes. Since most of those oscillations do not contribute to the damage, we have removed all local extremes that are paired into cycles with ranges below 10% of the maximum range. For the Wöhler exponents $\beta = 3, 4, 5, 6, 7, 8$, the damage for the original signal and the rainflow filtered is basically identical despite the reduction of the number of cycles by a factor 7.

From the MCTP model defined by the Markov matrix it is possible to theoretically compute the expected rainflow matrix, and then the expected damage by means of the software WAFO in Matlab, available free of charge, WAFO Group [237, 238], Brodtkorb *et al.* [41]. (Some comments on the algorithms can be found in Section 6.8.5.) In Table 6.8, rows 3 and 4, the computed expected damage is given and we can see that the two MCTP give the same expected damage. Obviously, the "narrow band" MCTP estimated from the filtered signal contains much fewer cycles per km of driving. The estimated values of damage for the signal are given in the second row. We can see that the agreement is excellent. Finally, in Figure 6.14, the simulated sequences of two MCTP are shown. One can see that the first sequence is extremely broad banded, while the second one has an irregularity factor close to one. However, the damage is almost the same.   □

In the following, we give only a very brief overview of this theory and try to avoid technical mathematical details, which is not always possible. The examples are somewhat artificial and have been chosen in order to illustrate the concepts. For a more detailed presentation of Markov models, we refer to the literature (se e.g. Rychlik [200], Frendahl and Rychlik [95] and Johannesson [119, 120]), or one can attempt to use the existing software in the toolbox WAFO, WAFO Group [237, 238], of MATLAB routines.

6.8.1 Markov Chains*

A stochastic process $Z_0, Z_1, \ldots$ with values $1, 2, \ldots, N$, is a Markov chain if there is a matrix P, $p_{ij} \geq 0$ for all $i, j \in S$, such that

$$\mathbf{P}(Z_k = j | Z_{k-1} = i, Z_{k-2} = z_{k-2}, \ldots, Z_0 = z_0) = p_{ij}. \tag{6.85}$$

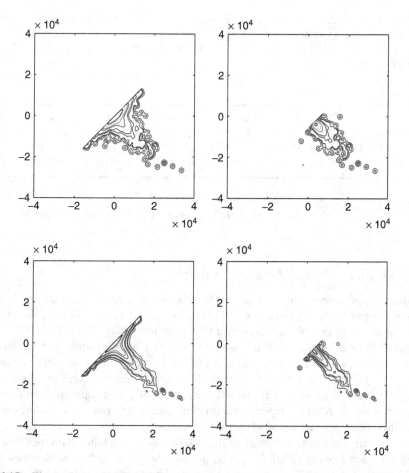

Figure 6.15 Illustration to Example 6.21. Top left – contour lines of the Markov matrix estimated from the sprung force, see Figure 6.8. Top right – contour lines of the Markov matrix estimated from the rainflow filtered sprung force (all ranges below 10% of the maximal range were removed). Bottom – the computed corresponding rainflow matrices

We assume that all values are essential, namely for all i, $\mathbf{P}(Z_k = i) > 0$ for some k and that the states are positive recurrent. It means that the expectation of the least $k > 0$ with $Z_k = i$ given that $Z_0 = i$ is finite and equal $1/\pi_i$, where $\pi_i > 0$. The row vector π is a probability vector, i.e. $\sum_{i=1}^{N} \pi_i = 1$. The probabilities π are solutions to the following equation system $\pi = \pi P$, viz.

$$\pi_i = \sum_{j \in S} \pi_j p_{ji}. \tag{6.86}$$

The row vector π is called a stationary probability of Z_k, since $\lim_{k \to \infty} \mathbf{P}(Z_k = i) = \pi_i$ and, if $\mathbf{P}(Z_0 = i) = \pi_i$ then $\mathbf{P}(Z_k = i) = \pi_i$ for all k. In the following, we will consider stationary Markov chains Z_k, i.e. $\mathbf{P}(Z_k = i) = \pi_i$. Finally, denote the joint probabilities of Z_k, Z_{k+1} by $F = [f_{ij}]$, where

$$f_{ij} = \mathbf{P}(Z_k = i, Z_{k+1} = j) = \pi_i p_{ij}. \tag{6.87}$$

Table 6.8 Row 1 – the Wöhler exponent β; Row 2 – the observed damage for the simulated force for 45 km driving as described in Example 6.6. Row 3 – the expected damage $\mathbf{E}[d(\beta)]$ computed using the MCTP with Markov transition matrix estimated from the simulated record of the force discussed in Example 6.1. Row 4 – the same as Row 3 but the Markov transition matrix is estimated from the rainflow filtered load when all cycles with ranges below 10% of the maximum range are removed

	$\beta = 3$	$\beta = 4$	$\beta = 5$	$\beta = 6$	$\beta = 7$	$\beta = 8$
observed damage	$5.5 \cdot 10^{15}$	$1.4 \cdot 10^{20}$	$4.7 \cdot 10^{24}$	$1.9 \cdot 10^{29}$	$8.4 \cdot 10^{33}$	$4.1 \cdot 10^{38}$
broad band MCTP	$5.2 \cdot 10^{15}$	$1.4 \cdot 10^{20}$	$4.7 \cdot 10^{24}$	$1.9 \cdot 10^{29}$	$8.6 \cdot 10^{33}$	$4.2 \cdot 10^{38}$
narrow band MCTP	$5.4 \cdot 10^{15}$	$1.4 \cdot 10^{20}$	$4.7 \cdot 10^{24}$	$1.9 \cdot 10^{29}$	$8.6 \cdot 10^{33}$	$4.2 \cdot 10^{38}$

If F is symmetrical, then the chain Z_k is time reversible. Only time reversible chains will be considered.

6.8.2 Discrete Markov Loads – Definition

The MC Z_k has a very simple dependence structure and can seldom be used to model variability of real loads. Therefore, we will introduce a wider class of loads, named Markov loads, that are more flexible and may fit a variety of real loads.

We give three examples of Markov loads. The first one has already been mentioned, namely Markov chains of turning points, which is a very useful model for irregular loads. However, if the load has a longer memory, for example, narrow band oscillations, where large cycles come in groups, or for slowly varying signals with superimposed faster vibrations, then a simple Markov dependence structure cannot describe these features well. One way to introduce "longer memory" is to use multi-step Markov chains of turning points, where transition probabilities depend on the two (or more) preceding extremes (not only on the last one), see Rychlik [200]. However, the complexity of such models grows very fast with the number of "steps", and numerical aspects limit the applicability of the models. The third class of Markov models, studied by Johannesson [119, 120], is the switching Markov chains, or hidden Markov models. The model contains MCTP as a special case, but allows to switch, at random points, between different chains. An example of a possible application is to model the load on a vehicle driving, loaded or unloaded, on a gravel road or on a highway, and driven by different drivers (see Johannesson [119, 120]). We turn next to the definition of Markov loads.

Markov loads are time series which are functions of Markov chains, viz. $Y_k = G(Z_k)$, where Z_k is a stationary MC taking values in $1, 2 \ldots, N$, $N \geq J$. (Here, as before, J is the number of values the discretized load can take.)

Remark 6.2 *Markov loads form a very general class of models and theoretically it can model any finite length sequence X_k, $k = 1, \ldots, n$. One simply chooses a very large $N = n \cdot J$. (The so-called Markov Chain Monte Carlo MCMC-algorithm could be used to generate the sequence.) However, Markov loads $Y_k = G(Z_k)$ where Z_k taking large number of values N are numerically prohibitive.*

We next give two very simple examples of Markov loads.

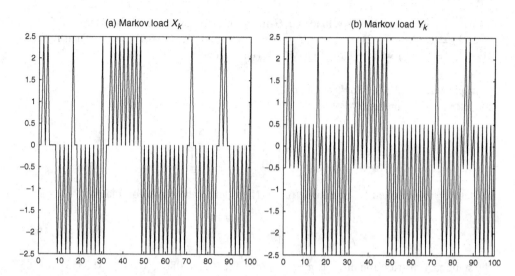

Figure 6.16 Markov loads defined in Example 6.22; (a) $X_k = G(Z_k)$ with $G(1) = 2.5$, $G(2) = G(3) = 0$, $G(4) = -2.5$; (b) $Y_k = G(Z_k)$ with $G(1) = 2.5$, $G(2) = -0.5$, $G(3) = 0.5$, $G(4) = -2.5$

Example 6.22 (Simple Markov Load 1)
Let Z_k be an MC taking four values $1, 2, 3, 4$ and having the transition matrix and stationary measure π

$$\boldsymbol{P} = \begin{bmatrix} 0 & 0 & 0.9 & 0.1 \\ 0 & 0 & 0.1 & 0.9 \\ 0.9 & 0.1 & 0 & 0 \\ 0.1 & 0.9 & 0 & 0 \end{bmatrix}, \qquad \pi = \begin{bmatrix} 0.25 & 0.25 & 0.25 & 0.25 \end{bmatrix}$$

Let discrete time series $X_k = G(Z_k)$ where $G(1) = 2.5$, $G(2) = G(3) = 0$, $G(4) = -2.5$. Note that X_k is not an MC. However for $Y_k = G(Z_k)$ where $G(1) = 2.5$, $G(2) = -0.5$ $G(3) = 0.5$, $G(4) = -2.5$ then X_k is an MCTP. In Figure 6.16, simulations of 100 values of Markov loads X_k and Y_k are presented. □

6.8.3 *Markov Chains of Turning Points*

Markov chains of turning points (MCTP) are Markov loads defined by the Markov matrix introduced in Section 3.1.5. For MCTP, there are explicit formulas for the expected rainflow matrix, and hence for the expected damage, see Rychlik [200] or Frendahl and Rychlik [95]. In addition, MCTP are particularly convenient for simulating loads for fatigue tests. The MCTP X_k taking values $(x_1 > x_2 > \ldots > x_J) = \mathbf{x}$ will be defined next.

Let $\boldsymbol{P}^{Mm}$, $\boldsymbol{P}^{mM}$ be $(J - 1, J - 1)$ dimensional upper and lower triangular matrices with transition probabilities from $x_1, \ldots, x_{J-1}$ to $x_2, \ldots, x_J$ and $x_2, \ldots, x_J$ to $x_1, \ldots, x_{J-1}$, respectively. Let Z_k be a stationary MC having the transition matrix

$$\boldsymbol{P} = \begin{bmatrix} \mathbf{0} & \boldsymbol{P}^{Mm} \\ \boldsymbol{P}^{mM} & \mathbf{0} \end{bmatrix}, \tag{6.88}$$

where

$$p_{ij}^{mM} = p_{(J-i)(J-j)}^{Mm}, \quad i, j = 1, \ldots J - 1. \tag{6.89}$$

In order to define the Markov load the G-function also has to be given. Here let $G(1) = x_1$, $G(k) = G(J + k - 2) = x_k$, for $2 \leq k \leq J - 1$, and $G(2(J - 1)) = x_J$. Then the time series $X_k = G(Z_k)$ will be called an MCTP (with range x).

Example 6.23 (Simple Markov Load 2)
Consider a $(3, 3)$-matrix P^{Mm}

$$P^{Mm} = \begin{bmatrix} 0.5 & 0.25 & 0.25 \\ 0 & 0.75 & 0.25 \\ 0 & 0 & 1 \end{bmatrix} \tag{6.90}$$

and let P^{mM} be defined by Equation (6.89). Next define the transition matrix

$$P = \begin{bmatrix} 0 & P^{Mm} \\ P^{mM} & 0 \end{bmatrix}. \tag{6.91}$$

The stationary distribution of Z_k is $\pi = [0.1765, 0.2353, 0.0882, 0.0882, 0.2353, 0.1765]$ and hence the $F = [\pi_i \cdot p_{ij}]$ matrix is

$$F = \begin{bmatrix} 0 & 0 & 0 & 0.0662 & 0.1176 & 0.0441 \\ 0 & 0 & 0 & 0 & 0.1765 & 0.0441 \\ 0 & 0 & 0 & 0 & 0 & 0.1765 \\ 0.1765 & 0 & 0 & 0 & 0 & 0 \\ 0.0441 & 0.1765 & 0 & 0 & 0 & 0 \\ 0.0441 & 0.1176 & 0.0662 & 0 & 0 & 0 \end{bmatrix}$$

As expected F is symmetrical and hence Z_k is time reversible.

Using the construction of MCTP $Y_k = G(Z_k)$, we will choose the G-function as follows: $G(1) = 3$, $G(2) = G(4) = 1$, $G(3) = G(5) = -1$, and $G(6) = -2$. However, it is not the only way of defining a Markov load from Z_k such that it is a sequence of turning points. Another way is to let $X_k = G(Z_k)$ be defined by $G(1) = 3$, $G(2) = 2$, $G(3) = 1$, $G(4) = -1$, $G(5) = -2$, and $G(6) = -3$. The sequence X_k is a Markov chain but the transition matrix P^{Mm} is not given by Equation (6.90). Simulation of both time series are presented in Figure 6.17.   □

6.8.4 Switching Markov Chain Loads

There are situations where the loading environment changes abruptly, for example, when a truck is driving on a highway and then on a gravel road, etc. Suppose that loads, under each of these conditions, could be adequately modelled as a Markov load. As Markov loads are convenient for generation of load sequences and for estimation of expected damage, one would like to include switching between different loading environment, also called regimes, into a more complex Markov load model. This is done by enlarging the state space. Modelling and properties of loads with Markov regimes are discussed in depth in Johannesson [119, 120], and here we give two examples of such models where we consider only two regimes.

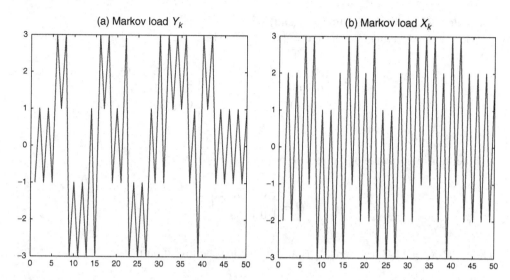

Figure 6.17 Markov loads defined in Example 6.23; (a) $Y_k = G(Z_k)$ with $G(1) = 3$, $G(2) = G(4) = 1$, $G(3) = G(5) = -1$, and $G(6) = -2$; (b) $X_k = G(Z_k)$ with $G(1) = 3$, $G(2) = 2$, $G(3) = 1$, $G(4) = -1$, $G(5) = -2$, and $G(6) = -3$.

Suppose that one has $r = 2$ regimes. Under each of the regimes the variability of the load can be well modelled by a Markov load $X_k^{(s)}$, $s = 1, 2$, having the same range x (some states may be non-essential). Consider a special case when both Markov loads $X_k^{(s)} = \tilde{G}(Z_k^{(s)})$, where $Z_k^{(s)}$ are an MC with (N, N)-transition matrices $P^{(s)}$. Introduce a new transition matrix

$$P = \begin{bmatrix} P^{(1)} & 0 \\ 0 & P^{(2)} \end{bmatrix} \tag{6.92}$$

and the function $G(i) = \tilde{G}(\mathrm{mod}(i, N) + 1)$, where the modulo-operation $\mathrm{mod}(i, N)$ finds the remainder of division of the first number by the second. The MC Z_k having transition matrix P is not positive recurrent, as, if it starts in regime s, it will stay in it for ever.

Suppose that regimes switch at random times. Let R_k indicate the regime at time k, i.e. it takes value 1 if load is at time k in regime "1" and value "2" otherwise. Suppose that R_k is an MC with transition matrix

$$Q = \begin{bmatrix} q_{11} & q_{12} \\ q_{21} & q_{22} \end{bmatrix}. \tag{6.93}$$

Now let Z_k be an MC having the transition matrix

$$P = \begin{bmatrix} q_{11}P^{(1)} & q_{12}P^{(2)} \\ q_{21}P^{(1)} & q_{22}P^{(2)} \end{bmatrix}. \tag{6.94}$$

and the function $G(i) = \tilde{G}(\mathrm{mod}(i, N) + 1)$. Then $X_k = G(Z_k)$ is the switching Markov load.

The transitions in this model occur according to the following scheme: At time k, process $R_k = s$, is first updated according to transition matrix Q. Suppose that the new value $R_{k+1} = w$. Then the new Z_{k+1} is updated using the transition matrix $P^{(w)}$.

Example 6.24 (Switching Markov Load)

Consider a load taking 3 values $x = \{1, 0, -1\}$. There are two regimes with transition matrices

$$P^{(1)} = \begin{bmatrix} 0 & 0.9 & 0.1 \\ 0.9 & 0 & 0.1 \\ 0.95 & 0.05 & 0 \end{bmatrix}, \qquad P^{(2)} = \begin{bmatrix} 0 & 0.05 & 0.95 \\ 0.1 & 0 & 0.9 \\ 0.1 & 0.9 & 0 \end{bmatrix},$$

and the transition matrix of the regime process R_k is

$$Q = \begin{bmatrix} 0.95 & 0.05 \\ 0.1 & 0.9 \end{bmatrix}.$$

The SMC $X_k = G(Z_k)$, where Z_k is an MC having the transition matrix

$$P = \begin{bmatrix} 0 \cdot 0.95 & 0 \cdot 0.05 & 0.9 \cdot 0.95 & 0.05 \cdot 0.05 & 0.1 \cdot 0.95 & 0.95 \cdot 0.05 \\ 0 \cdot 0.01 & 0 \cdot 0.9 & 0.9 \cdot 0.1 & 0.05 \cdot 0.9 & 0.1 \cdot 0.1 & 0.95 \cdot 0.9 \\ 0.9 \cdot 0.95 & 0.1 \cdot 0.05 & 0 \cdot 0.95 & 0 \cdot 0.05 & 0.1 \cdot 0.95 & 0.9 \cdot 0.05 \\ 0.9 \cdot 0.01 & 0.1 \cdot 0.9 & 0 \cdot 0.1 & 0 \cdot 0.9 & 0.1 \cdot 0.1 & 0.9 \cdot 0.9 \\ 0.05 \cdot 0.95 & 0.1 \cdot 0.05 & 0.95 \cdot 0.95 & 0.9 \cdot 0.05 & 0 \cdot 0.95 & 0 \cdot 0.05 \\ 0.05 \cdot 0.01 & 0.1 \cdot 0.9 & 0.95 \cdot 0.1 & 0.9 \cdot 0.9 & 0 \cdot 0.1 & 0 \cdot 0.9 \end{bmatrix}.$$

and the G-function is defined as follows $G(1) = G(2) = 1$, $G(3) = G(4) = 0$, and $G(5) = G(6) = -1$. (Note that X_k is not an MC.) The simulated values of Z_k and X_k processes are presented in Figure 6.18.

Finally, the stationary distribution of Z_k is

$$\pi = [0.2432, \ 0.0380, \ 0.2783, \ 0.1765, \ 0.0518, \ 0.2121]$$

and we can see that all states are essential. The probability mass function of Z_0, Z_1 is $F = [\pi_i \cdot p_{ij}]$

$$F = \begin{bmatrix} 0 & 0 & 0.2079 & 0.0006 & 0.0231 & 0.0116 \\ 0 & 0 & 0.0034 & 0.0017 & 0.0004 & 0.0325 \\ 0.2379 & 0.0014 & 0 & 0 & 0.0264 & 0.0125 \\ 0.0016 & 0.0159 & 0 & 0 & 0.0018 & 0.1430 \\ 0.0025 & 0.0003 & 0.0468 & 0.0023 & 0 & 0 \\ 0.0011 & 0.0191 & 0.0202 & 0.1718 & 0 & 0 \end{bmatrix}.$$

Since F is asymmetrical, Z_k is not time reversible.   □

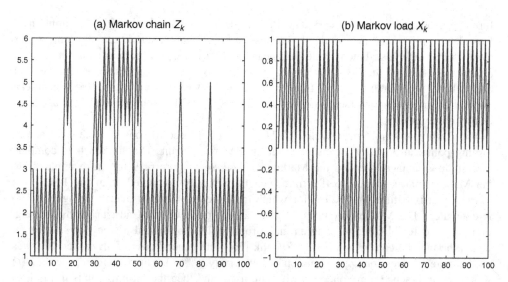

Figure 6.18 (a) The Markov chain Z_k used in Example 6.24; (b) the SMC $X_k = G(Z_k)$ with $G(1) = G(2) = 1$, $G(3) = G(4) = 0$, $G(5) = G(6) = -1$

6.8.4.1 Switching Markov Chain of Turning Points

The *Switching Markov Chains of Turning Points* (SMCTP) belong to the class of SMC. However, as this special case has found many applications in fatigue, we will define it here.

Again, consider two regimes $r = 2$ and suppose that under a regime $s = 1, 2$ the loads $X_k^{(s)}$ are MCTP. Then there are $(J - 1, J - 1)$ dimensional upper and lower triangular matrices $\mathbf{P}^{Mm1}$, $\mathbf{P}^{Mm2}$ and $\mathbf{P}^{mM1}$, $\mathbf{P}^{mM2}$ such that the $\mathbf{P}^{(s)}$ matrices are given by

$$\mathbf{P}^{(1)} = \begin{bmatrix} \mathbf{0} & \mathbf{P}^{Mm1} \\ \mathbf{P}^{mM1} & \mathbf{0} \end{bmatrix}, \qquad \mathbf{P}^{(2)} = \begin{bmatrix} \mathbf{0} & \mathbf{P}^{Mm2} \\ \mathbf{P}^{mM2} & \mathbf{0} \end{bmatrix}.$$

Now the rest is the same as for the general SMC given above.

6.8.5 Approximation of Expected Damage for Gaussian Loads

For stationary Gaussian loads, the Markov matrix can be accurately computed when the PSD $S(\omega)$ is known see Rychlik [199]; Krenk and Gluver [138] proposed a related method. Then the turning points in Gaussian loads can be approximated using an MCTP, and thus an approximation to the rainflow matrix is derived.

Lindgren and Broberg [143] have studied the accuracy of the Markov chain approximation for the dependence of the sequence of turning points in Gaussian loads. They found that the method is less accurate for Gaussian processes with narrow band or multi-modal PSD.

Bishop and Sherratt [29] have also used MCTP to approximate the rainflow cycle distribution for Gaussian loads. In their approach, the Markov matrix, i.e. the PDF for min-to-max cycles, was approximated using an explicit formula, the so-called Kowalewski PDF, viz.

$$f(u, v) = \frac{v - u}{4\alpha_2^2 \lambda_0 \sqrt{2(1 - \alpha_2^2)}} \exp\left(-\frac{(v - u)^2 + 4vu\alpha_2^2}{8\alpha_2^2(1 - \alpha_2^2)\lambda_0}\right) \tag{6.95}$$

for $v \geq u$, otherwise $f(u, v) = 0$. The PDF is derived on the basis of an assumption in Kowalewski [137], and independently in Sjöström [214], that the amplitude of a min-to-max cycle is Rayleigh distributed while the mean of the cycle has a normal distribution with parameters based on the first four spectral moments. Both the variables are independent of each other. The assumption is not true for non-degenerated Gaussian loads. However, it can serve as an approximation (see Lindgren and Broberg [143] for a discussion of the accuracy of the approximation).

The density of min-to-max cycles $f^{mM}(u, v)$ is often used to construct the sequence of turning points in $X(t)$ by means of a one-step Markov chain. This approach to compute the rainflow damage is called the Markov method, see Figure 6.19 for an example. Here the Markov matrix (Figure 6.19a) has been theoretically computed from the PSD. This Markov matrix defines an MCTP, for which the expected rainflow matrix in Figure 6.19b is computed. The Markov method has also been applied to non-Gaussian loads, see e.g. Frendahl and Rychlik [95]. Lindgren and Broberg [143] presented a study of spectra for which the method can be applied. In Rychlik [200] the general case of an m-step MC was used, where a regression approximation of the density f^{mMm} for the Gaussian process $X(t)$ was used. In practice, only one-step MCs are used, and then the turning points in the load are approximated by an MCTP.

6.8.6   Intensity of Interval Upcrossings for Markov Loads*

There are explicit formulas to compute $\mu^+(u, v)$ when the Markov load is specified. However, in other situations the Monte Carlo method can always be used. Here we will discuss

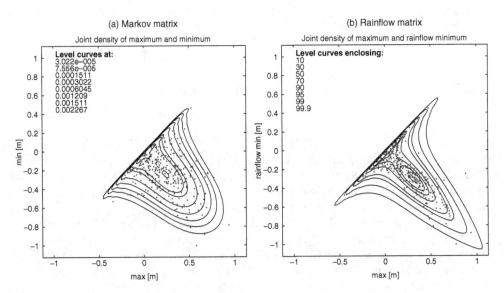

Figure 6.19   (a) Contour lines of the Markov matrix computed for the PSD used in Example 6.11. The dots represent the observed min-to-max cycles in a 30-minute long simulation of the load. (b) The rainflow matrix computed for the MCTP with Markov matrix from the left plot. The dots represent the observed rainflow cycles in the simulated load

computations of $\mu^+(x_i, x_j)$ (note that for $x_{i+1} < u < x_i$ and $x_j < v < x_{j-1}$, $\mu^+(u, v) = \mu^+(x_i, x_j)$). For Markov chains computations of $\mu^+(u, v)$ are particularly simple. However, Markov loads are often not an MC. Consequently, we will express $\mu^+(x_i, x_j)$ using crossings of intervals of Z_k. In order to do this, a new concept of oscillations intensity $\mu^+(U, V)$, between sets U and V, has to be introduced, so that $\mu^+(u, v; X_k) = \mu^+(U, V; Z_k)$, for $U = \{k \in S : G(k) < u\}$ and $V = \{k \in S : G(k) > v\}$. Characterization of load variability by means of intensity of oscillations between sets was introduced in Beste *et al.* [25]. The formula for computing $\mu^+(U, V)$ is explicit and involves only standard matrix operations, but it requires some additional notation and will thus not be given here. However, in the following subsection we will give a "physical" algorithm, presented in Rychlik [203], to evaluate $\mu^+(U, V)$.

6.8.6.1 Crossing Intensities for Markov Chains – Currents in Electric Circuits*

Let S be a set of terminals which are connected with wires, for example $S = \{1, 2, \ldots, J\}$. For two terminals $k, l \in S$, the resistance of a wire connecting k with l is R_{kl} (if $R_{kl} = +\infty$, there is no connection between k and l). An *electric circuit* is defined by its resistance matrix $\boldsymbol{R}$. Equivalently, one can use the conductance matrix $\mathbf{C}$, $C_{kl} = 1/R_{kl}$ with convention that $1/ + \infty = 0$. In the following, we assume that $C_{kl} = C_{lk}$, $C_{kk} = 0$, $\sum_{k,l} C_{kl} < +\infty$ and $\sum_l C_{kl} > 0$ for all k.

For completeness, we state Ohm's and Kirchhoff's Laws:

1. (Ohm) If k is at voltage v_k and l is at voltage v_l, then the current from l to k is $(v_k - v_l)C_{kl}$.
2. (Kirchhoff) The sum of all currents flowing into a given terminal is 0.

If an outside source is attached to a certain terminal, then the Kirchhoff Law does not apply unless account is taken of current flowing out from the outside source. But for all terminals k which are not attached to any outside source Ohm's and Kirchhoff's Laws imply

$$\sum_{k,l} (v_k - v_l)C_{kl} = 0. \tag{6.96}$$

Assume that we have an outside source with two terminals at voltage 0 and 1. Choose two disjoint subsets U, V of S and connect all terminals $k \in V$ to the 1 voltage of the source and all terminals $l \in U$ to the voltage 0. Now it can be shown that with conductance matrix $\boldsymbol{C} = \boldsymbol{F}$ ($C_{ij} = F_{ij}$), $\mu^+(U, V) = I(U, V)$, i.e. the crossing intensity from set U to V is equal to the current flowing out of the source and can be computed by means of Ohm's and Kirchhoff's Laws.

Obviously, knowing $\mu^+(U, V; Z_k)$, one also knows $\mu^+(u, v; X_k)$, and the expected damage for the Markov load can be obtained by means of Equation (6.78). The second use of the formulas for $\mu^+(U, V)$ is to solve the inverse problem, that is, given the rainflow matrix, to find a suitable Markov load that has the same rainflow matrix.

6.9 Summary

This chapter is devoted to modelling the variability of load signals. A load measurement may be considered as a deterministic load signal, which it is, once it has been measured. However, before the signal is measured, we only know the typical characteristics that govern the load signal, e.g. type of road, truck specification, pay load, driver behaviour, and so on. This is manifested by the fact that if the 'same signal' is measured twice, the resulting load signals will become somewhat different, but they can be described by common characteristic properties. It is these characteristic properties that the random process tries to capture, for example, a Gaussian load is characterized by its PSD, and a Markov load by its Markov transition matrix. Once we have a model of a random load, it is possible to evaluate mean properties and variability of the load, but also to simulate synthetic loads.

We have presented two classes of random loads:

Gaussian loads, which are based on modelling the frequency content and correlation in the load signals. Therefore, these are well suited as input to systems with frequency dependent behaviour.

Markov loads, where, for example, the sequence of turning points of the load signal is modelled by a Markov chain. Therefore, these are well suited for modelling fatigue loads, since the fatigue damage process is rate independent, and only the turning points are of importance.

The Gaussian load model gives a more complete description of the properties of the load signal, since it models the complete correlation structure, whereas the Markov load model is specialized to the fatigue application, since it models exactly what is important for the fatigue damage, namely the turning points of the signal. The generality of the Gaussian model also implies some disadvantages compared to the more simple Markov model, namely more complicated procedures for estimation, for generation of signals, and for theoretical calculations, as well as less accurate reproduction of the damage, when the frequency content is not relevant, but only the turning points. An advantage of the Gaussian load model is that it has a natural extension to the multi-input case, while for the Markov load model the extension is quite complicated. The bottom line is that both the Gaussian load model and the Markov load model are appropriate for modelling fatigue loads, but

- when the frequency content of the signal is of importance for the fatigue damage, e.g. in case of resonance phenomena, the Gaussian load model should be used, since it models exactly the frequency information in a stationary signal, and
- when the frequency content of the signal is *not* important for the fatigue damage, e.g. in the case of a quasi-static analysis, the Markov load model should be considered, since it models exactly what is important for the fatigue damage, namely the turning points of the signal.

A large part of this chapter is dedicated to computing the expected pseudo damage for different model assumptions. Empirical modelling of the range-pair distribution is exemplified in Section 6.3 based on a measured load signal. It is also possible to model the load more explicitly as in Section 6.4, where a road with pot holes is modelled as a standard Gaussian process with added irregularities, representing pot holes, at random locations. Different approaches to approximatie the expected damage for Gaussian loads are presented in Section 6.5. For non-Gaussian loads, the upcrossing intensity can be used to approximate the damage, see Section 6.6. The modelling of Markov loads, and the computation of their expected damage is treated in Section 6.8.

Another important topic is the variability of the measured damage. As discussed above, if we make a second measurement the load signal, it will become somewhat different, and consequently also the calculated damage. Section 6.7 presents methods for estimating the coefficient of variation of the measured damage depending on the limited measurement period. The uncertainty represents the precision in the measured damage value, and hence is a useful tool for determining how long we should measure, which is also discussed in Section 8.3.

7

Load Variation and Reliability

7.1 Modelling of Variability in Loads

Traditional truck fatigue design is based on a comparison between "the most severe customer" and "the weakest vehicle configuration". Such comparisons are, of course, quite vague and strongly dependent on subjective judgements. In consequence, it often leads to components being overdesigned with respect to their fatigue performance. Therefore, "the most severe customer" needs to be replaced with a proper statistical description of the severity of the customer population. Such a **load** description should also be compared to a statistical representation of the population of vehicle **strengths**, see Figure 7.1.

This chapter treats the problem of describing load variability and also deals with the load interaction with fatigue strength, the load-strength reliability problem.

Many advanced reliability methods have been developed, either based on advanced mathematical solvers or on extensive Monte Carlo simulations. However, they depend on data input that is usually not available in truck fatigue applications. In practice, one knows two numbers at the most for each input variable that represent its nominal value and its variation, and the use of the advanced methods forces the user to complete the lack of required information with advanced guesses.

It is desirable to use a reliability method that

- makes use of statistical tools, in order to improve the simple worst-case approach,
- is simple enough to be controlled by the skilled engineer, and
- is able to take full advantage of existing knowledge.

One method that should fulfil these requirements in the truck application is in the scientific literature denoted: **the first-order, second-moment reliability method**. The term "first-order" relates to the fact that the failure function, in our case the logarithm of the fatigue life, is approximated as a linear function of the load, i.e. a first-order approximation. The term "second-moment" denotes that it is based only on expected values and variances of the random variables involved.

Based on the simple **Gauss approximation formula**, the method combines scatter, originating from natural variation in physical variables, and uncertainty, which could be unknown

Guide to Load Analysis for Durability in Vehicle Engineering, First Edition. Edited by P. Johannesson and M. Speckert.
© 2014 Fraunhofer-Chalmers Research Centre for Industrial Mathematics.

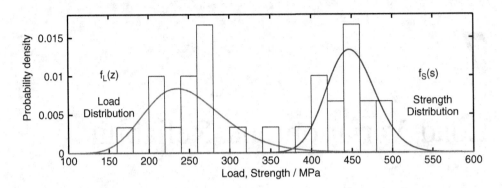

Figure 7.1   An illustration of the load-strength interaction

errors in necessary computations, errors in mathematical models and uncertainties due to small sample sizes.

This chapter will explain why this type of model is used, by rough comparisons to other established methods, and further describe an application of the method for fatigue problems related to load variability.

7.1.1   The Sources of Load Variability: Statistical Populations

A certain vehicle will be subjected to loads from a large number of road irregularities, curves, and obstacles. It will be handled by a certain number of different drivers and be exposed to different environments, pay loads and traffic regulations. For the individual vehicle, the variability of loads with time (or distance), that is, the load-time history, can be seen as a random process that is influenced by curves, pot holes, drivers, etc. Statistical models of this variability are treated in Chapter 6.

There is also a variability between different vehicle load processes since any two vehicles will experience different drivers, environments and events. For reliability calculations, we are interested in the population of vehicles. Such a population will vary both with respect to different load processes, built up from the variational sources given above, and with respect to the vehicle strength, see illustration in Figure 7.2.

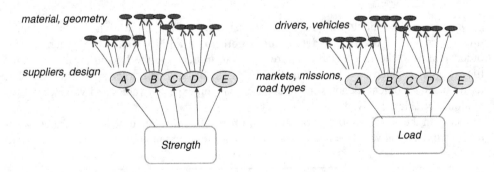

Figure 7.2   Some sources of variability

7.1.2 Controlled or Uncontrolled Variation

Some load variability sources can be regarded as *uncontrolled*, such as driver behaviour, road irregularities, or environment influences, since they cannot be predicted beforehand for a certain vehicle. Other sources may be *controlled*, such as different missions, or different vehicle specifications. Uncontrolled variability can only be taken into account by statistical considerations, while controlled variability may be handled by either classifications or statistics. Sometimes it is possible to choose if a controlled variability source should be a basis of classification, or be regarded as uncontrolled and be subject to statistical reliability analysis. The basis for the choice must be a trade-off between the efforts for proper classification and the necessity of larger safety factors for variability. For instance, it may be easy to control the difference between an off-road vehicle and a motorway transport vehicle. If this variation is a basis of classification, the off-road vehicle can be strengthened according to its more severe loads. By contrast, if this variability is regarded as uncontrolled, the motorway vehicle must also be made stronger to avoid population failure rate, which, of course, will increase the costs.

All sources of load variability should be classified with respect to the ability to control them, and sources that are possible to use as a basis of classification should be omitted from the reliability investigation.

7.1.3 Model Errors

The predicted reliability of a vehicle depends not only on the variability in load and strength; it is also highly influenced by the modelling process and possible errors in it.

In case of verification tests there are systematic differences between the test environment, the test track or the laboratory rig test equipment, and the service environment in field. Additionally, fatigue and durability tests always need to be accelerated to reduce time and cost.

In the case of computer-aided design, a large number of mathematical and physical models must be introduced in order to predict the fatigue damage during a target life.

The uncertainty introduced by these models is usually treated by the use of safety factors based on experience and judgements. A reliability method that aims at rational control of uncertainty should include these possible model errors. One possibility is to regard model errors as random variables and include them as components in the variability analysis.

In order to distinguish between different types of uncertainty components we will use the following definitions:

- *Uncertainty* is the overall term for the basis of the reliability measure. It will usually include a combination of both pure random variation and unknown errors.
- *Scatter or physical uncertainty* denotes variables that are random with respect to each vehicle population or subpopulation, i.e. the value of such a random variable will differ between any two vehicles both with regard to strength and loads. Driver influence, material scatter, manufacturing variation due to tolerances, and road irregularities are scatter components.
- *Statistical uncertainty* denotes errors in estimated parameters. For instance, if a component test yields a Wöhler curve, two parameters will be estimated from the test results. Due to

a limited number of specimens, these parameters are not true for the whole population of components, but they have a certain estimation error, which is here called statistical uncertainty. The parameter uncertainty can often be estimated from the test, and decreases with increasing number of test results. This type of uncertainty differs from the scatter type since it will be the same for all vehicles whose strength is estimated from that particular test.

- *Model uncertainty*. Physical and mathematical models always introduce unknown errors, whose possible magnitude may be estimated by judgements, by comparisons of models or by failure experience. In an accelerated test, one may introduce corrections to the resulting strength by means of a safety factor. In that case the factor should be used, but the uncertainty about the factor must be regarded as a model uncertainty component. Model errors are systematic in the sense that their unknown errors will be the same for all vehicles, until a new model is used.

In Section 7.6, an example with uncertainty components of all the types defined above is evaluated.

7.2 Reliability Assessment

There are several methods for reliability assessment, and the choice of model complexity can be described in two ways. From a statistical modelling point of view, one may choose a first moment, a second moment or a full probabilistic model. From the physical modelling point of view, it is of interest to study the model response to the variation in the input variables. The choice is between a first-order, that is a linear model, or second- or higher-order models.

7.2.1 The Statistical Model Complexity

A statistical model of a reliability problem is a combination of a certain number of random variables, each representing an identified source of uncertainty. The different complexity of the statistical model is related to the details of the description of each random variable.

7.2.1.1 First-Moment Methods

A first-moment reliability method is only based on a *single number* for each variable. This number may be the expected value or a certain percentile in the variable distribution. In the truck application, this method is represented by the use of a "severe customer" versus the "weakest truck". Making this concept more detailed means that a number of *single numbers* are combined into an overall safety factor, or by the application of **partial safety factors**.

7.2.1.2 The Second-Moment Reliability Index

The natural extension of the first-order methods is to also include the variances of the actual variables in the reliability calculation. Such an approach is based on the fundamental statistical property that the variance of the sum of two random variables equals

$$\mathrm{Var}[X_1 + X_2] = \mathrm{Var}[X_1] + \mathrm{Var}[X_2] + 2\mathrm{Cov}[X_1, X_2]. \tag{7.1}$$

This property can be generalized to several variables, which means that if the fatigue life can be represented as a linear function of a number of random variables, the reliability calculation reduces to

$$\text{Var[``fatigue life'']} = c_1^2 \text{Var}[X_1] + c_2^2 \text{Var}[X_2] + \cdots + \text{Covariances}, \qquad (7.2)$$

where the c-values are sensitivity coefficients for the different variables influence on fatigue life. In case the covariances are negligible, the final variance of the fatigue life can be obtained by a simple quadratic summation. Since this method takes both the mean value and the variability into account in a rational way, it is a large step towards a relevant probabilistic result compared to the first-order method. Still, it is simple enough to be used even for very complex physical relationships. It is therefore the basis of the load-strength method, which will be the main theme in this chapter.

7.2.1.3   The Full Probabilistic Model

A further extension of the probabilistic approach means that higher moments of the variables are included or a full representation of each variable is used, i.e. its individual probability distribution. Such an approach would be the ideal for the reliability calculation, as it would give a complete description of the life distribution. With today's computer resources, it is also possible to perform the necessary calculations, either by using numerical algorithms for integration of the multidimensional probability distribution, or by making Monte Carlo simulations. **Unfortunately, there is a serious limitation on the use of such methods: The input distributions are usually unknown**. Regardless of the great popularity of these methods, they are not recommended unless the input information is good enough. Without proper knowledge, the input distributions must be constructed by guesses and the output will give a result with false accuracy hiding uncertainties within a complex mathematical framework.

In the case of the present problem, reliability of heavy vehicles, the knowledge about the input distributions is far too weak to take advantage of the greater complexity in the full probabilistic model.

7.2.2   The Physical Model Complexity

The development of computational resources has made it possible to obtain very accurate descriptions of local stresses and strains in sensitive parts of vehicle components. This is done by multibody simulations and finite element calculations using outer load inputs and digitized drawings of the structure. For fatigue assessment, there is a need for a post-processing of the stress/strain data in order to obtain a damage measure. To this end, the computer resources are not very helpful, as the complex damage accumulation process highly depends on data that are not included in the drawing. For instance, there are non-metallic inclusions within the material, scratches or other surface irregularities, and defects in weldments, i.e. on the microscale, where physical models exist, there is a lack of information. The fatigue assessment must therefore be based on quite simple empirical models, and the result of calculations includes, regardless of the advanced numerical methods, considerable model uncertainties. This fact together with the uncertainty regarding the multidimensional input load process makes predictions of absolute fatigue life very unreliable. Even with a fully deterministic load, predicted life within a factor 2 of the subsequently observed life is considered world class with the present state-of-the-art. Still, the advanced mathematical calculations on the structure are extremely valuable for comparisons, and for the present problem such comparisons can be made in order to study the influence of input variations on predicted fatigue life. Controlled variation of input loads, variation according to material specifications or geometric variation within tolerances can easily be studied and used as a part of reliability calculations. This part must then be completed by making a corresponding analysis of the uncertainty in input loads and post-processing.

To find the influence of variations on the local stress/strain behaviour of a structure, the numerical mathematical procedure needs to be run with varying inputs. This may be very time-consuming and if the study is performed by Monte Carlo simulations, the time spent on calculations may not be acceptable. In such cases, it is desirable to simplify the model to, for instance, a quadratic multidimensional polynomial in the input variables. One popular technique is called the response surface method and one example of this is described in Section 7.3.1.

A rational treatment of different levels of approximations is to be found in Madsen *et al.* [151], Ditlevsen and Madsen [74], Melchers [159], where the two main approximation levels are denoted as the FORM (a First-Order Reliability Method) and the SORM (a Second-Order Reliability Method). In Section 7.5 a second-moment FORM for fatigue assessments is presented.

7.3 The Full Probabilistic Model

The whole load-strength reliability problem can be mathematically formulated in a single formula,

$$P(\text{``failure''}) = P(\Lambda > \Sigma) = \int_0^\infty (1 - F_\Lambda(x)) f_\Sigma(x) \, dx, \tag{7.3}$$

where Λ and Σ are the load and strength random variables, as before, $F_\Lambda(x)$ is the cumulative distribution function for the load variable, and $f_\Sigma(x)$ is the probability density function for the strength variable. The integration should be performed over the whole space of possible load values x. In practice, this formula can usually not be used, because of the lack of knowledge about the two statistical distributions. This gives two possibilities:

1. Make some initial assumptions about the distribution types, for instance some standard forms as normal, log-normal and Weibull, and estimate the parameters of these distributions from data. Then it is possible to solve the integral above and find the reliability by means of the probability of failure. An immediate problem with this approach is that the distributions should reflect all sources of scatter and uncertainty, which makes them hard to estimate. Another problem is that, because we are interested in very small probabilities, the result is very sensitive to the subjective choices of distribution types. A third problem is that both load and strength are complicated functions of several sources, and their final distributions are hard to obtain.
2. The last problem can be solved by simulations: Use the distributions of the sources of scatter and uncertainty and make a large number of numerical trials in order to find the

distributions for load and strength. Then combine them numerically, and find the proper probability of failure from the numeric empirical sample. Unfortunately, the same main problem as mentioned above is inherent in this method as well: the lack of knowledge about the input distributions, which will be reflected in the important but sensitive tails of the output distributions.

Nevertheless, stochastic simulations can be most valuable for studying the behaviour of a complicated system caused by variation in input variables, particularly when the system response is highly non-linear within the space of input variation. Therefore, the technique of Monte Carlo simulations is treated next.

7.3.1 Monte Carlo Simulations

The development of computer resources has made it possible to make extremely many calculations on a certain mathematical model. This has enabled the substitution of advanced mathematical tools for reliability with so-called Monte Carlo simulations, which means that a complex reality is modelled by statistical distributions. Using a huge set of random outcomes from this artificial reality as input makes it possible to find a full statistical distribution of the output. The procedure is as follows:

1. For each input variable, find its statistical distribution. This is done by first identifying the population of outcomes that is relevant for the application. For instance, in the case of the variable "road type", the variable values must first be defined by a certain classification. Next, the distribution of road types for a certain vehicle class must be established by measurements, interviews or judgements. From these data, an analytical expression for the full distribution must be established, for instance: "a normal distribution with expectation μ and variance σ^2". This is the most critical part of the whole procedure and must usually include a certain amount of guesswork.
2. In many cases, there are several input variables that are correlated. This fact must be taken into account by using multidimensional statistical distributions for these input variables. In the case of the "road type" distribution, one can expect that this is influenced at least by the "mission" for the vehicle, and the "market". This can be solved by making one Monte Carlo simulation for each pair of "mission", "market", or by investigating the three-dimensional random variable

$$\{\text{"roadtype", "mission", "market"}\}. \tag{7.4}$$

If correlations are neglected, there may be great errors in the output. The simple case of the variance of two random variables illustrates the problem: Assume that we have two random variables, X and Y with variances σ_X^2 and σ_Y^2, respectively, and the correlation coefficient ρ. Then we have the following result

$$\text{Var}[X + Y] = \sigma_X^2 + \sigma_Y^2 + 2\sigma_X\sigma_Y\rho = \begin{cases} \sigma_X^2 + \sigma_Y^2 & \text{if } \rho = 0 \\ (\sigma_X + \sigma_Y)^2 & \text{if } \rho = 1 \\ (\sigma_X - \sigma_Y)^2 & \text{if } \rho = -1 \end{cases}. \tag{7.5}$$

This shows that when neglecting correlation, the output variance will be underestimated if the variables are positively correlated. Positively correlated means that there is a tendency that Y is large when X is large and Y is small when X is small. If the variables are negatively correlated, we overestimate the variance. To make the problem even clearer, we can look at the special case when the two variances are identical, $\sigma_X = \sigma_Y$,

$$\text{Var}[X + Y] = \begin{cases} 2\sigma_X^2 & \text{if } \rho = 0 \\ 4\sigma_X^2 & \text{if } \rho = 1 \\ 0 & \text{if } \rho = -1 \end{cases} . \tag{7.6}$$

It is clear that in case there are substantial correlations, these must be handled. This can be done by using conditional distributions, i.e. fix one variable at each value and make repeated simulation studies, or by using multivariate input distributions.

3. When the statistical distributions are defined, a random number generator is used to draw one set of input variables at a time, calculate the interesting property with this input and store the result. This procedure is then repeated as many times as necessary to get a proper output distribution (usually thousands). Random number generators can be found in many software packages. They may include several distribution types including multidimensional normal distributions. If the type required is not implemented, it is easy to get univariate random numbers from any distribution type by the following transformation of a uniform random number: Assume that the wanted distribution is $F(x)$ and that we have a random number U that is uniformly distributed between 0 and 1. Then we generate a number U_i and find the number required simply by taking the inverse $X_i = F^{-1}(U_i)$. In cases with multivariate distributions, the procedure is more complicated and is not treated here.

4. The result from the simulations is a large set of output values. If the input distributions are correct, these output values are a sample of the true output distribution and can be further analysed by taking the mean and variance, or finding empirical percentiles such as the "95% customer severity". These latter numbers are found by sorting the resulting output values and choosing the percentile required in the following way: The output values are collected in a vector $Y = \{Y_1, Y_2, \ldots, Y_N\}$, numbered by the order of calculation and representing N simulations. The vector is sorted by its values to $Y_{sort} = \{Y_{(1)}, Y_{(2)}, \ldots, Y_{(N)}\}$, numbered by their sizes. Then the empirical 95% percentile can be found by

$$y_{0.95} = Y_{(\text{round}(0.95 \cdot N))}. \tag{7.7}$$

7.3.1.1 Response Surface Approximation

In case the calculations in the Monte Carlo simulations are too time-consuming, the mathematical model may be simplified within the interesting area for simulation. A general method for doing this is the so-called *response surface method*. This fits a polynomial function to the more complicated mathematical model.

Assume that we have a numerical procedure that takes the input variables x_1 and x_2 to calculate the property $y = g(x_1, x_2)$. A second-order polynomial model is assumed to be

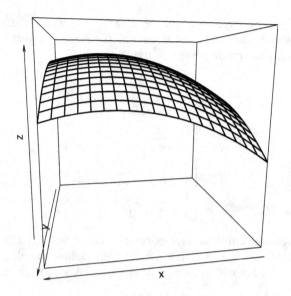

Figure 7.3 A second-order response surface approximation

a good approximation in the area of input variable variation and the following simplified model is used,

$$y = a_0 + a_1x_1 + a_2x_2 + a_{12}x_1x_2 + a_{11}x_1^2 + a_{22}x_2^2 + \varepsilon, \qquad (7.8)$$

where the extra term ε represents the error in the approximation. An efficient way to estimate the new model parameters $a_0, a_1, a_2, a_{11}, a_{22}, a_{12}$ is to make a second-order design plan for the computation. One such plan is the following:

run	x_1'	x_2'	
1	-1	-1	y_1
2	$+1$	-1	y_2
3	-1	$+1$	y_3
4	$+1$	$+1$	y_4
5	$-\sqrt{2}$	0	y_5
6	$+\sqrt{2}$	0	y_6
7	0	$-\sqrt{2}$	y_7
8	0	$+\sqrt{2}$	y_8
9	0	0	y_9

$$(7.9)$$

where the variables are transformed according to

$$x_1' = \frac{x_1 - \mu_1}{\sigma_1}, \qquad x_2' = \frac{x_2 - \mu_2}{\sigma_2} \qquad (7.10)$$

with μ and σ equal to the mean and variance of each variable, and the y-values in the right most column are the results from the calculation. The parameter vector a' is estimated by a simple matrix operation on the design matrix X

$$X = \begin{bmatrix} 1 & -1 & -1 & 1 & 1 & 1 \\ 1 & +1 & -1 & -1 & 1 & 1 \\ 1 & -1 & +1 & -1 & 1 & 1 \\ 1 & +1 & +1 & 1 & 1 & 1 \\ 1 & -\sqrt{2} & 0 & 0 & 2 & 0 \\ 1 & +\sqrt{2} & 0 & 0 & 2 & 0 \\ 1 & 0 & -\sqrt{2} & 0 & 0 & 2 \\ 1 & 0 & +\sqrt{2} & 0 & 0 & 2 \\ 1 & 0 & 0 & 0 & 0 & 0 \end{bmatrix} \qquad (7.11)$$

which is constructed with columns corresponding to the variables in Equation (7.8), the first column with ones on each line, and to the following equal to the value of the variables x_1', x_2', $x_1'x_2'$, $x_1'^2$, $x_2'^2$. Together with the output vector y, the least squares solution is given by

$$\hat{a}' = (X^T X)^{-1} X^T y. \qquad (7.12)$$

and using the transformation (7.10), the corresponding estimated parameter vector $\hat{a}$ can be determined. The adequacy of the model can be checked by studying the residual vector r,

$$r = y - X\hat{a}'. \qquad (7.13)$$

This residual vector represents the model error ε, which should be compared to the variance of y caused by the different input variables to see if the approximation is reasonable. For instance, one can compare the mean square error

$$MSE = \frac{1}{n} \sum_{i=1}^{n} r_i^2 \qquad (7.14)$$

with the contribution from the largest influential variable,

$$\max_i \left\{ \left(\frac{\partial y}{\partial x_i} \right)^2 \sigma_{x_i}^2 \right\}. \qquad (7.15)$$

As a rule of thumb, if the mean square error is less than 4% of the maximum variable contribution, then the model is accurate enough. In case of large mean square errors, the model should be refined to higher degree polynomials or be replaced by a model more related to physics than the polynomial form. However, all rules of thumb must be handled with care; non-linearities such as thresholds outside the domain of the response surface may still violate the approximation and physics must be more carefully considered in the modelling work.

The example shown here is a common type of design when making experiments. The design may be extended to more variables by the following construction: First construct a two-factor basic design corresponding to the four first rows above, then add two lines for each variable, where all variables but the actual variable are kept at zero and the variable in question is put at $\pm\alpha$. Complete the design with a centre point, where all the transformed variables are zero. The number α is in the standard procedure chosen to give a rotational design, which in the two-dimensional case results in $\alpha = \sqrt{2}$.

7.3.2 Accuracy of the Full Probabilistic Approach

As discussed at the beginning of this chapter, the use of complete probabilistic distributions must usually be based on some guesses about the type of distribution. This is because of the few observations that are possible to obtain from tests and field measurements. The output from Monte Carlo simulations must be handled with care and the use of the results as reliability measures needs additional safety considerations. For mean values, variances and moderate percentiles, say, between 1% and 99%, the procedure is not so sensitive to the subjective choices, and the results are indeed useful. However, if the reliability measure is fairly linear within the limits of input variation, then these results can normally be found with much less effort, namely by the second-moment approach, which is described in Section 7.5. Comparisons of the two approaches are presented in Lorén et al. [150], Lorén and Svensson [149].

7.4 The First-Moment Method

In structural safety applications for buildings, bridges and similar constructions, the partial safety factor method is used by the definition of *design values* for load, strength (resistance) and geometry, denoted q_{id}, r_{jd}, and l_{kd}, respectively. The design values may be written as

$$q_{id} = \gamma_i q_{ic}$$
$$r_{jd} = \varphi_j r_{jc} \qquad (7.16)$$
$$l_{kd} = \varsigma_k l_{kc}$$

where q_{ic}, r_{jc}, and l_{kc} are characteristic values, γ_i are load factors, φ_j are resistance factors, and ς_k are geometrical factors. The structure is verified by using the design values in the load-strength calculation. Characteristic values and partial safety factors are often regulated in standards like the Eurocode, and the characteristic values are typically the mean value for dead load and for geometrical parameters, the 98th percentile for annual maxima of loads, and the 5th percentile for strength parameters.

The advantage of this type of standardized method is that it may be constructed to be unambiguous and therefore suitable for constructions that are highly regulated by authorities. The drawbacks are that several worst-case factors may result in overdesign, since the probability of simultaneous events is much smaller than reflected in the calculation method. This problem can be handled by empirical corrections, which however may make the method complex and difficult to handle.

The simple method of comparing the worst-case customer with the weakest configuration is a first step with this method. Variants of the application include classification of customers in different classes and choosing the most severe customer in each class. Possibly, it would be fruitful to extend this simple method by using the partial safety factor method described above in order to get a more detailed picture of the sources of uncertainty. However, it may be quite complex and demand calibrations, and the advantages of such an approach is hard to see in comparison to the second-moment method. Therefore, we here concentrate on the second-moment approach and evaluate it in some detail below.

7.5 The Second-Moment Method

For truck development problems, the knowledge of scatter and uncertainty is often limited, but some knowledge is usually available, for example, from previous experience. That is, the expected load and strength are estimated and the variation can be roughly judged for most of the uncertainty sources. This overall picture points out the second-moment approach as the road to improvement of the simple safety factor practice. In this chapter, we will present the theoretical basis for the second-moment formulation by presenting Gauss approximation formula and further evaluating the approach by means of a second-moment, first-order reliability index.

7.5.1 The Gauss Approximation Formula

For a function of several variables, the variance can be found by the Gauss approximation formula. Assume that the log fatigue life can be described as the function $f(x_1, x_2, \ldots, x_n)$, where $x_1, x_2, \ldots, x_n$ are variables influencing the life. If x_i has the mean μ_i and the standard deviation σ_i, then the Gauss approximation formula says

$$\text{Var}[f(x_1, x_2, \ldots, x_n)] \approx \sum_{i=1}^{n} c_i^2 \sigma_i^2 + \sum_{i \neq i} c_i c_j \text{Cov}[x_i, x_j] \qquad (7.17)$$

where the c-values are the *sensitivity coefficients* for the different variables influencing the log fatigue life. Often the covariances are negligible and then the final variance of fatigue life can be obtained by the simple quadratic summation. The sensitivity coefficients are the partial derivatives of the function with respect to the different variables, evaluated at the mean values. However, it can easily be obtained for a complicated function numerically by differentiation. Then the sensitivity coefficients are given by

$$c_i = \left. \frac{\partial f}{\partial x_i} \right|_{x_i = \mu_i} \approx \frac{f(\mu_i + 2\sigma_i, \mu_{\neq i}) - f(\mu_i - 2\sigma_i, \mu_{\neq i})}{4\sigma_i}, \qquad (7.18)$$

where $\mu_{\neq i}$ denotes the vector of expected values for the variables with other indices than i. Note that the step size in the given differential quotient is chosen to be as large as two standard deviations. This choice is robust against possible numerical problems with small steps, and can also be justified by the fact that our goal is to study the behaviour at fairly small probabilities and not in the centre of the distributions; Svensson and de Maré say:

> This formula produces a measure of the variance of the function with only two function evaluations per variable. If a Monte Carlo simulation is based only on the knowledge about input mean and variance, then it gives no more information than this simple procedure in case of a linear response.
>
> – Svensson and de Maré [229].

7.6   The Fatigue Load-Strength Model

The second-moment approach is here formulated as a special reliability index for fatigue life assessments, based on the load-strength formulation. For non-failure, we want the strength to be greater than the load during a certain design target time or distance of usage,

$$\Sigma > \Lambda(t) \qquad \text{for} \qquad t < T_d, \tag{7.19}$$

where the strength Σ is modelled as a fixed value and the load $\Lambda(t)$ is modelled as a growing value that represents the damage accumulation. For the reliability calculation we want to assess the condition at the target time of usage by investigating if

$$\Sigma > \Lambda(T_d). \tag{7.20}$$

In the load-strength model, strength, Σ, and load, Λ, are random variables. The reliability (or the failure risk) may be assessed with a suitably defined reliability index, γ. Such an index is defined and discussed in Section 7.6.2. The **reliability index** will be used directly for comparisons and be regarded as a **measure of distance to the failure mode**.

This distance may be a combination of expected values, statistical measures, model uncertainties and engineering judgements, where the model uncertainties and judgement parts can be estimated and updated based on experience or studies of similar constructions.

A special problem for reliability assessments in fatigue is to establish proper measures of the load and the strength. These measures should be scalars in order to manage their statistical distributions in the reliability calculations. This implies a projection of the multidimensional load and strength features to one-dimensional spaces which will introduce model errors. These must be taken into account by being included in the model uncertainty.

7.6.1   The Fatigue Load and Strength Variables

In fatigue, the service load can be seen as a random process, a time-varying load signal acting on the structure to be analysed. The fatigue strength of the structure is best described as a random variable; it varies between nominally identical samples. In reality, the strength of a certain individual truck (drawn at random from a population of nominally identical trucks) is statistically independent of its usage, i.e. the severity of the driver and transport mission (drawn at random from the population of drivers and missions relevant for the studied truck population). Thus, strength and load are statistically independent.

To facilitate statistical reliability studies, one-dimensional measures of load and strength are sought. To reduce the time-varying load to a single-number representation, a damage-accumulation model is needed. The most frequently used model of this type is the Palmgren-Miner damage accumulation hypothesis. To get a one-dimensional strength description, one point on the Wöhler curve is selected. The rest of the Wöhler curve is then described by this point and the exponent in Basquin's equation.

Combining Basquin and Palmgren-Miner to get the desired one-dimensional load description introduces an artificial coupling between strength description and load description. The numerical value of the load measure, often labelled "equivalent load", will depend heavily on the fatigue exponent in the Basquin equation. But the exponent is a strength characteristic!

The subtle difference between actual load and strength (statistically quite independent) and their one-dimensional representations (connected via the exponent) must be kept in mind when analysing load-strength models.

As stated earlier in this book, the damage-accumulation theory assumes that the fatigue life of a structure subjected to a variable amplitude load can be calculated by comparison of the load cycle count and the strength of the structure at the corresponding constant amplitude loads. Many tests have demonstrated that this assumption is not universally valid, and different approaches have been presented for improvements.

1. *The relative Miner rule.* Variable amplitude tests can be characterized by their "Miner sums" at failure, i.e. their observed damage measure D, Equation 3.24. In future predictions, the failure criterium: $D > 1$ may then be replaced by the experimentally observed damage measure. For instance, in some weldment standards, there are recommendations such as the criterion $D > 0.5$ for broad-band loading and $D > 1$ for narrow-band loading. Thus, for a certain application, it is possible to use the relative Miner rule in the following way. Find the constant amplitude fatigue strength by test, make additional spectrum tests and adjust the criterion according to the mean value of the spectrum test results.
2. *The Gassner line.* If the spectrum load for future usage is well known, one may use scaled realizations of this very spectrum for tests of the relevant structure and define the strength by means of a Gassner line. This is a line in a log-log-diagram describing the log life as a function of the log max load in the spectrum. In fact, this method avoids the trust in the Palmgren-Miner rule and only assumes that scaling preserves the sequence damaging behaviour.
3. *The beta-norm load approach.* The two methods above may be generalized by representing the fatigue strength as a line, which is the log life as a function of the beta-norm load. Here, the beta-norm load is defined using the Palmgren-Miner rule as will be explained in Example 7.1.

The choice between the different approaches above can be seen as a trade-off between accuracy demands and testing resources, see Figure 7.4. The Gassner line approach avoids trusting the damage accumulation theory, but demands one specific strength curve for each spectrum type that can be expected to occur in future use. The accuracy is high because of small model errors, but the required testing efforts are great. The established approach using constant amplitude tests for predicting variable loads in service has the opposite position: Heavy trust in theory gives large model uncertainties, but needs only comparably cheap constant amplitude tests. The beta-norm load approach can be adjusted to any trade-off between accuracy and test efforts. In fact all the other approaches are special cases of this generalization, that is, at a constant amplitude test the beta-norm load is equal to the amplitude and in case of a load spectrum that is known beforehand, the beta-norm load approach is the same as the Gassner line approach. Further, the method given by the relative Miner is nothing else than a combination of two-spectrum types, the constant amplitude and the spectrum used for adjustment, which is also a special case of the beta-norm load approach.

7.6.2 Reliability Indices

For the second-moment approach, we will here present a reliability index that is based only on expected values and variances of the scatter and uncertainty components. Reliability

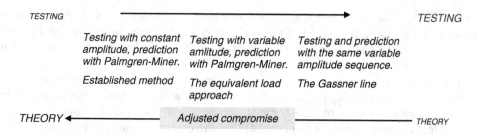

Figure 7.4 Illustration of the trade-off between trust in theory and testing efforts

assessments may be performed using a suitably defined **reliability index**, γ, being a function of the (stochastical) strength and final load measures, Σ and Λ, respectively. The index is used for direct comparisons and could be regarded as a measure of distance to the failure mode or as a measure of the "safety margin". With the index suggested below, $\gamma = 0$ indicates 50% failure probability, positive values of γ indicate lower failure probability. The required or acceptable value of γ, i.e. the required or acceptable margin of safety, is a matter of judgement. Thus, the acceptable value should be defined according to the application and the studied structure. Further, actual operational experience should be monitored and the set "acceptable" value should be updated – increased or decreased – on the basis of experience and company policy. Acceptable values are discussed in Section 7.6.8, "usage of the reliability index". One such index is the Cornell reliability index, which is based on the difference between strength and load, $\Sigma - \Lambda$

$$\gamma = \frac{\mu_\Sigma - \mu_\Lambda}{\sqrt{\tau_\Sigma^2 + \tau_\Lambda^2}}, \tag{7.21}$$

where μ_Σ and μ_Λ are the expected values of the strength and load, respectively, and τ_Σ^2 and τ_Λ^2 are measures of their uncertainties. Such an index is very convenient to use if the underlying variables have normal distributions. In the case of fatigue load and strength, this is not the case, but taking the logarithm of the variables gives a good approximation to normality. For fatigue, the method to obtain a scalar measure is, as described before, using the Palmgren-Miner rule. This can be done by using the abstract damage measure D or by formulating equivalent loads based on the accumulation theory. The latter approach gives numbers that are easier to interpret for the engineer, since the equivalent load can be expressed in stress or force units. Therefore, we will here formulate a reliability index based on the logarithm of the equivalent strength and load.

7.6.3 The Equivalent Load and Strength Variables

We here use a variant of the Basquin equation that is formulated with respect to an endurance limit for the component or material,

$$N = n_e \left(\frac{S}{\alpha_e} \right)^{-\beta}. \tag{7.22}$$

The number n_e is a suitably chosen cycle number corresponding to the endurance limit α_e, and S is the load amplitude. A proper choice for steel constructions is $n_e = 10^6$. However, any value of n_e may be chosen. This formulation has the advantage that each property in the formula, except the exponent, has a physical interpretation; the parameter α_e has the same unit as the stress value and can be interpreted as an endurance limit, and the parameter n_e is a cycle number.

The exponent β is assumed to be a component characteristic and is estimated by tests or by experience. The property α_e is regarded as a random variable that varies between specimens, reflecting different microstructural features, geometry and manufacture variation, etc. The total effect of these variations is manifested by the observed scatter in fatigue tests. The random variable α_e is defined as the **equivalent strength**, denoted S_{eq}, and is modelled as a log-normal variable, where the parameters represent a certain component type or a specific material.

The strength of a component is estimated from laboratory test results. For spectrum tests, the applied load consists of a variable amplitude time history that for damage analysis is described by its amplitude spectrum based on the rainflow cycle counting procedure giving the number of M applied load amplitudes, $\{S_k;\ k = 1, 2, \ldots, M\}$. The applied load amplitudes are not necessarily different, and thus the constant amplitude case is also covered. According to the Palmgren-Miner cumulative damage assumption using Equation (7.22), we obtain

$$D = \sum_{k=1}^{M} \frac{1}{N_k} = \sum_{k=1}^{M} \frac{S_k^\beta}{n_e \alpha_e^\beta} = \frac{1}{n_e \alpha_e^\beta} \sum_{k=1}^{M} S_k^\beta, \qquad (7.23)$$

where N_k is the life according to Equation (7.22) for the load amplitude S_k, and M is the total number of counted cycles.

Failure occurs when D equals unity, and the value of α_e for a specific specimen is

$$S_{eq} = \left(\frac{1}{n_e} \sum_{k=1}^{M_S} S_k^\beta \right)^{1/\beta}, \qquad (7.24)$$

which is an observation of the random **equivalent strength variable**. Note that the observed random variable on the right-hand side is M_S, the number of cycles to failure for the specific load sequence applied to the test specimen.

For the application of the load-strength model we need a comparable property for a service load spectrum. This is constructed in a similar way on the basis of the calculated damage for the spectrum,

$$D_{T_d} = \sum_{k} \frac{1}{N_k} = \frac{T_d}{T} \sum_{k=1}^{M_L} \frac{L_k^\beta}{n_e \alpha_e^\beta} = \frac{K_L}{n_e \alpha_e^\beta} \sum_{k=1}^{M_L} L_k^\beta. \qquad (7.25)$$

Here we sum over all M_L counted amplitudes with values L_k for the driven distance T, and T_d is the design target life by means of, for instance, driving distance. As an abbreviation we also introduce the scaling factor $K_L = T_d/T$.

We now define the **equivalent load variable** L_{eq} as

$$L_{eq} = \left(\frac{K_L}{n_e} \sum_{k=1}^{M_L} L_k^\beta \right)^{1/\beta}, \qquad (7.26)$$

and it is modelled as a log-normal random variable. It can be seen from Equation 7.25 that

$$D_{T_d} = \frac{L_{eq}^{\beta}}{S_{eq}^{\beta}}. \tag{7.27}$$

This quotient represents a safety factor and equals unity when L_{eq} equals S_{eq}. The distance between the logarithms of these two random properties represents a safety margin, which we will use for the reliability index,

$$\gamma = \frac{\mathbf{E}[\ln S_{eq} - \ln L_{eq}]}{\sqrt{\mathrm{Var}[\ln S_{eq} - \ln L_{eq}]}} = \frac{m_S - m_L}{\sqrt{\tau_S^2 + \tau_L^2 + \tau_{\beta}^2}} = \frac{\delta}{\tau}, \tag{7.28}$$

where the nominator, δ, contains the average distance between the log-values of the observed equivalent strengths and loads, respectively, and the denominator, τ, is the second-moment approximation of the uncertainty in the distance between the logarithmic equivalent strengths and loads. Since the uncertainty in the fatigue exponent β affects both the load and strength variable, it is treated specially, which will be shown later.

Example 7.1 (Steering Arm)

The reliability index will here be illustrated by a case study of a steering arm, a safety component with an assumed target life of one million km. The case study has been presented in Svensson *et al.* [230]. In total, 9 steering arms were tested until failure. The steering arms were tested with a variable amplitude load with three different spectra. The spectra had the same shape, but differed in scaling, see the range-pair plot in Figure 7.5. In all cases, the load ratio for the applied force-time history is $R = -1$, and for all spectra, including those from field measurements, load ranges below 10% of the largest one have been omitted.

The laboratory tests were compared to field measurements from several expeditions. In total, 17 measurements of the force acting on the steering arm were performed in 6 different European countries as well as in Ethiopia, covering a large part of the applications and road conditions that the component may face.

We first estimate the strength. In the present case the exponent is not known beforehand, but must be estimated from the test. This is achieved by estimating, (see Johannesson *et al.*

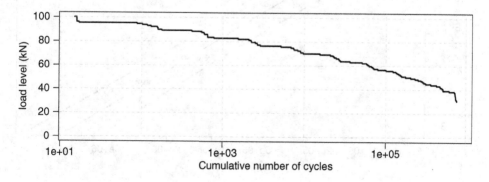

Figure 7.5 One of the load spectra for the laboratory tests

[124]), the parameters in the equation:

$$N = 10^6 \left(\frac{|S|_\beta}{\alpha_e}\right)^{-\beta}, \qquad |S|_\beta = \left(\frac{1}{M}\sum_{k=1}^{M} S_k^\beta\right)^{1/\beta}, \qquad (7.29)$$

where $|S|_\beta$ is denoted by *the equivalent beta-norm load*, which is the constant amplitude load applied M times that gives the same damage as the applied load spectrum. For the actual data, we obtained the estimates: $\widehat{\alpha}_e = 48.0$ kN, $\hat{\beta} = 6.5$, and the standard deviation for the logarithmic life, $s_l = 0.20$. The estimated Wöhler line is illustrated in Figure 7.6.

The observed equivalent strengths can now be calculated using Equation (7.24),

$$S_{eq,i} = \left(\frac{1}{10^6}\sum_{k=1}^{M_{S,i}} S_{i,k}^{6.5}\right)^{1/6.5}, \qquad (7.30)$$

giving the observations {48.9, 47.8, 48.3, 45.9, 47.9, 46.7, 46.6, 47.4, 50.4} in unit kN, and their logarithmic average is $m_S = \overline{\ln\ S_{eq}} = 3.87$.

The equivalent load is estimated from the seventeen service load observations. Each load spectrum from a field measurement is used to calculate its equivalent load stress amplitude, according to Equation (7.26),

$$L_{eq,i} = \left(\frac{K_{L,i}}{n_e}\sum_{k=1}^{M_{L,i}} L_k^\beta\right)^{1/\beta} = \left(\frac{T_d}{10^6 \cdot T_i}\sum_{k=1}^{M_{L,i}} L_k^{6.5}\right)^{1/6.5} \qquad (7.31)$$

using the target life $T_d = 10^6$ km. The resulting average of the logarithmic equivalent load was $m_L = \overline{\ln\ L_{eq}} = 2.70$. These estimates can now be inserted in the nominator of the reliability index, Equation (7.28),

$$m_S - m_L = \overline{\ln\ S_{eq}} - \overline{\ln\ L_{eq}} = 3.87 - 2.70 = 1.17. \qquad (7.32)$$

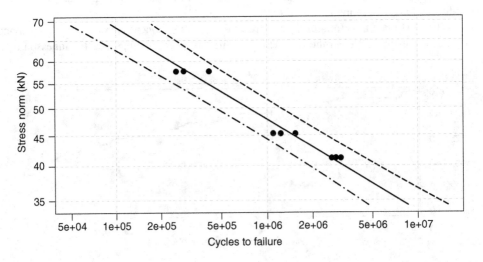

Figure 7.6 Fatigue test results with the estimated Wöhler line and 95% prediction limits

This observed distance needs to be scaled by the prediction uncertainty. The example will be continued after the introduction of the theory of uncertainty measures. □

7.6.4 Determining Uncertainty Measures

In order to find the denominator of the reliability index, Equation (7.28), we need three terms representing uncertainties in the load, the strength, and the fatigue exponent. We first give some common rules for finding prediction uncertainties and then treat the specific problems related to the three actual uncertainty components.

7.6.4.1 Scatter and Uncertainty

In the reliability assessment it is desirable to include not only variability due to true random variation, but also uncertainty originating from lack of knowledge. Such uncertainties may be statistical uncertainties about parameters in empirical models or uncertainties about possible model errors. It is important to distinguish between these types as uncertainties are often systematic and therefore, (1) are possible to eliminate by further investigations, and (2) do not vary when different worst-case scenarios are studied.

7.6.4.2 Statistical Methods for Estimating Scatter

Scatter has its origin in a natural variation of a property and is best estimated by using a random sample of the actual property. Having the samples $x_1, x_2, \ldots, x_n$, the standard deviation is estimated as

$$s = \sqrt{\frac{1}{n-1} \sum_{i=1}^{n} (x_i - \overline{x})^2},$$ (7.33)

where $\overline{x}$ is the average value of the observations. Since the reliability measure is related to assessment of future values of the different sources of uncertainty, it is good to use the prediction uncertainty. A future prediction of the variable will contain two errors, namely the error in the estimated mean value, and an additional random deviation, giving the prediction standard deviation

$$s'_p = \sqrt{s^2 + \frac{s^2}{n}} = s\sqrt{1 + \frac{1}{n}}.$$ (7.34)

Further, there is an uncertainty in the estimated standard deviation which should be considered. A simple method to do this is to use the statistical t-distribution, and adjust the estimated prediction uncertainty with a factor that depends on the number of observations used for the standard deviation estimate. This gives the final estimate of the X contribution to the uncertainty:

$$s_p = \frac{t_{0.975, n-1}}{1.96} s\sqrt{1 + \frac{1}{n}},$$ (7.35)

where the number 1.96 in the denominator normalizes the correction factor to a property corresponding to standard deviation, since the t-value approaches 1.96 when n is large, see

Table 7.1 Values from the t-distribution. In the table, "df" means the number of degrees of freedom, which in the case of standard deviation estimates as above is equal to n − 1, the number of observations minus 1

df	2	3	4	5	6	7	8	9	10	11	18	30	∞
$t_{0.025}$	4.3	3.2	2.8	2.6	2.4	2.4	2.3	2.3	2.2	2.2	2.1	2.0	1.96

Table 7.1. Using these adjustments, the resulting standard deviation measure is a combination of scatter and statistical uncertainty. It will reflect the uncertainty due to few observations, which is advantageous when one wants to find the balance between test efforts and safe predictions: A large number of observations will reduce the prediction variance component and makes it possible to decrease the safety distance.

Example 7.2 (Weld Toe)
The radius of the weld toe at the critical spot for a component is a parameter in a computer code calculating the fatigue strength. The radius of ten components has been measured with the mean value 1.1 mm and the standard deviation 0.3 mm. The prediction standard deviation for this influential factor is

$$s_{r_p} = \frac{t_{0.975,n-1}}{1.96} s \sqrt{1 + \frac{1}{n}} = \frac{2.3}{1.96} \cdot 0.3 \cdot \sqrt{1 + \frac{1}{10}} = 0.36.$$

□

7.6.4.3 Methods for Statistical Modelling of Model Uncertainties

Model uncertainties may be treated as random variables in the actual context. This means that possible errors in model assumptions will be included in the reliability measure, thus avoiding extra safety distances because of model assumptions. It also makes it possible to compare these possible errors with scatter and statistical uncertainties to find the weakest link in the reliability chain and the right priorities.

Of course, model uncertainties are very difficult to estimate, because a statistical estimate would demand the identification of the population of possible models with a correct mean value. One possible approximation to this is to use calculation Round Robin projects. In such a project, a well-defined problem is distributed to a number of investigators, each using his own favourite model to calculate a given property. The result can be regarded as a random sample from the space of possible model errors and the method above can be used to estimate the mean and variance of the error.

Another method is to use the judgement of a skilled engineer. It is often possible for an experienced fatigue designer to assess the possible error in his method. Assume that the engineer can say that the error is 10% as the most. Such a judgement can then be transformed into a statistical measure by, for example, using the uniform distribution:

$$s_m = \frac{0.10}{\sqrt{3}} = 0.058,$$

which can be added to other scatter and uncertainty sources in the safety index.

7.6.4.4 The Combination of Scatter and Uncertainty

For the reliability index, we will use the Gauss approximation formula repeatedly. For the strength variable there will, for instance, be variance contributions from at least material scatter, model errors, supplier variation, and manufacturing tolerances. We can usually assume that these variables are uncorrelated and the uncertainty of the equivalent strength may be written,

$$
\tau_S = \sqrt{c_{test}^2 \cdot \sigma_{test}^2 + c_{model}^2 \cdot \sigma_{model}^2 + c_{suppl}^2 \cdot \sigma_{suppl}^2 + c_{man}^2 \cdot \sigma_{man}^2}
$$
$$
= \sqrt{\tau_{test}^2 + \tau_{model}^2 + \tau_{suppl}^2 + \tau_{man}^2}, \tag{7.36}
$$

where we have included each sensitivity coefficient in the contributing uncertainty measure τ.

Example 7.3 (Weld Toe, Cont.)

Assume now that our fatigue evaluation method includes both material properties and weld characteristics. Then extra contributions from uncertainties due to weld characteristics must be added in Equation (7.36). In Example 7.2, the weld toe radius was found to have the average 1.1 mm and prediction standard deviation $s_{r_p} = 0.36$ mm. In order to find the contribution uncertainty measure for the weld toe radius, the sensitivity coefficient must be found. This was done by using a computer code that calculates the fatigue strength for different radii. The code was used for two calculations, one with the radius $1.1 - 2 \cdot 0.36 = 0.38$, and one with $1.1 + 2 \cdot 0.36 = 1.82$. All other parameters in the calculation were held at their nominal values. The resulting fatigue strengths were 180 kN and 230 kN, respectively. The sensitivity coefficient is

$$
c_r = \left| \frac{\ln\left[f\left(0.38, \mu\right)\right] - \ln\left[f\left(1.82, \mu\right)\right]}{4 \cdot 0.36} \right| = \left| \frac{\ln\left(230\right) - \ln\left(180\right)}{1.44} \right| = \left| \frac{5.19 - 5.44}{1.44} \right| = 0.17,
$$

and the contributing uncertainty measure is $\tau_r = c_r \cdot s_{r_p} = 0.17 \cdot 0.36 = 0.06$. That is, the variation in the weld toe radius contributes to the fatigue strength uncertainty by approximately 6%. □

7.6.5 The Uncertainty due to the Estimated Damage Exponent

A crucial parameter in the load-strength interaction in fatigue is the Wöhler exponent. This parameter is in industrial practice often fixed at a specific value, for instance, $\beta = 3$ for welds, $\beta = 5$ for vaguely defined structures, or $\beta = 8$ for high-strength components with smooth surfaces. Experience and historical data may contain information about the uncertainty in such fixed values; the weld exponent may, for instance, be regarded as uncertain within the limits {2.5; 3.5}. In other cases, the exponent is estimated from specific tests and the uncertainty can be determined directly from a statistical analysis of the test result. It can be shown that the sensitivity coefficient with respect to β is

$$
\frac{\partial(\ln S_{eq} - \ln L_{eq})}{\partial \beta} = \frac{1}{\beta}\left(\frac{1}{\beta}\ln\frac{n_T}{\tilde{N}} + \bar{\xi}_S - \bar{\xi}_L\right), \tag{7.37}
$$

where $\tilde{N}$ is the geometric average of the fatigue lives of the reference specimens, n_T is the target life by means of fatigue cycles, and $\bar{\xi}_S$ and $\bar{\xi}_L$ are the averages of numbers representing **spectrum type**. The exponent sensitivity depends on the difference between the spectrum type numbers for the reference test and for the usage, respectively.

The spectrum type measure ξ is a function of the spectrum of load amplitudes,

$$\xi = \frac{\sum S_k^\beta \ln S_k}{\sum S_k^\beta} - \frac{1}{\beta} \ln \left(\frac{1}{M} \sum S_k^\beta \right), \tag{7.38}$$

where the sums are evaluated over all M counted cycles. The spectrum type measure has the nice properties that it is zero for a constant amplitude load and it is scale invariant.

For some typical load spectra used in fatigue assessment, see Figure 7.7, the spectrum type measure is

Spectrum type	Omission level (%)	ξ
Constant amplitude	–	0
Gaussian	10	0.21
Gaussian	20	0.12
Linear	10	0.25
Linear	20	0.10

For the omission level of 10%, all cycles below 10% of the maximum magnitude are omitted.

Example 7.4 (Steering Arm, Cont.)
To find the exponent uncertainty for the given example, we need the variance of the estimated exponent and the sensitivity coefficient Equation 7.37. The exponent is in the actual case

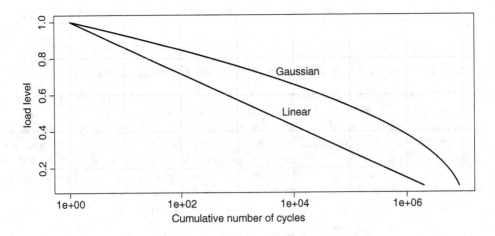

Figure 7.7   The shapes of the Gaussian and linear spectra

estimated as $\hat{\beta} = 6.5$ and, based on the regression, its standard deviation is calculated, see Johannesson *et al.* [124],

$$s_\beta = s_l \sqrt{\frac{1}{\sum_{i=1}^9 (a_i - \bar{a})^2}}, \qquad a_i = \left\{ \frac{\sum S_k^\beta \ln S_k}{\sum S_k^\beta} \right\}_i. \tag{7.39}$$

In this case, we obtain $s_\beta = s_l \sqrt{\frac{1}{0.18}} = 0.2 \cdot 2.4 = 0.48$. Just like the other uncertainty measures, this one should be adjusted for the uncertain s_l estimate by the t-distribution: $s'_\beta = s_\beta \frac{t_{0.975,7}}{t_{0.975,\infty}} = 0.48 \cdot \frac{2.36}{1.96} = 0.58$. This uncertainty actually corresponds to a subjective uncertainty interval based on judgement of about 6.5 ± 1.0, since a uniform distribution assumption then would give $s''_\beta = \frac{1.0}{\sqrt{3}} = 0.58$.

The sensitivity coefficient is calculated according to Equation 7.37. Here, β is replaced by its estimate, and the average type measures are calculated by Equation 7.38 for the load and strength spectra. The geometrical average of the reference tests is found through the actual lives, giving $\tilde{N} = 5.45 \cdot 10^5$. The target life in cycles must be estimated from the load measurements, and is here taken as the average number of counted cycles per km times the target distance, one million km, giving $n_T = 4.45 \cdot 10^8$. The resulting sensitivity is

$$c_\beta = \frac{1}{\hat{\beta}} \left(\frac{1}{\hat{\beta}} \ln \frac{n_T}{\tilde{N}} + \bar{\xi}_S - \bar{\xi}_L \right) = \frac{1}{6.5} \left(\frac{1}{6.5} \ln \frac{4.45 \cdot 10^8}{5.45 \cdot 10^5} + 0.25 - 0.70 \right) = 0.09,$$

and the uncertainty component from the fatigue exponent is

$$\tau_\beta = c_\beta \cdot s'_\beta = 0.09 \cdot 0.58 = 0.052.$$

$\square$

7.6.6 The Uncertainty Measure of Strength

In Equation (7.24) we defined the equivalent strength for a certain specimen,

$$S_{eq} = \left(\frac{1}{n_e} \sum_{k=1}^{M_S} S_k^\beta \right)^{1/\beta}, \tag{7.40}$$

where the random property is the accumulated damage to failure, while the spectrum is supposed to be known and the value n_e is a fixed constant. As mentioned before, we work with the logarithms and calculate the mean and the standard deviation of the observed logarithmic equivalent strengths,

$$m_S = \frac{1}{n} \sum_{i=1}^n \ln S_{eq,i}, \qquad s_S = \sqrt{\frac{1}{n-1} \sum_{i=1}^n (\ln S_{eq,i} - m_S)^2}, \tag{7.41}$$

where n is the number of fatigue tests.

From this estimated standard deviation, we want to obtain a measure of prediction uncertainty for the reliability index. We will do that by using a proper 95% prediction interval for a future assessment of the strength, given the test. If we assume that the logarithm of the equivalent strength is normally distributed, such a prediction interval can be calculated from the Student-t distribution,

$$m_S \pm s_S \cdot t_{0.025,n-1} \cdot \sqrt{1 + \frac{1}{n}}. \tag{7.42}$$

Here, the t-value compensates for the uncertainty in the standard deviation estimate and the one over n term under the root sign takes the uncertainty in the mean value m_S into account. Since the t-value approaches approximately 1.96 when the number of tests increases to infinity, we finally define the actual uncertainty component for the reliability index,

$$\tau_{S,1} = \frac{s_S \cdot t_{0.025,n-1}}{1.96} \cdot \sqrt{1 + \frac{1}{n}}, \tag{7.43}$$

which thereby is a measure corresponding to a standard deviation of the predicted equivalent strength.

This initial uncertainty measure includes the scatter of the specimens, the uncertainty in the estimated mean, and the uncertainty in the variance estimate. The uncertainty due to the exponent β, treated in the previous section, influences both the equivalent strength uncertainty and the equivalent load uncertainty and will be added separately. The influence of the exponent uncertainty on the calculated standard deviation is ignored.

However, when comparing the strength with the load in future there may be additional variation and model error sources involved. For instance, one must take into account that the tested specimens are taken from a single supplier, while several suppliers may be used in future. Such a source of variation is difficult to quantify, but by experience one might be able to assess its influence by, say, 10%. Since we here work with natural logarithms, such a judgement can be quantified directly as a standard deviation contribution, $\tau_{S,2} = 0.10$.

Apart from variation sources like suppliers there are also other uncertainty sources. One is due to the lack of equivalence between laboratory reference tests and the service situation. Multidimensional loads, environment, multiaxiality, residual stresses or mean values are different in a laboratory test compared to service and such model errors should be represented in the reliability index. By experience, specially designed tests, and judgements, these sources may also be quantified by one or more percentage variation resulting in estimates, $\tau_{S,j}$, where the index $\{j = 3, \ldots, m\}$ indicates different identified sources. By putting more emphasis on test design and using components or structures instead of material specimens, these uncertainties can be reduced and balanced to the cost of the additional work.

The damage-accumulation theory including the Basquin equation and the formulation of the equivalent load is an approximation and will contribute to the uncertainty in life prediction. Just like the problem above, contributions such as this may be estimated by experience, literature data, specially designed tests and so forth, adding more percentage variation components $\tau_{S,j}$, $\{j = m + 1, \ldots, p\}$. By performing a laboratory test using load spectra with similar properties as the ones in service, these model uncertainties can be reduced at the cost of more advanced tests.

In total, the uncertainty measure of strength is calculated by a simple addition of the corresponding variances of log strength,

$$\tau_S = \sqrt{\sum_{j=1}^{p} \tau_{S,j}^2}. \tag{7.44}$$

The quadratic summation results in an emphasis of the largest contributions and in practice all components that are smaller than 20% of the largest one can be neglected.

Example 7.5 (Steering Arm, Cont.)
For the given example, the uncertainty in the strength variable is calculated as follows. We had the observed equivalent strengths, S_{eq} (unit kN):

$$\{48.9, 47.8, 48.3, 45.9, 47.9, 46.7, 46.6, 47.4, 50.4\}.$$

Using Equation (7.41) on these values we obtain,

$$m_S = 3.87, \qquad s_S = \sqrt{\frac{1}{n-1} \sum_{i=1}^{n} (\ln S_{eq,i} - m_S)^2} = 0.028,$$

and adjusting by the t-distribution gives,

$$\tau_{S,1} = \frac{s_S \cdot t_{0.025,n-1}}{1.96} \cdot \sqrt{1 + \frac{1}{n}} = \frac{0.028 \cdot t_{0.025,8}}{1.96} \cdot \sqrt{1 + \frac{1}{9}} = \frac{0.028 \cdot 2.3}{1.96} \cdot \sqrt{\frac{10}{9}} = 0.035.$$

Since the tests are made in a laboratory set-up, there will probably be some severity deviations between the conditions at test and in service. This adds an uncertainty to the reliability assessment, which is here judged to be 5% at the most. Further, the tested components are sampled from one batch for one single supplier. An extra variation in future batches is also judged to add 5% deviation at the most. Using the simple uniform distribution for these judgements gives,

$$\tau_{S,2} = \tau_{S,3} = \frac{0.05}{\sqrt{3}} = 0.03.$$

Other uncertainties in strength are judged to be negligible and the final strength component for the denominator in the reliability index is

$$\tau_S = \sqrt{0.035^2 + 2 \cdot 0.03^2} = 0.055.$$

Note that in this particular case, the exponent beta was estimated from the test, and the correct denominator in the standard deviation calculation should therefore be $n - 2$, and not $n - 1$. This would change the estimate of τ_S from 0.055 to 0.059, which is ignored here in order to avoid confusion with the common case when β is given from other sources.   □

7.6.7   The Uncertainty Measure of Load

The load variation by means of reliability originates from the population of users. These may be identified as customers, markets, missions or owners depending on the application. In the case of trucks, it is desirable to distinguish between populations with common missions and environments, such as timber transport, city distribution or highway usage. For each population, different reliability indices can be calculated. The important thing is that it is made clear what population the reliability index relates to.

Once the population has been defined, the mean and the variance of the corresponding load need to be estimated. This is often a rather difficult task. In the automotive industry there is often an established relation between the assumed population and the company-specific test track. In military air space applications, special predefined missions are established, and in other industrial practices, target loads are specified by experience and sometimes given as standardized load spectra. These load specifications are rather rough approximations of the reality and usually only represent the mean user or a certain extreme user.

Because of the extremely fast development of information technology, many companies now work with direct measurements of loads in service, which makes it possible to obtain the properties needed for a proper reliability assessment. However, this usually demands large measurement campaigns, where the sampling of customers, environments and markets is a subtle matter, demanding great care when planning the campaigns, see Chapter 8. From such measurements in a specific population, it is possible to calculate equivalent loads according to Equation (7.26) for each sample and find the mean and the standard deviation of the logarithmic transformation of the population,

$$m_L = \frac{1}{n} \sum_{i=1}^{n} \ln L_{eq,i}, \qquad s_L = \sqrt{\frac{1}{n-1} \sum_{i=1}^{n} (\ln L_{eq,i} - m_L)^2}. \tag{7.45}$$

If the number of samples is small, there will be an uncertainty in the estimated standard deviation and our uncertainty measure must be adjusted as in the strength case, again assuming a normal distribution for our logarithmic property,

$$\tau_{L,1} = s_L \frac{t_{0.975,n-1}}{1.96} \sqrt{1 + \frac{1}{n}}. \tag{7.46}$$

If the load measurements have been made on a test track or if the measurement campaign is known not to be a random sample of the actual population, there may be an unknown bias in the estimates and more uncertainty components should be introduced,

$$\tau_L = \sqrt{\tau_{L,1}^2 + \tau_{L,2}^2 + \tau_{L,3}^2 + \cdots}, \tag{7.47}$$

where each different component $\tau_{L,i}$ is estimated as the coefficient of variation of equivalent load, i.e. based on a percentage judgement.

Example 7.6 (Steering Arm, Cont.)
According to the formulae above, we can calculate the prediction uncertainty due to the field measurements.

$$\tau_{L,1} = s_L \frac{t_{0.975,16}}{1.96} \sqrt{1 + \frac{1}{17}} = 0.46 \frac{2.1}{1.96} \sqrt{\frac{18}{17}} = 0.51.$$

The strain gauge measurement that found the basis of comparison was made on the link rod, which is the part that introduces the force in the steering arm. Due to the uncertainties in calibration of the force sensor and the possible deviations due to variation in the angle at which the force is applied, the calculated force is judged to be uncertain up to 10% from the actual steering arm force. As before, we use the uniform distribution assumption on this engineering judgement and find,

$$\tau_{L,2} = \frac{0.10}{\sqrt{3}} = 0.058.$$

Other uncertainties are assumed negligible and the final load uncertainty component is

$$\tau_L = \sqrt{0.51^2 + 0.058^2} = 0.51.$$

□

7.6.8 Use of the Reliability Index

The Cornell reliability index, which is the basis of the actual second-moment reliability method, is primarily presented in the area of structural reliability, that is structures within building technology. In this discipline, a great deal of work has been done for the application of second-moment reliability indices, for instance, by the organization JCSS, the Joint Committee on Structural Safety. One problem that is treated is how one should interpret the resulting index and how it should be judged with respect to safety. As pointed out before, there is always little knowledge about the tail in the distribution, and calculated low probabilities of failure are highly uncertain. Figures 7.8 and 7.9 illustrate the problem. The steering arm data from Example 7.1 has been used for the illustrations. The nine observed equivalent strengths have been fitted to six different statistical distribution types, namely 1) log-normal, 2) normal, 3) two-parameter Weibull, 4) two-parameter Weibull on the inverse, 1/strength, and 5) three-parameter Weibull. The seventeen equivalent loads have been fitted to 1) log-normal, 2) normal, 3) two-parameter Weibull on the inverse, 4) two-parameter Weibull, and 5) three-parameter Weibull on the inverse. The resulting probability density functions are plotted in Figure 6.8, showing a large dependence on the choice of distribution type, particularly in the important part where the tails interact. A formal calculation of the probability of failure based on the different estimated distributions gives the following values: 1) log-normal 0.0055, 2) normal, $5.6 \cdot 10^{-6}$, 3) two-parameter Weibull, 0.056, 4) two-parameter Weibull-inv, $1.2 \cdot 10^{-5}$, and 5) three-parameter Weibull, 0. Since the choice of distribution type must usually be regarded as subjective, the calculations of small probabilities to failures are indeed dubious.

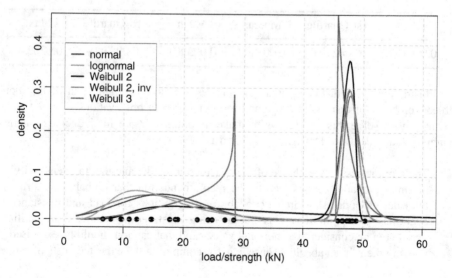

Figure 7.8 Five different distribution types fitted to the same two sets of data

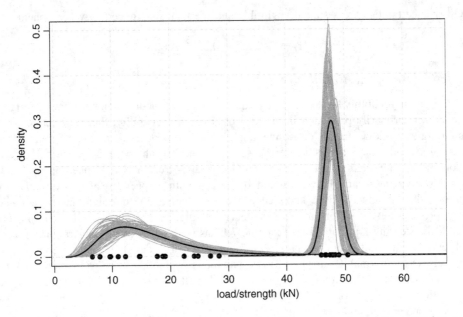

Figure 7.9   The bootstrap technique is used to simulate 100 random choices from our data set

If we believe in a certain distribution type, the small number of observations will still introduce a large uncertainty in the tails, the statistical uncertainty. To illustrate this fact we use the bootstrap technique and simulate 100 random choices from our data set. For each randomly drawn set of nine strengths and seventeen loads we fit a log-normal distribution. The result is seen in Figure 7.9 and the corresponding calculated probabilities for failure were distributed as follows:

Min.	1st Quartile	Median	Mean	3rd Quartile	Max.
0.000132	0.002104	0.004537	0.005901	0.008790	0.020250

The illustrations show that calculated probabilities of failure are highly sensitive to both the subjective choice of distribution type and to the number of observations.

One way to solve this problem is to make a common frame for judgements about the reliability index demand. Der Kiureghian and Ditlevsen say:

> The arbitrariness in the choice of the distribution model and the 'tail-sensitivity' of small probabilities has lead to the recommendation that probabilistic structural design codes standardize probability distributions for load and resistance quantities. One point of view is that in such a construct the computed probabilities should be considered as notional values and that caution should be exercised in using them in an absolute sense, e.g. for computing the expected costs of rare events.
>
> – Der Kiureghian and Ditlevsen [71].

Table 7.2 Table from JCSS [118]. Tentative target reliability indices β (and associated failure rates) related to a one-year reference period and ultimate limit states. Observe that the number β here corresponds to our γ_d

1 Relative cost of safety measure	2 Minor consequences of failure	3 Moderate consequences of failure	4 Large consequences of failure
Large (A)	$\beta = 3.1 (p_F \approx 10^{-3})$	$\beta = 3.3 (p_F \approx 5 \cdot 10^{-4})$	$\beta = 3.7 (p_F \approx 10^{-4})$
Normal (B)	$\beta = 3.7 (p_F \approx 10^{-4})$	$\beta = 4.2 (p_F \approx 10^{-5})$	$\beta = 4.4 (p_F \approx 5 \cdot 10^{-6})$
Small (C)	$\beta = 4.2 (p_F \approx 10^{-5})$	$\beta = 4.4 (p_F \approx 5 \cdot 10^{-5})$	$\beta = 4.7 (p_F \approx 10^{-6})$

In Table 7.2, a proposal for such a standardized decision matrix is shown. Here, rough judgements of the consequences of failure are combined with the relative cost of safety achievements, giving the demand on the reliability index. The numbers in parentheses represent the probability of failure, assuming a normal distribution (log-normal in our case).

Example 7.7 (Steering Arm, Cont.)
For our example the reliability index is

$$\gamma = \frac{m_S - m_L}{\sqrt{\tau_S^2 + \tau_L^2 + \tau_\beta^2}} = \frac{3.87 - 2.70}{\sqrt{0.046^2 + 0.51^2 + 0.05^2}} = \frac{1.17}{0.51} = 2.3,$$

which should be larger than a demanded value, γ_d. The resulting index is far from the lowest demand in the presented table, and the basis of the construction must be re-considered. An immediate conclusion from inspection of the formula above is that the largest uncertainty source is the extreme variation in the observed usage of the vehicles. Can a variation of 50% among users be acceptable? Is the target life of one million km reasonable for all the usages that are represented in the sample? Could the construction be strengthened by design or better material? Is the table of tentative target reliability indices relevant for the actual application? These questions need to be considered in combination with truck engineering knowledge and especially experience from similar constructions. □

7.6.9 Including an Extra Safety Factor

As an alternative way of interpreting the reliability index, we will here introduce a concept that is a mixture of the probabilistic interpretation and a deterministic safety distance. We rewrite the demand in the following way:

$$m_S - m_L > S_d + 2\tau, \qquad \tau = \sqrt{\tau_S^2 + \tau_L^2 + \tau_\beta^2}. \tag{7.48}$$

Here the target is a sum of two terms. The second term is based on the calculated prediction uncertainty, and by the multiplication by 2 it corresponds to an approximate 95% prediction interval. This interpretation is based on the assumption that there is an approximate normal distribution of the actual difference and defines a limit that corresponds to 2.5% probability of failure. The first term, S_d, should be interpreted as a *deterministic extra safety distance* to the failure mode, which corresponds to an extra safety factor in the original unit of stress or force.

As indicated above, the concept can also be formulated in terms of safety factors, namely

$$\phi = \phi_d \cdot \phi_s \quad \text{with} \quad \phi_d = e^{S_d} \quad \text{and} \quad \phi_s = e^{2\tau} \tag{7.49}$$

where ϕ is the total safety factor, which can be expressed as a product of a deterministic and a statistical safety factor, ϕ_d and ϕ_s, respectively. Note that we have used central safety factors defined as the quotient between the median equivalent strength and the median equivalent load.

The reason for this construction is that, using a proper transformation as the logarithm in this case, the normal distribution assumption can be justified in the central part of the distribution, that is, within 95% probability. This part is therefore considered to be in statistical control, meaning that we have the statistical tools needed to reduce this part by more laboratory tests, more field measurements, or better mathematical and physical models. This can be done in the same manner for all applications, regardless of the level of safety demands. Safety considerations can be handled with the extra safety factor and tables corresponding to the one in Table 7.2 can be established by initial rough probability judgements, eventually completed and updated by experience.

Example 7.8 (Steering Arm, Cont.)
Our running example is a safety critical component for a truck and the target reliability for these types of parts is decided in the actual company to be the extra safety factor of $\phi_d = 1.5$. This means that the deterministic part in Equation 7.48 is

$$S_d = \ln (1.5) = 0.41.$$

The second term in Equation 7.48 is

$$2\tau = 2\sqrt{\tau_S^2 + \tau_L^2 + \tau_\beta^2} = 2\sqrt{0.046^2 + 0.51^2 + 0.05^2} = 2 \cdot 0.51 = 1.02$$

and the demand is

$$m_S - m_L > 0.41 + 1.02 = 1.43.$$

In our case the difference is only $m_S - m_L = 3.87 - 2.70 = 1.17$, and the target is not reached.

In Figure 7.10, the situation is illustrated by a symbolic normal density function for the property $\ln S_{eq} - \ln L_{eq}$. The distance to failure mode, i.e. when the property is zero, is split into i) a statistical distance of two standard deviations, and ii) a deterministic extra safety distance.

In terms of safety factors, the demand is

$$\phi > \phi_d \cdot \phi_s = 1.5 \cdot e^{2\tau} = 1.5 \cdot e^{1.02} = 1.5 \cdot 2.77 = 4.16$$

whereas the obtained safety factor is $e^{1.17} = 3.22$, which is less than the demand, and therefore the safety margins for the construction are too narrow.

Just as in the previous example, using the established table from JCSS, we find that the basis of the construction needs to be re-considered.   □

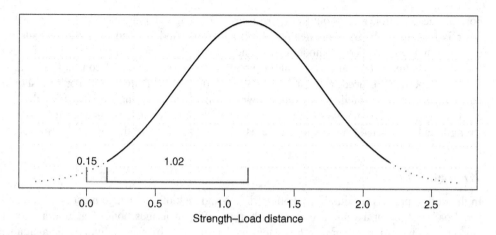

Figure 7.10 A symbolic statistical density for the difference between logarithmic strength and logarithmic load, $\ln S_{eq} - \ln L_{eq}$

This formulation of the reliability index does not solve the problem of the limited knowledge of the tails in the load and strength distributions, but it avoids the false interpretations of probability assessments as failure frequencies. In addition, in cases when failures mainly originate from extreme events like human mistakes, it should be correct to separate these events from the reliability assessment based on specified conditions.

7.6.10 Reducing Uncertainties

The sources of uncertainty with respect to failure probability can be classified as scatter, statistical uncertainty, and model uncertainty. Scatter refers to natural variation that can be regarded as random. Such uncertainties can be reduced by changing to a material quality with less scatter, by narrowing specification limits or by adjusting vehicles to different classes of usage profiles, but they cannot be reduced by more measurement or experiments. In the basic formula (7.35), the scatter contribution is represented by the estimated standard deviation s.

Statistical uncertainties are present because of a limited amount of reference data for estimating parameters, and are represented in two places in Equation (7.35), namely the number of tests, n under the root sign, and the t-factor, which is also n-dependent. The immediate method to diminish these uncertainties is therefore to increase the number of tests. Another possibility is to identify populations that are assumed to have a common standard deviation and combine test results from different occasions for the estimate. This will increase the precision in the s-estimate and reduce the t-factor. Other possibilities are to optimize test plans with respect to information and construct models with fewer parameters. This last method will reduce statistical uncertainty, but also increase model uncertainty, and an optimal trade-off must be searched for, see Svensson [227], Johnson and Svensson [128] for a study on fatigue life prediction.

Model errors and lack of knowledge about population differences can be diminished by designing special tests or analysing field failures. For instance, in case the uncertainty due to an accelerated test is predominant in a reliability assessment, tests that investigate the possible acceleration bias should be performed. The original judgement about a possible error is then replaced by a mathematical extrapolation model based on the tests. Such a model contains new possible model errors and statistical uncertainty, but the sum of these two variances may be smaller than the originally judged uncertainty. If constant amplitude fatigue properties are used, a possible model error due to the Palmgren-Miner rule should be included in the reliability measure. However, making additional tests can result in a correction factor (the relative Miner rule, see page 266) which can be used on the average strength and completed with a smaller statistical uncertainty from the test evaluation.

7.7 Summary

In the design process for durability both variation and lack of knowledge must be accounted for. Loads acting on a vehicle vary according to different usages both over time and over customers. The variation can to some extent be determined experimentally, but the available data needs to be completed by judgements based on experience, interviews, and other more or less uncertain sources. The description of load variation and uncertainty must be completed by a corresponding description of the strength, which needs experimental test results as well as mathematical and physical models for its relation to outer loads. This means that also on the strength side, both variation and uncertainties must be considered.

Methods for the assessment of the reliability of a vehicle include many different levels of complexity, both regarding the mathematical modelling and the statistical modelling. The statistical methods can be classified with respect to the basic statistical properties used, as first- or second-moment methods or full probabilistic approaches. The first-moment methods are extensions of the classic safety factor approach and lack immediate possibilities to take advantage of new knowledge, such as more experiments, more field measurements or better mathematical models. Second-moment methods include simple measures of the variability and make it possible to combine them in a rational way, keeping track of the different contributions of the sources. The full probabilistic approach includes detailed information about the uncertainty sources, which are usually non-existent.

Among second-moment methods, the Cornell reliability index is a powerful tool for treating fatigue reliability problems, and the use of this index is evaluated in some detail in this chapter in order to demonstrate how one can proceed to obtain and combine the necessary variation measures of different typical sources, both for variation by means of random scatter and for possible systematic errors, for example, model errors, sampling biases, and statistical uncertainties.

Part Three

Load Analysis in View of the Vehicle Design Process

8

Evaluation of Customer Loads

8.1 Introduction

When a truck is to be designed, the loads it is expected to meet have to be estimated. The design specification, detailing the minimum performance to be demonstrated in tests and calculations, is based on the expected vehicle usage and pertinent load estimation. The load estimation is based on several different sources. There is experience of the performance of previous designs which has been verified on test grounds, in rig tests and/or through calculations. There have also been measurements on trucks in service. The process seems satisfactory since most trucks function well and failures are rare. The problem is rather that some parts of the trucks may become unnecessarily strong and heavy, which causes costs both for the producer in terms of material, and for the consumer in terms of permissible pay load reduction. Since trucks are tested on proving grounds, it is possible that they pass tests on the proving ground of their producers but fail on the proving ground of their competitors. This test practice tends to produce over-designed trucks. A design that is adequately strong for the world's most demanding customer will be over-designed for everybody else, and a design just fulfilling the median customer's expectations would be under-designed/unsatisfactory for every other customer. A truly unique truck design for every customer is economically impossible; more than one experimental truck is needed to verify a completely new design. By "grouping" similar customers, an economically feasible design (without too much over-design, with reasonable production volume) may be found. The key is to understand and describe customer loads and their variation, that is, the variation between customers, markets, transport tasks, etc.

When discussing the evaluation of customer loads there are two scales. On the small scale the problem is to evaluate a specific customer or to compare the severity of two (or more) individual customers. On the large scale the problem is to evaluate the severity of a population of customers, or to compare the severity of two different populations of customers (e.g. European customers and Brazilian customers). These two problems need to be treated separately, and with different methods. In this chapter we will concentrate on the large scale, the evaluation of populations of customers. Methods for the evaluation of a specific customer or for measurements are treated primarily in Chapter 3 but also in Chapters 4–6.

Guide to Load Analysis for Durability in Vehicle Engineering, First Edition. Edited by P. Johannesson and M. Speckert.
© 2014 Fraunhofer-Chalmers Research Centre for Industrial Mathematics.

The important problem of selecting which customers to measure is treated in Section 8.2 by a discussion of the principles of survey sampling. In Section 8.3 the uncertainty in the measured load severity is evaluated. The assessment of the customer distribution by random sampling of customers is the topic of Section 8.4. Another strategy is treated in Section 8.5, where the customer distribution is built up by a model of the customer usage in combination with road classification. In Section 8.6 vehicle independent load descriptions are discussed, followed by discussion and summary in Section 8.7.

8.2   Survey Sampling

A natural approach to obtain load specifications is to measure loads on trucks in service. Unless loads are measured on all trucks all the time, specific vehicles must be selected for measurements. Questions to be answered are:

- What trucks (where in the world) should be measured?
- How many measurements are needed?
- How long should the measurements be?

In practice, the amount of time, money, equipment and personnel available is limited. Thus, two additional questions of great interest are:

- How should a measurement campaign be planned to get the best possible, i.e. most reliable answers from the limited resources available?
- Just how reliable are those answers?

In order to answer those questions, it is necessary to specify what information we are looking for. It is clear that each kind of truck has its own pattern of usage and that it varies between regions. Let's say we are interested in the load history during the design life of all trucks of a specific kind in use today. In that case we have specified the population we are looking for. The population is the collection of load histories of these trucks.

In the future it is not entirely impossible that all trucks of a certain kind will be equipped with an on-board logging system reporting loads to the producer. In this case measurements from the whole population of interest are available. But today we have to choose a subpopulation to measure in some way.

A common approach is to choose a sample from the hauliers who have special connections with the producer. Often these hauliers have load profiles of special interest and the advantage is that measurement efforts are directed to crucial customers. The drawback is that there is no assurance that the measured load profiles represent the whole population of profiles. To be able to draw conclusions about the whole population, there are only two possibilities. Either the whole population is measured or a randomly selected subpopulation.

8.2.1   Why Use Random Samples?

From a common sense point of view, random sampling seems both impractical and risky. If, for example, the outcome of a complete survey of all customers of groceries is to be predicted, a random sample can be very skewed and also difficult to obtain. The people in the sample can live anywhere and can therefore be difficult to reach. Also there is nothing to prevent all people in the random sample from being male and then the sample will not be representative of the population. An alternative way to sample is the quota sample, which guarantees that exactly the right proportions of males and females, young and old, people in the countryside and people in the suburbs are included in the sample.

The following example will show how quota sampling can go wrong. In the 1948 presidential election in the USA, a quota sample of 50 000 interviews was taken. For example, in St Louis, an interviewer "was required to interview thirteen subjects of whom exactly six were to live in the suburbs, and seven in the central city; exactly seven were to be men, and six women. Of the men, exactly three were to be under forty years old, and four over forty; exactly one was to be black, and six white. The monthly rental to be paid by the six men were specified also: one was to pay $44.01 or more; three were to pay $18.01 to $44.00; two were to pay $18.00 or less", Freedman *et al.* [94]. The estimate based on this vast quota sample predicted that the Republican candidate Dewey would get 50% of the votes and be the next president. In fact the Democrat Truman won and Dewey got only 45% of the votes; an error of 5%. A simple random sample of this size would have given a random error of less than 0.5%!

After this election the Gallup institute started using random samples instead of quota samples. The explanation for the vast error in the quota sample is the bias, that is, systematic error, caused by the subjective choice of each interviewer. The Republicans tend to be a little more well-educated, a little wealthier and a little friendlier in each quota. The answer would have been to do a stratified design and choose the subjects in each stratum randomly, in a scientific sense.

When a systematic non-random sampling is done there will be a bias which can be difficult to estimate. A random sample has ideally no bias but a random error whose distribution is possible to estimate. The effects of known variables can be balanced by stratification and the effects of unknown variables can be estimated in the random setting.

There are several different ways to select a subpopulation at random. We will discuss the following procedures:

1. simple random sample
2. stratified random sample
3. cluster sample
4. sampling with unequal probabilities

These procedures can be combined and sampling with unequal probabilities is often a consequence of (2) or (3).

8.2.2 Simple Random Sample

Let's say we have a list of all trucks running today and that there are $M = 10\ 000$ trucks. We want to measure $n = 40$ load histories. Then there are $1.1 \cdot 10^{112}$ possible ways to choose subsets of 40 trucks out of 10 000. Simple random sampling means that all these subsets have the same probability of occurring as our sample. An estimate which is based on a simple random sample of size n usually has a precision which is proportional to n. It is also in general possible to obtain estimates with no bias in this case. That the estimator is unbiased means that, on average, it has no deviation from the entity of interest.

Example 8.1 (Simplified Example)
Let the population consist of just $M = 4$ trucks from three different hauliers; a large one with two trucks, and two smaller ones with one truck each. The measured values are 8 and 10 from haulier A, 2 from haulier B, and 4 from haulier C. Let the sample size of the simple random sample be two, i.e. $n = 2$. Then there are six random samples of size two. The six mean values of these six possible random samples are 3, 5, 6, 6, 7 and 9. The population mean is $(2 + 4 + 8 + 10)/4 = 6$ and thus the deviation of the sample means are $-3, -1, 0, 0, 1$ and 3 and the variance of the sample mean are $(9 + 1 + 0 + 0 + 1 + 9)/6 = 3.3$. The precision which is the reciprocal of the variance is 0.3. □

8.2.3 Stratified Random Sample

If the population we are interested in has subpopulations which are quite different from each other with respect to load profiles, it can be wise not to use simple random sampling. It could be advantageous to partition the population into subpopulations, called strata, and then do simple random sampling within each stratum. In this way it is possible to increase the precision without increasing the total sample size.

Example 8.2 (Simplified Example, Continued)
We divide the population of hauliers into the two subpopulations: large hauliers (A), and small hauliers (B and C). We want to include at least one truck from each subpopulation. We now have four possible stratified random samples of size 2; either we choose 2 or 4 and either 8 or 10. The four different sample means are then 5, 6, 6 and 7. The corresponding deviations from the population mean 6 are $-1, 0, 0$ and 1. The variance of the estimator is now $(1 + 0 + 0 + 1)/4 = 0.5$ and the precision has increased to $1/0.5 = 2$ compared to 0.3 in the random sample case. □

8.2.4 Cluster Sample

Performing random sampling can be impractical. Sometimes it is easier to choose hauliers at random in the first stage and then do a random sample from each haulier's truck fleet.

Example 8.3 (Simplified Example, Again)
We choose two out of our three hauliers at random with equal probability. From each chosen haulier we choose one truck at random. We will obtain the following samples $\{2, 4\}$ with probability 1/3, $\{2, 8\}$ with probability 1/6, $\{2, 10\}$ with probability 1/6, $\{4, 8\}$ with probability 1/6 and at last $\{4, 10\}$ with probability 1/6.

It is not immediately obvious how to estimate the population mean from each sample. Note that haulier A has twice as big a fleet as hauliers B and C. The conclusion is that when one of his trucks appears in the sample it should be given a bigger weight than the one-truck hauliers. To obtain an unbiased estimate the weight of each truck is the reciprocal to the probability that the truck will appear in the sample, corrected with the total number of trucks. For example, the estimated mean from sample $\{2, 8\}$ is $[2/(2/3) + 8/(1/3)]/M = 6.75$, since $M = 4$. In the same way the weighted mean from $\{2, 4\}$ is $[2/(2/3) + 4/(2/3)]/4 = 2.25$ (and not 3!).

The weighted means of the four samples above are 2.25, 6.75, 8.25, 7.5 and 9. The deviations from the population means 6 are -3.75 (with probability 1/3), 0.75, 2.25, 1.5 and 3 with probability 1/6, respectively. The variance of this weighted mean estimate is 7.5 and its precision 0.13. □

8.2.5 Sampling with Unequal Probabilities

A common situation is that some customers are of crucial importance and we want to be sure to include them in our sample.

Example 8.4 (Simplified Example, Last Time)
We have one haulier with a big fleet and two with small fleets. We want to be sure to include the big fleet in our sample. Therefore, we will make sure that haulier A will be included. We will choose either B, $\{2\}$ or C, $\{4\}$ and one of the trucks from haulier A, $\{8, 10\}$. The possible samples are now $\{2, 8\}$, $\{2, 10\}$, $\{4, 8\}$ and $\{4, 10\}$, all appearing with probability 1/4. To calculate the weighted means of each sample we have to calculate the probability for each truck to be included and then compensate for the total number of trucks, exactly as in Example 8.3. The total number of trucks is $M = 4$ as before and the probability for each truck to be included is one half. Thus the weighted mean of the first sample is $[2/0.5 + 8/0.5]/4 = 5$. We will obtain the same numbers as in Example 8.2. □

We have seen that different ways of sampling give rise to different precisions of the estimate of the quantity of interest. The precisions range from 0.13 up to 2.0 in our examples where we have four trucks.

One way to get a high precision is to use stratified sampling with homogeneous strata, where great differences between trucks occur as differences between strata but which do not affect the variance. For truck applications, "haulage" may be selected as one stratum and "construction" as another. The load variance inside each group (stratum) is expected to be smaller than the variance between "haulage" and "construction" usage.

If cluster sampling is used, it is important that the big differences occur within clusters and that the differences between clusters are small.

When sampling with unequal probabilities, it is interesting to note that an unbiased estimator of the population mean is obtained when the actual value to be measured is divided by its inclusion probability[1] (as well as the population size). A small variance is then obtained if large values are divided by large inclusion probabilities and small values are divided by small inclusion probabilities.

Example 8.5 (Simplified Example, Very Last Time)
Inclusion probabilities are ideally proportional to the entity of interest. As before, let the population consist of the following values: 2, 4, 8 and 10. Let the sample size be 2, which then is the sum of the inclusion probabilities and we ideally obtain the following probabilities: 1/6, 2/6, 4/6 and 5/6, respectively. These probabilities occur if the sample $\{2, 8\}$ has probability

---

[1] The inclusion probability of a measurement is the probability it will occur in the sample.

1/6, {4, 10} probability 2/6, and {8, 10} probability 3/6. The corresponding weighted means are $[2/(1/6) + 8/(4/6)]/4$, $[4/(2/6) + 10/(5/6)]/4$, and $[8/(4/6) + 10/(5/6)]/6$ which all equal 6 and the variance is 0, i.e. infinite precision.   □

The sampling plan in Example 8.5 is called proportional-to-size sampling but is not achievable in practice since it is assumed that the quantities to be estimated are known in advance. However, it is not uncommon that there is a known auxiliary variable which roughly follows the variable of interest. Then the sampling can be carried out with probabilities proportional to the auxiliary variable. A typical auxiliary variable is the result from a previous investigation. It can be measurements from another country, from a previous measurement campaign, from calculations based on complete vehicle simulations or from questionnaires used as a planning tool.

Another difficulty which can arise is that the sizes to be estimated vary in a way that makes the probability proportional to size larger than one, which is not an admissible probability. This can happen if the sample size is bigger than one. Either the too large probabilities are replaced by one or stratification is used where the large quantities are put in a separate stratum.

An attractive feature of the sampling plan with probabilities proportional to size is that units with higher values are more likely to occur in the sample than the ones with smaller values. The more problematic a customer is expected to be, the bigger chance that the customer will be included in the sample.

This feature is attractive only if a unit with a big value is more interesting than a unit with a smaller value. For example, if strength is the entity of interest, components with lower strengths are more critical than those with higher strength. In this case it can be useful to invert the data to obtain high values for weak components and vice versa.

8.2.6 An Application

In some applications where service loads are measured, special interest is focused on the distribution of an entity which is related to the fatigue life of the vehicle component studied. One such entity is the damage intensity v. Assume that the component meets the load amplitudes $S_1, S_2, \cdots, S_k$, when the truck is driven $t = 1000$ km. To calculate the damage $D(t)$, use the Wöhler curve $N(S) = \alpha S^{-\beta}$ and obtain

$$D(t) = \sum_{i=1}^{k} \frac{1}{N(S_i)} = \sum_{i=1}^{k} \frac{S_i^{\beta}}{\alpha} = \frac{d(t)}{\alpha}. \tag{8.1}$$

For this component the expected pseudo damage is assumed to develop according to $E[d(t)|truck] = vt$ which defines the damage intensity v for the studied component installed in a specific truck. We are now interested in the distribution of v for our truck population. The procedure is now to choose a sample of trucks according to Section 8.2.5, measure the service loads under $t = 1000$ km and calculate

$$v \approx \sum_{i=1}^{k} \frac{S_i^{b}}{t}. \tag{8.2}$$

The measurements from our n trucks form an estimate of the distribution of v.

Using the theory indicated in the previous section we can get unbiased estimates of the probabilities of different events. We can, for example, consider the event that we have large damage intensities in our population by studying the variable z, which is 1 if the truck in question has a damage intensity exceeding a given threshold and zero otherwise. In a simple random sample the average of the outcomes of the variable z is an unbiased estimate of the probability that the damage intensity will exceed the threshold.

When our sample is random with unequal inclusion probabilities, we use a weighted average of the observations of z to obtain an unbiased estimate of the exceeding probability.

In our discussion we have assumed that we can measure the damage intensity v exactly without any error. Of course, this is a simplification. In fact, the statistical precision of the measurement of the damage intensity is proportional to the measurement distance t. If $t = 1000$ km is enough or not depends on the kind of mission the truck has.

If $t = 1000$ km contains more than 5–10 repetitions of the transport task it should be enough, but if the truck is involved in long distance deliveries then $t = 1000$ km can be too short a distance to cover a single transport task and the different types of loads the truck will meet during one mission.

The number of trucks which are required depends on the precision we want to achieve and also on which procedure we use to obtain a sample.

8.2.7 Simple Random Sampling in More Detail

Simple random sampling of n trucks out of a population of M trucks means that every sub-sample of size n has the same chance to occur. Each truck will appear in the sample with probability n/M. Simple random sampling is used when there is no information available as to where to measure and when there is no immense extra cost involved in measuring individually selected trucks. The strength of simple random sampling is that the analysis is simple and hence transparent and convincing. The sample mean is an unbiased estimate of the population mean and the sample variance is an unbiased estimate of the corrected population variance.

We will assume that the trucks are numbered $1, 2, 3, \ldots, M$ and to each number i there is assigned a value y_i. We obtain a population of values, $y_1, y_2, y_3, \ldots, y_M$, with population mean

$$\bar{y}_U = \sum_{i=1}^{M} y_i/M \tag{8.3}$$

and corrected variance

$$S^2 = \sum_{i=1}^{M} (y_i - \bar{y}_U)^2/(M-1). \tag{8.4}$$

Let us denote a random sample of size n by Z, its sample mean by

$$\bar{y} = \sum_{i \in Z} y_i/n \tag{8.5}$$

and its sample variance by

$$s^2 = \sum_{i \in Z} (y_i - \bar{y})^2/(n-1). \tag{8.6}$$

The variance of the sample mean is

$$\mathrm{Var}(\overline{y}) = \frac{S^2}{n}\left(1 - \frac{n}{M}\right) \tag{8.7}$$

and an unbiased estimate of this quantity is

$$\mathrm{Var}^*(\overline{y}) = \frac{s^2}{n}\left(1 - \frac{n}{M}\right). \tag{8.8}$$

The corrected variance S^2 is introduced since we study sampling without replacement, which implies a weak dependence between the observations. In this sampling scheme the expectation of the sampling variance s^2 is the corrected variance S^2 and not the population variance $\sigma^2 = \frac{(M-1)}{M}S^2$.

A question that should be asked is how big a sample is needed. To answer this question we will study the corrected sample deviation $SE(\overline{y})$ of the estimate $\overline{y}$ of the population mean $\overline{y}_U$. We have

$$SE(\overline{y}) = \sqrt{\mathrm{Var}(\overline{y})} = \sqrt{\frac{S^2}{n}\left(1 - \frac{n}{M}\right)} \approx S/\sqrt{n} \approx s/\sqrt{n} \tag{8.9}$$

when $n << M$, which is proportional to the length of the confidence interval of the population mean $\overline{y}_U$. In fact the length of the confidence interval

$$[\overline{y} - 2s/\sqrt{n}; \overline{y} + 2s/\sqrt{n}] \tag{8.10}$$

is $4s/\sqrt{n}$ for a coefficient of confidence of approximately 95%. Increasing the sample size one step from n to $n+1$ decreases the average width with the factor $\sqrt{(1 - 1/(n+1))}$. For instance, for the sample size $n = 2$, an extra sample decreases the confidence interval width to 82% of its former value, i.e. an improvement of 18%. But already for the sample size $n = 7$ the gain is only about 6% by one extra measurement. This calculation is relevant when the distribution of the y:s is not too skew.

Let v denote the damage intensity of a particular truck as in Section 8.2.6 above and introduce a new variable z such that $z = 1$ when $v > v_0$ and zero else. Then the sample mean of z estimates the probability that a randomly chosen truck will have a damage intensity above the threshold v_0. If the threshold is high, most z:s in the population are zeroes and there is a great chance that the sample mean will also be zero, which will give only little information about the tail of the distribution. In this situation the number of non-zero observations is relevant rather than the original sample size.

8.2.8   Conclusion

There are many different ways to obtain random samples and it is possible to include some hauliers or trucks with probability one and still obtain unbiased estimates. The important point is that the population of interest is well defined and that every member of that population can be included in the sample with a positive probability, known to the investigator. For further reading, see Särndal et al. [207] and Lohr [148].

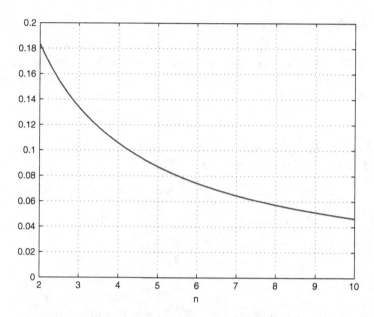

Figure 8.1 The effect on confidence interval length by an extra observation when original sample size is n, i.e. $n = 2$ shows the gain going from 2 to 3 samples.

8.3 Load Measurement Uncertainty

The aim here is to evaluate the uncertainty in a measured load severity, e.g. pseudo damage or equivalent load amplitude. This information is needed in the load-strength model presented in Chapter 7, and in the following sections in this chapter. The uncertainty should also be considered when comparing different measurements in order to ensure that the difference is not only random scatter. We will discuss the following topics:

Precision in load severity. How statistically reliable is the measurement? How many laps should be measured in order to get a certain statistical precision? Here confidence intervals are useful and they can be calculated by splitting the load signal into shorter load segments.

Comparison of two drivers. Is a measured difference statistically significant or can it be due to random scatter? Here we can use a confidence interval for difference in load severity.

Comparison of many drivers. Can we detect a difference between the test drivers? Here we can use ANalysis Of VAriance (ANOVA), which is a statistical technique for estimating the difference in mean between several individuals.

8.3.1 Precision in Load Severity

An estimate is always attached with a statistical uncertainty; in our case we are interested in the pseudo damage, d, the equivalent load amplitude, A_{eq}, or the equivalent mileage, M_{eq}. Quantifying the uncertainty is one of the topics in Chapters 6 and 7.

8.3.1.1 Distribution Assumption

Reasonable distributions for the load severity are normal, Weibull and log-normal. The log-normal distribution will be used here since it agrees well with observations and is practical (mathematically convenient) to work with. It is also in accordance with the load-strength model in Section 7.6. Thus, for the pseudo damage, d, we assume

$$\ln d \sim N(\mu_d, \sigma_d) \tag{8.11}$$

and for the pseudo damage intensity (i.e. the pseudo damage normalized with respect to the measured distance L), $\tilde{d} = d/L$, we have

$$\ln \tilde{d} = \ln d - \ln L \sim N(\mu_d - \ln L, \sigma_d). \tag{8.12}$$

Further, the equivalent amplitude, see Equation (3.28), is defined as

$$A_{\text{eq}} = \left(\frac{K}{N_0}d\right)^{1/\beta} \tag{8.13}$$

which means that

$$\ln A_{\text{eq}} = \frac{1}{\beta}(\ln K - \ln N_0 + \ln d) \sim N((\mu_d + \ln K - \ln N_0)/\beta, \sigma_d/\beta). \tag{8.14}$$

Finally, for the equivalent mileage, $M_{\text{eq}} = d/\tilde{d}_{\text{ref}}$, we get

$$\ln M_{\text{eq}} = \ln d - \ln \tilde{d}_{\text{ref}} \sim N(\mu_d - \mu_{\tilde{d}_{\text{ref}}}, \sqrt{\sigma_d^2 + \sigma_{\tilde{d}_{\text{ref}}}^2}) \tag{8.15}$$

where $\tilde{d}_{\text{ref}} = d_{\text{ref}}/L_{\text{ref}}$, and hence by using Equation (8.12)

$$\ln \tilde{d}_{\text{ref}} \sim N(\mu_{d_{\text{ref}}} - \ln L_{\text{ref}}, \sigma_{d_{\text{ref}}}) = N(\mu_{\tilde{d}_{\text{ref}}}, \sigma_{\tilde{d}_{\text{ref}}}). \tag{8.16}$$

As we can see from the calculation above, all our different measures of load severity follow a log-normal distribution.

8.3.1.2 Splitting into Subloads

In order to evaluate the statistical precision in the load severity measurement, the parameters in the distributions need to be estimated. The mean of the logarithmic damage is simply estimated through the observed damage

$$\mu_d^* = \ln d \tag{8.17}$$

while the standard deviation, σ_d, can be estimated using the methods for estimating the coefficient of variation described in Section 6.7. Note that, for small values of σ_d, it is approximately equal to the coefficient of variation, $R(d)$. More precisely

$$R(d) = \sqrt{e^{\sigma_d^2} - 1} \approx \sigma_d, \qquad \text{for small } \sigma_d \tag{8.18}$$

and

$$\sigma_d = \sqrt{\ln(R(d)^2 + 1)} \approx R(d), \qquad \text{for small } R(d) \tag{8.19}$$

As a rule of thumb, $\mathbf{R}(d) < 0.25$, can be regarded as "small", since

$$\sqrt{\ln(0.25^2 + 1)} \approx 0.246.$$

We recall that a measured load signal can be split into N subloads, giving damage values $d_1, d_2, \ldots, d_N$, and the coefficient of variation can be estimated as

$$\mathbf{R}(d) \approx \frac{1}{\sqrt{N}} \cdot \frac{s}{\overline{d}} \tag{8.20}$$

with

$$\overline{d} = \frac{1}{N} \sum_{i=1}^{N} d_i, \qquad s^2 = \frac{1}{N-1} \sum_{i=1}^{N} (d_i - \overline{d})^2 \tag{8.21}$$

see Section 6.7.1 for details.

Example 8.6 (How Many Laps Do We Need to Measure?)

A series of measurements has been performed on the test track, where three drivers were instructed to drive 5 laps each according to specific driving instructions. As one would expect, the calculated severities of the measured laps are all different, see Figure 8.2.

Now we will examine the uncertainty in the measured pseudo damage. Here the division is that a subload represents one lap, and for driver one we get

$$\overline{d} = \frac{1}{5}(3.83 + 4.13 + 3.60 + 4.26 + 4.86) \cdot 10^{23} = 4.14 \cdot 10^{23} \tag{8.22}$$

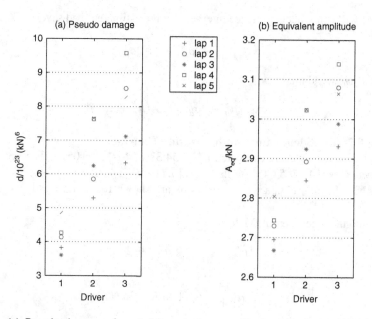

Figure 8.2 (a) Pseudo damage, d, and (b) equivalent amplitude, A_{eq}, for each driver and lap, using damage exponent $\beta = 6$

$$s^2 = \frac{1}{5-1} \sum_{i=1}^{5} (d_i - \bar{d})^2 = 2.31 \cdot 10^{45} = (4.81 \cdot 10^{22})^2 \qquad (8.23)$$

and

$$\mathbf{R}(d) \approx \frac{1}{\sqrt{5}} \frac{s}{\bar{d}} = \frac{4.81 \cdot 10^{22}}{\sqrt{5} \cdot 4.14 \cdot 10^{23}} = 0.0519. \qquad (8.24)$$

The estimated standard deviation of the damage becomes

$$\sigma_d^* = \sqrt{\ln(\mathbf{R}(d)^2 + 1)} = 0.0519 \qquad (8.25)$$

and the estimated standard deviation in the equivalent amplitude becomes

$$\sigma_{A_{eq}}^* = \sigma_d^*/\beta = 0.0519/6 = 0.0086. \qquad (8.26)$$

Now assume that our goal is to estimate the equivalent amplitude with a precision of $\sigma_{A_{eq}} \leq 1\%$; how many laps do we need to measure?

$$\frac{1}{\sqrt{n}} \frac{s}{\bar{d}} / \beta \leq 0.01 \quad \Rightarrow \quad n \geq \frac{s^2}{(0.01 \cdot \bar{d} \cdot \beta)^2} = 3.7. \qquad (8.27)$$

Thus, we need to measure at least 4 laps.

In order to reduce the uncertainty even more, we need to measure more laps. For example, to reduce the uncertainty by a factor $k = 2$, we have to increase the number of measured laps by a factor k^2, i.e. in our case from 4 to 16 laps. □

Now we will construct confidence intervals for the estimated load severity. A 95% interval for $\ln d$ is

$$[\mu_d^* - 2\sigma_d^*; \mu_d^* + 2\sigma_d^*] \qquad (8.28)$$

and for the pseudo damage

$$[\exp (\mu_d^* - 2\sigma_d^*); \exp (\mu_d^* + 2\sigma_d^*)]. \qquad (8.29)$$

Example 8.7 (Confidence Interval for Pseudo Damage)
For driver one we have the estimates $\mu_d^* = 54.37$ and $\sigma_d^* = 0.0519$. The 95% confidence interval becomes [54.27; 54.48] for $\ln d$, and [3.71 $\cdot$ 10^{23}; 4.57 $\cdot$ 10^{23}] for the pseudo damage in unit (kN)^6. In Figure 8.3a,b these intervals are shown for all three drivers. □

A 95% confidence interval for $\ln A_{eq}$ is

$$[\mu_{A_{eq}}^* - 2\sigma_{A_{eq}}^*; \mu_{A_{eq}}^* + 2\sigma_{A_{eq}}^*] \qquad (8.30)$$

with

$$\mu_{A_{eq}}^* = \mu_d^* + \ln K - \ln N_0 \qquad (8.31)$$

and

$$\sigma_{A_{eq}}^* = \sigma_d^*/\beta \qquad (8.32)$$

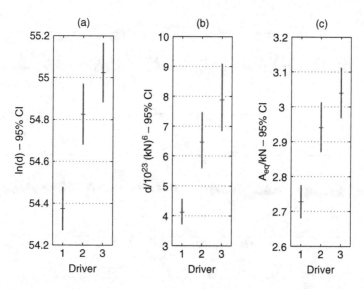

Figure 8.3 Confidence intervals for (a) logarithmic pseudo damage, (b) pseudo damage, and (c) equivalent amplitude, using damage exponent $\beta = 6$

and for the equivalent load amplitude A_{eq}

$$[\exp(\mu^*_{A_{eq}} - 2\sigma^*_{A_{eq}}); \exp(\mu^*_{A_{eq}} + 2\sigma^*_{A_{eq}})]. \tag{8.33}$$

Example 8.8 (Confidence Interval Equivalent Amplitude)
For driver one we have the estimates $\mu^*_{A_{eq}} = 7.91$ and $\sigma^*_{A_{eq}} = 0.0086$, and using $K = 1000$ and $N_0 = 10^6$, the 95% confidence interval becomes [7.89; 7.93] for $\ln A_{eq}$. For the equivalent amplitude, the confidence interval is [2.68; 2.77] in unit kN, see Figure 8.3c. □

Example 8.9 (Uncertainty in Load Severity for Service Loads)
The main example in Chapter 3 was the three measured service loads in Figure 3.1 showing the load signal on different road types. We will here investigate the uncertainty in the estimated load severity for these three signals. We employ the technique above, by splitting the load signals into 5 subloads each, see Figure 8.4. First we calculate the pseudo damage for each signal using exponent $\beta = 5$, and estimate the pseudo damage intensity (per km) to $\tilde{d}_1 = 0.00261$, $\tilde{d}_2 = 0.00164$, $\tilde{d}_3 = 0.00112$ for the city, highway and country road, respectively. The uncertainty in the estimated pseudo damage intensity for the city is calculated using Equations (8.20, 8.21)

$$\bar{d} = \frac{1}{5}(0.0033 + 0.0158 + 0.0173 + 0.0156 + 0.0018) = 0.0108 \tag{8.34}$$

$$s^2 = \frac{1}{5-1} \sum_{i=1}^{5} (d_i - \bar{d})^2 = 0.0000572 = (0.00756)^2 \tag{8.35}$$

and

$$\mathbf{R}(\tilde{d}) = \frac{1}{\sqrt{N}} \frac{s}{\bar{d}} = \frac{0.00756}{\sqrt{5} \cdot 0.0108} = 0.313 \tag{8.36}$$

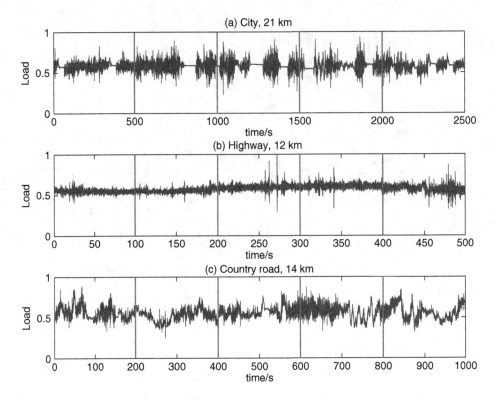

Figure 8.4 Measured service loads, vertical wheel forces

The results for all roads are $\mathbf{R}(\tilde{d}_1) = 0.313$, $\mathbf{R}(\tilde{d}_2) = 0.537$, and $\mathbf{R}(\tilde{d}_3) = 0.250$. These results will be used in Example 8.16.   □

8.3.1.3 Resampling from Rainflow Matrix

Both the estimation procedure and the relation between the rainflow matrix and the wanted final estimate A_{eq} are complicated functions. Therefore, the uncertainty is very hard (almost impossible) to compute analytically. Consequently, the bootstrap method comes in handy. The bootstrap procedure is based on resampling the measured rainflow matrix, which means that we construct new artificial rainflow matrices from the measured one. The resampling is performed by picking cycles at random, with frequencies specified by the measured rainflow matrix. The resampled rainflow matrices contain the same number of cycles as the measured rainflow matrix. From each resampled rainflow matrix, we can calculate a A_{eq}. The scatter in the resampled A_{eq}s now resembles the scatter in the estimation of A_{eq}. We can therefore construct a confidence interval for A_{eq}. One possibility is to construct the confidence intervals by using the normal distribution, giving an approximate 95% confidence interval of $[\mu^*_{eq} - 2\sigma^*_{eq}; \mu^*_{eq} + 2\sigma^*_{eq}]$ where $\mu^*_{eq} = \ln A^*_{eq}$ is calculated from the measured rainflow matrix, and σ^*_{eq} is the standard deviation computed from the resampled $\ln A_{eq}$s.

We should also note that the resampling procedure of the rainflow matrix, that is, by picking cycles at random, is not quite correct, since there might be a correlation between the cycles. This correlation structure has not been taken into account, since it is unknown. For general information about the bootstrap technique, see, for example, Hjorth [114], Davison and Hinkley [65], and Efron and Tibshirani [88].

8.3.2 Pair-wise Analysis of Load Severity

Here we will compare two measurements (e.g. representing different drivers or different operating conditions), and ask the following question: Can we, with certain statistical confidence, state that there is a difference, or could the difference be explained by random scatter? This question can be answered by calculating the confidence intervals for the difference in load severity. If the confidence interval does not cover the value zero, we can establish with 95% confidence that there is a difference. If not, the difference could be explained by random scatter.

A 95% confidence interval for $\ln d_2 - \ln d_1$ can be constructed in the following way

$$\left[\mu_{d_2}^* - \mu_{d_1}^* - 2\sqrt{\sigma_{d_2}^{*2} + \sigma_{d_1}^{*2}}; \ \mu_{d_2}^* - \mu_{d_1}^* - 2\sqrt{\sigma_{d_2}^{*2} + \sigma_{d_1}^{*2}} \right] \tag{8.37}$$

and for the ratio of the pseudo damage values d_2/d_1, we have the following

$$\left[e^{\mu_{d_2}^* - \mu_{d_1}^* - 2\sqrt{\sigma_{d_2}^{*2} + \sigma_{d_1}^{*2}}}; \ e^{\mu_{d_2}^* - \mu_{d_1}^* + 2\sqrt{\sigma_{d_2}^{*2} + \sigma_{d_1}^{*2}}} \right]. \tag{8.38}$$

The confidence interval for the difference in logarithmic equivalent amplitude is constructed in a similar way.

Example 8.10 (Confidence Interval for Difference in Load Severity)
For drivers one and two we have the estimates $\mu_{d_1}^* = 54.37$, $\mu_{d_2}^* = 54.83$, $\sigma_{d_1}^* = 0.0519$ and $\sigma_{d_2}^* = 0.0726$. The 95% confidence interval for $\ln d_2 - \ln d_1$ becomes [0.273; 0.630] and for the ratio d_2/d_1 it becomes [1.31; 1.88]. The confidence interval for the difference in logarithmic equivalent amplitude between drivers one and two is [0.046; 0.105] and the ratio $A_{eq,2}/A_{eq,1}$ has the following interval [1.05; 1.11]. Since the interval for the ratio does not cover unity (or equivalently that the interval for the logarithmic difference does not cover zero) we can draw the conclusion with 95% confidence that there is a difference between drivers one and two. More precisely, we can say with (at least) 95% confidence that driver two is more severe than driver one. For all the differences between the three drivers, see Figure 8.5. □

8.3.3 Joint Analysis of Load Severity

There seems to be a difference between the drivers. Thus, the following question arises: Is there a statistically significant difference between the three drivers? If so, how large is the scatter between the drivers, and how large is the scatter within a driver, i.e. the scatter between different laps driven by the same driver? These questions can be answered by the statistical technique ANalysis Of VAriance (ANOVA), see, for example, Searle *et al.* [213], Box *et al.* [35] and Montgomery [162].

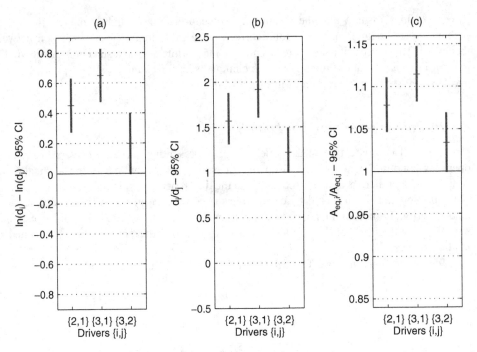

Figure 8.5 (a) 95% confidence intervals for the difference in logarithmic pseudo damage values between different drivers. (b) 95% confidence intervals for the ratio in the pseudo damage values between different drivers. (c) 95% confidence intervals for the ratio in the equivalent amplitude between different drivers

We now describe the statistical model behind ANOVA where the goal is to distinguish between different variance components; in our case the variance between and within drivers. Denote by y_{ij} the measured quantities for driver i on lap j, which in our case is chosen as the log-severity $y_{ij} = \ln A_{\text{eq},ij}$, see Figure 8.2. The log-severity y_{ij} is assumed to follow the model

$$y_{ij} = \mu + x_i + e_{ij} \tag{8.39}$$

where

μ = The overall mean log-severity (for the population of drivers). This is a constant.

x_i = The systematic deviation from the mean log-severity μ, for driver i. This is a random quantity, if the driver is randomly chosen from the population of drivers. The driver influence x_i is assumed to follow a normal distribution, $N(0, \sigma_x)$.

e_{ij} = The random deviation from the mean log-severity $\mu + x_i$, for driver i, lap j. This is a random quantity. The within-drivers variation e_{ij} is also assumed to be independent and to follow a normal distribution, $N(0, \sigma_e)$.

Example 8.11 (Difference Between Drivers?)

We will test whether there is a between-drivers effect, and also estimate the between- and within-drivers variances, σ_x^2 and σ_e^2, respectively. We use the method of Analysis of Variance

Table 8.1   Estimates of σ_x and σ_e, with 90% confidence intervals

Standard deviation	σ_x	σ_e
Estimate	0.0544	0.0247
Confidence interval	[0.0258; 0.3488]	[0.0186; 0.0374]

(ANOVA) which is explained in some detail in Section 8.6. Here we will merely report the result of the analysis. The analysis shows that it is very unlikely that the observed driver effect is caused by chance.

In fact, the probability of observing such a large estimate of the driver variance if there is no difference between the drivers is less than 0.00005. This value is given by a standard statistical program under the heading of the p-value of the hypothesis of no driver effect.

We conclude that there is an actual difference in pseudo damage caused by the different drivers which cannot be explained by chance. The estimates and confidence intervals of σ_x and σ_e are presented in Table 8.1. In the case of our three test drivers (Figure 8.2), a significant drivers-effect was found, and the within- and between-drivers scatter was in the same order. □

8.4   Random Sampling of Customers

8.4.1   Customer Survey

In order to estimate the distribution of load severity of real customers, measurements need to be carried out. In the following examples we will use a survey where 36 drivers were instructed to drive a specific route, including several types of roads. The rainflow matrix for the vertical load direction of each customer was then extrapolated to a full design life, and the corresponding equivalent amplitude was calculated, see Figure 8.6. Further, these 36 customers are supposed to represent a random sample of European customers.

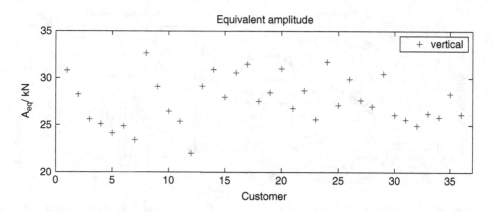

Figure 8.6   Equivalent amplitude for 36 customers

8.4.2 Characterization of a Market

Here we want to characterize a population of customers by a statistical distribution. Several proposals for the type of distribution is found in the literature, e.g. the Weibull distribution is used in Olsson [178] and Samuelson [206], the Normal distribution is used to model the equivalent amplitude in Bignonnet [27] and Thomas et al. [232], and the log-normal distribution is used in Svensson et al. [230] and Karlsson et al. [133]. All these distribution assumptions seem reasonable. However, it should be checked by making a probability plot.

Here the customer distribution is modelled by a log-normal distribution, which is in accordance with the load-strength model formulated in Chapter Chapter 7. The log-normal probability plot in Figure 8.7 shows that the distribution assumption seems reasonable. Thus, $\ln A_{eq}$ is characterized by its mean μ, and its standard deviation σ. From these we can calculate the target customer A_q, defined as a quantile in the distribution. When analysing a customer survey it is important also to present the uncertainties in the results. Uncertainty analysis is, for example, needed in order to decide how large a random sample of customers would be needed in order to estimate the target customers with sufficient accuracy. A good way to evaluate uncertainties is through confidence intervals for the quantities of interest. Below we present the results on calculating estimates and confidence intervals of the above quantities.

The customer distribution is estimated from the observations $A_{eq,1}, \ldots, A_{eq,n}$ as

$$\mu^* = \frac{1}{n}\sum_{i=1}^{n}\ln A_{eq,i}, \qquad s^2 = \frac{1}{n-1}\sum_{i=1}^{n}(\ln A_{eq,i} - \mu^*)^2 \qquad (8.40)$$

with 95% confidence intervals

$$I_{\mu} = [\mu^* - 2s/\sqrt{n}; \mu^* + 2s/\sqrt{n}] \qquad (8.41)$$

$$I_{\sigma} = \left[s\sqrt{\frac{n-1}{b}}; s\sqrt{\frac{n-1}{a}}\right] \qquad (8.42)$$

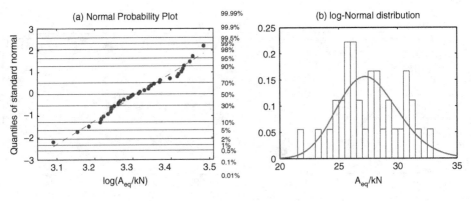

Figure 8.7   (a) Normal probability plot for $\ln A_{eq}$. (b) Histogram and estimated log-normal density for A_{eq}

where $a = \chi_{0.025}^2(n - 1)$ and $b = \chi_{0.975}^2(n - 1)$ are quantiles of the χ^2-distribution with $n - 1$ degrees of freedom, and the factor 2 in the interval for μ is approximately the 97.5% quantile of the normal distribution. The 95% confidence interval for the median of the A_{eq}, $\theta = e^\mu$, is

$$I_\theta = [e^{\mu^* - 2s/\sqrt{n}}; e^{\mu^* + 2s/\sqrt{n}}]. \tag{8.43}$$

Example 8.12 (Estimation of The Customer Distribution)
For the vertical loads in our specific case with $n = 36$ customers, the parameters are estimated at $\mu^* = 3.31$ and $s = 0.09$ with 95% confidence intervals

$$I_\mu = [3.28; 3.34], \qquad I_\sigma = [0.08; 0.12]$$

using $a = \chi_{0.025}^2(36 - 1) = 20.6$ and $b = \chi_{0.975}^2(36 - 1) = 53.2$. The estimate and 95% confidence interval for the median of the A_{eq} is

$$\theta^* = 27.5 \text{ kN}, \qquad I_\theta = [26.6; 28.3] \text{ kN}. \qquad \square$$

To obtain intervals with other levels of confidence, the values of a, b and the factor 2 are replaced by appropriate quantiles of the χ^2-distribution and the normal distribution, respectively. The quantiles can be obtained easily, for example, from tables or by using Matlab or Excel.

The target customer A_q is defined as a quantile in the distribution, namely $\ln A_q = \mu + z_q \cdot \sigma$, where z_q is a quantile in the standard normal distribution. There is no simple formula for the variance of the estimate, which is needed to construct the confidence interval. A useful approximation is found in Olofsson [177, Equation (4.34) on page 21]

$$\text{Var}[\ln A_q^*] \approx \sigma^2 \left(\frac{1}{n} + \frac{z_q^2}{2(n - 1)} \right). \tag{8.44}$$

Example 8.13 (Estimation of The Target Customer)
The target customer is defined as a quantile in the distribution, $\ln A_q = \mu + z_q \cdot \sigma$, and estimated at $\ln A_q^* = \mu^* + 1.64\sigma^* = 3.47$ for the 95% customer giving $A_{95\%}^* = 32.0$ kN. The variance for $\ln A_{95\%}^*$ is approximately 0.000574. A 95% confidence interval for the target customer is

$$I_{A_{95\%}} = [e^{\ln A_{95\%}^* - 2s_{95\%}/\sqrt{n}}; e^{\ln A_{95\%}^* + 2s_{95\%}/\sqrt{n}}] \text{ kN}$$

$$= [e^{3.42}; e^{3.51}] \text{ kN} = [30.5; 33.6] \text{ kN}.$$

The corresponding estimates and intervals for the 99% and 99.9% customer are

$$A_{99\%}^* = 34.1 \text{ kN}, \qquad I_{A_{99\%}} = [e^{3.47}; e^{3.59}] \text{ kN} = [32.1; 36.2] \text{ kN},$$

$$A_{99.9\%}^* = 36.7 \text{ kN}, \qquad I_{A_{99.9\%}} = [e^{3.53}; e^{3.68}] \text{ kN} = [34.0; 39.5] \text{ kN}.$$

An alternative approach is to compute the confidence interval using the bootstrap method. $\qquad \square$

The results of Examples 8.12 and 8.13 are summarized in Figure 8.8.

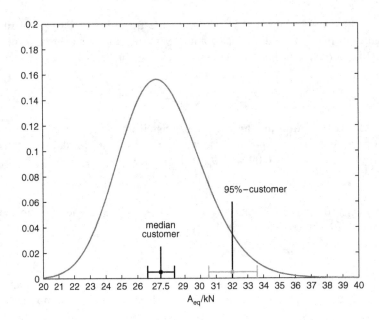

Figure 8.8 Estimated customer distribution with confidence intervals for the median customer and the 95%-customer

8.4.3 Simplified Model for a New Market

In order to determine the target customer in Brazil, a number of measurements have to be carried out. Suppose that the equivalent load A_{eq} is log-normally distributed and that $\ln A_{eq}$ has expectation μ and standard deviation σ. This can be formulated as $A_{eq} = \theta A$, with A being a dimensionless log-normal random variable, $\ln A$ having expectation 0 and standard deviation σ, and θ being the median of A_{eq}, i.e. $\theta = e^{\mu}$. The outcome of A can be interpreted as the normalized severity of the individual customer.

In the simplified model we assume that the standard deviation σ is known by experience, for example, being the same as for the European market, and therefore we only need to estimate the median θ for the new market.

We can estimate the median from direct measurements of A_{eq}, as in the previous section, but now the confidence intervals are different. The main advantage is that it is easy to estimate the median and its confidence interval also in the case of indirect measurements. The indirect measurements can be "easy to measure" quantities, like the speed profile, that are correlated to the interesting quantity A_{eq}. Consequently, an indirect measurement of A_{eq} is attached with a possibly large measurement error.

8.4.3.1 Direct Measurements

If the equivalent loads of n randomly selected customers are measured (without error), an interval with the coefficient of confidence of 95% for the median equivalent amplitude θ is given by

$$I_\theta = [e^{\mu^* - 2\sigma/\sqrt{n}}; \ e^{\mu^* + 2\sigma/\sqrt{n}}]. \tag{8.45}$$

where μ^* denotes the average of the logarithmic measured equivalent amplitudes. In the same manner a 95% confidence interval for the target customer $\ln A_q = \mu + z_p \cdot \sigma$ is obtained as

$$I_{A_q} = [e^{\mu^* + \sigma(z_p - 2/\sqrt{n})}; \ e^{\mu^* + \sigma(z_p + 2/\sqrt{n})}] = [A_q^* e^{-2\sigma/\sqrt{n}}; \ A_q^* e^{2\sigma/\sqrt{n}}] \qquad (8.46)$$

with $A_q^* = e^{\mu^* + z_p \cdot \sigma}$.

8.4.3.2 Indirect Measurements

It is expensive and impractical to equip a truck with advanced measurement devices. Therefore it is preferable if signals already recorded can be used. An example is the brake pressure which is available in the ABS system and which can be used as a substitute for a more accurately measured brake pressure and possibly also be used to estimate the brake torque.

Assume that the equivalent amplitude can be calculated from

$$\tilde{A}_{eq} = f(x) = A_{eq} \cdot \varepsilon \qquad (8.47)$$

where x is an easy to measure variable, for instance, velocity profile or ABS signals, $f(\cdot)$ is a deterministic function and ε is modelled as a random error, $\ln \varepsilon \sim N(0, \sigma_\varepsilon)$.

Let us assume that we also have m such indirect measurements on the independent random variables $\tilde{A}_{eq,1}, \tilde{A}_{eq,2}, \dots, \tilde{A}_{eq,m}$, where the independent measurement errors have the standard deviation σ_ε each. When trying to estimate the median equivalent amplitude θ, the variance of the measurement error will be added to the variance of the equivalent loads and consequently the estimate $\tilde{\mu}^*$ based on the average of these measurements will have the variance

$$\text{Var}[\tilde{\mu}^*] = \text{Var}\left[\frac{1}{m}\sum_{i=1}^{m} \ln \tilde{A}_{eq,i}\right] = \frac{1}{m}(\sigma^2 + \sigma_\varepsilon^2) = \frac{1}{m}\tilde{\sigma}^2. \qquad (8.48)$$

In this case an interval with the coefficient of confidence of 95% for the median equivalent amplitude θ is

$$I_\theta = \left[\tilde{\theta}^* e^{-2\tilde{\sigma}/\sqrt{m}}; \ \tilde{\theta}^* e^{2\tilde{\sigma}/\sqrt{m}}\right] \qquad (8.49)$$

with $\tilde{\theta}^* = e^{\tilde{\mu}^*}$.

Note that in the analysis above we have assumed that there is no systematic error in the indirect measurement. In the presence of a multiplicative systematic error δ we get

$$\tilde{A}_{eq} = f(x) = A_{eq} \cdot \varepsilon \cdot \delta \qquad (8.50)$$

and consequently the estimate of the median θ will become biased, which means that the systematic error will not diminish no matter how many observations are available. Thus, if one has many measurements, it is important to minimize the systematic error δ in the function $f(x)$, even if it implies that the magnitude of the random error ε increases.

8.4.3.3 Common Normalized Severity

A favourable situation would be if the normalized severity A has the same distribution on all markets and the different markets differ only with respect to the median equivalent amplitude θ. As has been defined above, the random variable A is log-normally distributed, with $\ln A$ having expectation 0 and standard deviation σ. In the case where the normalized customer severity A is the same for all markets, it is enough to determine the common σ, and the market specific median θ to calculate the customer population θA of the market in question.

8.4.3.4   Indirect Measurements and On-board Logging

One problem with customer surveys is getting a large enough sample, but also to get a random sample of customers. The on-board logging devices in a large population of customer trucks would here be most helpful. Even though it may not be possible to measure A_{eq} directly, many other available signals from the on-board logging system may be useful in the sense that they are correlated to the equivalent amplitude. Such an indirect measurement of A_{eq} is then attached to a larger random variation. We may model the indirect measurements of A_{eq} as $\ln \tilde{A}_{eq} \sim N(\mu, \tilde{\sigma})$, where μ and σ are the mean and the standard deviation, respectively, of the logarithmic customer distribution, and $\tilde{\sigma} = \sqrt{\sigma^2 + \sigma_\varepsilon^2}$ is the standard deviation of the measurement where σ_ε^2 is the extra variance contribution due to the indirect measurement. Consequently, the methods developed for confidence intervals may be used to efficiently analyse the on-board logging data.

8.4.4   Comparison of Markets

From two customer surveys from different markets, it should be interesting to compare their severities. Above all, can we say that one market is more severe than the other? This question can be addressed by a confidence interval for the quotient in the median severity

$$I_{\theta_2/\theta_1} = \left[\frac{\theta_2^*}{\theta_1^*} e^{-2s\sqrt{1/n_2 + 1/n_1}}, \ \frac{\theta_2^*}{\theta_1^*} e^{+2s\sqrt{1/n_2 + 1/n_1}} \right] \tag{8.51}$$

where θ_1^* and θ_2^* are the estimated medians, which are obtained from the means $\ln \theta_1^*$ and $\ln \theta_2^*$ estimated from n_1 and n_2, respectively, number of measurements, and s is an estimate of the standard deviation σ, which is here assumed to be the same for both markets. If this interval does not cover the value one, we can establish with 95% confidence that there is a difference. In the other case, when it covers one, we cannot detect a difference at the degree of confidence of 95%, but we get a 95% confidence interval for the possible difference. Compare Section 8.3.2.

8.5   Customer Usage and Load Environment

There are several ways to assess the customer distribution of load severity. One way is to directly measure the load severity in the population of interest, as has been described above. Another way is to make a survey of how the trucks are used for the population of interest. Then the load environment can be described using load influentials (see Section 1.3), e.g. road type, curve intensity, climate, and so on, where the road type is the most important one. In this section the load environment will be described by classifying the road types. The combination of customer usage and load environment then builds up the customer distribution.

Assume that the load environment has been classified into n classes, with damage intensities per km

$$\tilde{d} = (\tilde{d}_1, \tilde{d}_2, \ldots, \tilde{d}_n).$$ (8.52)

The pseudo damage of a given customer can be written as

$$d_{\text{life}} = L_1 \tilde{d}_1 + L_2 \tilde{d}_2 + \cdots + L_n \tilde{d}_n$$ (8.53)

where the L_k:s are the distances driven in the different classes during the service life $L = L_1 + L_2 + \cdots + L_n$ of the truck. It can be rewritten as

$$d_{\text{life}} = L(w_1 \tilde{d}_1 + w_2 \tilde{d}_2 + \cdots + w_n \tilde{d}_n)$$ (8.54)

where $w_k = L_k/L$, $\sum w_k = 1$, are weights describing the proportions driven in the different classes.

Example 8.14 (Load Severity of a Specific Customer)
A specific customer of a city distribution truck states that his usage profile is 60% city driving, 30% highway, and 10% country road. The load for the different road classes has been measured (Figure 8.4 on page 296) resulting in the pseudo damage intensities $\tilde{d} = (0.00261, 0.00164, 0.00112)$ per km. For the expected service life of $L = 600 \cdot 10^3$ km the pseudo damage becomes

$$d_{\text{life}} = L(w_1 \tilde{d}_1 + w_2 \tilde{d}_2 + w_3 \tilde{d}_3)$$
$$= 600 \cdot 10^3 \cdot (0.60 \cdot 0.00261 + 0.30 \cdot 0.00164 + 0.10 \cdot 0.00112)$$
$$= 600 \cdot 10^3 \cdot 0.00217 = 1.30 \cdot 10^3$$ (8.55)

which in terms of the equivalent amplitude becomes

$$A_{\text{eq}} = \left(\frac{d_{\text{life}}}{N_0} \right)^{1/\beta} = \left(\frac{1.30 \cdot 10^3}{10^6} \right)^{1/5} = 0.265.$$ (8.56)

□

We are interested in assessing the customer distribution in terms of pseudo damage or equivalent amplitude of a given population (e.g. city distribution trucks in Europe). As a simple example we may classify road types into city, highway, and country road. Within each class there can be expected to be some variation, for example, the damage on a highway may be somewhat different within the market in question, due to variation in road quality within the market. However, this scatter can be expected to be quite small, if the classification has been well performed. The population of customers may be described by the customer usage, i.e. the statistical distributions of total distance driven, L, and the proportions, w_k, driven in the different classes. The combination of the random variation in customer usage and the load environment defines the customer load distribution.

Both the customer usage and the load environment need to be estimated. When measuring the damage in the different road classes there is a statistical uncertainty in the estimated pseudo damage intensity, due to the limited length of the measurement. Customer usage can be assessed by making a survey asking the customers how they use their trucks. There is also the possibility of using on-board logging devices to make the customer survey.

On the basis of models for customer usage and load environment we will now demonstrate how the mean and the variance of the customer distribution can be computed. This is the necessary input to the load-strength model presented in Section 7.6, and can also be used to estimate, for example, the 95% customer.

8.5.1 Model for Customer Usage

First, we assume that the vector of the damage intensities $\tilde{d} = (\tilde{d}_1, \tilde{d}_2, \ldots, \tilde{d}_n)$ for the different classes is known, that is, there is no uncertainty in the estimated load environment. The different weights, w_k, are random variables with

$$\mathbf{E}[w_k] = \mu_k, \quad \text{Var}[w_k] = \sigma_k^2 \quad \text{and} \quad \text{Cov}[w_i, w_j] = \sigma_{i,j} \tag{8.57}$$

and represent the scatter in customer usage. The customer distribution can now be computed. In statistical terms it means that we want to compute the conditional expectation of the pseudo damage d_{life} given that we know the damage intensities $\tilde{d}$

$$\mathbf{E}[d_{\text{life}}|\tilde{d}] = L(\mathbf{E}[w_1]\tilde{d}_1 + \mathbf{E}[w_2]\tilde{d}_2 + \cdots + \mathbf{E}[w_n]\tilde{d}_n) \tag{8.58}$$
$$= L(\mu_1\tilde{d}_1 + \mu_2\tilde{d}_2 + \cdots + \mu_n\tilde{d}_n)$$

and the corresponding conditional variance

$$\text{Var}[d_{\text{life}}|\tilde{d}] = L^2(\text{Var}[w_1]\tilde{d}_1^2 + \text{Var}[w_2]\tilde{d}_2^2 + \cdots + \text{Var}[w_n]\tilde{d}_n^2 \tag{8.59}$$
$$+ 2\sum_{i<j}\text{Cov}[w_i, w_j]\tilde{d}_i\tilde{d}_j)$$

$$= L^2(\sigma_1^2\tilde{d}_1^2 + \sigma_2^2\tilde{d}_2^2 + \cdots + \sigma_n^2\tilde{d}_n^2 + 2\sum_{i<j}\sigma_{i,j}\tilde{d}_i\tilde{d}_j). \tag{8.60}$$

The coefficient of variation is

$$\mathbf{R}[d_{\text{life}}|\tilde{d}] = \frac{\sqrt{\text{Var}[d_{\text{life}}|\tilde{d}]}}{\mathbf{E}[d_{\text{life}}|\tilde{d}]}. \tag{8.61}$$

An often used distribution to model weights is the Dirichlet distribution, $w \sim Dir_n(\alpha_1, \alpha_2, \ldots, \alpha_n)$, where $w = (w_1, w_2, \ldots, w_n)$ with $\sum_{k=1}^n w_k = 1$. (In the case of only two weights, the Dirichlet distribution becomes the more well-known beta distribution.) The expectation for the different weights is

$$\mathbf{E}[w_k] = \frac{\alpha_k}{\alpha_0} = \mu_k \tag{8.62}$$

where $\alpha_0 = \sum_{k=1}^n \alpha_k$, and the variance and the covariance are

$$\text{Var}[w_k] = \frac{\alpha_k(\alpha_0 - \alpha_k)}{\alpha_0^2(\alpha_0 + 1)} = \sigma_k^2, \tag{8.63}$$

$$\text{Cov}[w_i, w_j] = -\frac{\alpha_i \alpha_j}{\alpha_0^2(\alpha_0 + 1)} = \sigma_{i,j}. \tag{8.64}$$

The coefficient of variation for the different weights is

$$\mathbf{R}[w_k] = \frac{\sqrt{\text{Var}[w_k]}}{\mathbf{E}[w_k]} = \frac{\sqrt{(\alpha_0 - \alpha_i)}}{\sqrt{\alpha_k(\alpha_0 + 1)}} = \frac{\sqrt{(1 - \mu_k)}}{\sqrt{\mu_k(\alpha_0 + 1)}} = r_k. \tag{8.65}$$

The expected values μ_k of the weights are not enough to determine the parameters $\alpha_1, \alpha_2, \ldots, \alpha_n$, since $\mu_n = 1 - \sum_{k=1}^{n-1} \mu_k$. However, an extra equation is available through Equation (8.65), and we find that the parameter α_0 is a function of the mean μ_k and the coefficient of variation r_k

$$\alpha_0 = \frac{1 - \mu_k}{\mu_k \cdot r_k^2} - 1. \tag{8.66}$$

Thus, all parameters of the Dirichlet distribution can be determined by the mean values $\mu_1, \mu_2, \ldots, \mu_{n-1}$ together with one of the coefficients of variations r_k.

Example 8.15 (Three Road Types)

This is a continuation of Example 3.20, where the load environment has been classified into city driving, highway and country roads. We will here assess the customer distribution of a city distribution truck, where the customer usage is estimated as 45% city driving, 30% highway and 25% country road, and the design life is $L = 600 \cdot 10^3$ km. Thus, the mean values of the weights are $\mu_1 = 0.45$, $\mu_2 = 0.30$ and $\mu_3 = 0.25$. The parameters of the Dirichlet distribution that will be used to model the weights can be estimated if we also know the coefficient of variation for one of the weights. Here the coefficient of variation for the city weight was found to be $r_1 = 0.60$. First α_0 is calculated using Equation (8.66)

$$\alpha_0 = \frac{1 - \mu_1}{\mu_1 \cdot r_1^2} - 1 = \frac{1 - 0.45}{0.45 \cdot 0.60^2} - 1 = 2.40. \tag{8.67}$$

From Equation (8.62) the parameters of the Dirichlet distribution are calculated; $\alpha_1 = 1.078$, $\alpha_2 = 0.719$, and $\alpha_3 = 0.599$. The variances of the weights are calculated from Equation (8.63) and are as follows: $\sigma_1^2 = 0.0729$, $\sigma_2^2 = 0.0619$, and $\sigma_3^2 = 0.0552$. The covariances are calculated from Equation (8.64) and are the following: $\sigma_{1,2} = -0.0398$, $\sigma_{1,3} = -0.0331$, and $\sigma_{2,3} = -0.0221$. The damage intensities for the three road classes are calculated from the measurements in Figure 8.4, $\tilde{d} = (0.00261, 0.00164, 0.00112)$, using damage exponent $\beta = 5$. For the design life, the expected damage is

$$\begin{aligned}
\mathbf{E}[d_{\text{life}} | \tilde{d}] &= L(\mu_1 \tilde{d}_1 + \mu_2 \tilde{d}_2 + \mu_3 \tilde{d}_3) \\
&= 600 \cdot 10^3 \cdot (0.45 \cdot 0.00261 + 0.30 \cdot 0.00164 + 0.25 \cdot 0.00112) \\
&= 600 \cdot 10^3 \cdot 0.00195 = 1.17 \cdot 10^3 \tag{8.68}
\end{aligned}$$

and the variance is

$$\begin{aligned}
\text{Var}[d_{\text{life}} | \tilde{d}] &= L^2(\sigma_1^2 \tilde{d}_1^2 + \sigma_2^2 \tilde{d}_2^2 + \sigma_3^2 \tilde{d}_3^2 + 2\sigma_{1,2} \tilde{d}_1 \tilde{d}_2 + 2\sigma_{1,3} \tilde{d}_1 \tilde{d}_3 + 2\sigma_{2,3} \tilde{d}_2 \tilde{d}_3) \\
&= (600 \cdot 10^3)^2 \cdot (0.0729 \cdot 0.00261^2 + 0.0619 \cdot 0.00164^2 + 0.0552 \cdot 0.00112^2
\end{aligned}$$

$$-2 \cdot 0.0398 \cdot 0.00261 \cdot 0.00164 - 2 \cdot 0.0331 \cdot 0.00261 \cdot 0.00112$$

$$-2 \cdot 0.0221 \cdot 0.00164 \cdot 0.00112)$$

$$= (600 \cdot 10^3)^2 \cdot (7.32 \cdot 10^{-7} - 6.15 \cdot 10^{-7}) = (600 \cdot 10^3)^2 \cdot 1.17 \cdot 10^{-7}$$

$$= 4.22 \cdot 10^4 = (205)^2. \tag{8.69}$$

The coefficient of variation for the damage is

$$\mathbf{R}[d_{\text{life}}|\tilde{d}] = \frac{\sqrt{\text{Var}[d_{\text{life}}|\tilde{d}]}}{\mathbf{E}[d_{\text{life}}|\tilde{d}]} = \frac{205}{1.17 \cdot 10^3} = 0.176. \tag{8.70}$$

In terms of equivalent load the result is

$$\mathbf{E}[A_{\text{eq}}|\tilde{d}] \approx \left(\frac{\mathbf{E}[d_{\text{life}}|\tilde{d}]}{N_0}\right)^{1/\beta} = \left(\frac{1.17 \cdot 10^3}{10^6}\right)^{1/5} = 0.259 \tag{8.71}$$

and the coefficient of variation

$$\mathbf{R}[A_{\text{eq}}|\tilde{d}] \approx \frac{\mathbf{R}[d_{\text{life}}|\tilde{d}]}{\beta} = \frac{0.176}{5} = 0.0352 \tag{8.72}$$

which represents the scatter in the load distribution. □

8.5.2   Load Environment Uncertainty

Now we will assume that the damage intensity for each class is also a random variable with expectation

$$\mathbf{E}[\tilde{d}_k] = \nu_k \tag{8.73}$$

and variance

$$\text{Var}[\tilde{d}_k] = (\nu_k \rho_k)^2 \tag{8.74}$$

where ρ_k is the coefficient of variation of $\tilde{d}_k$. We assume that there is no correlation between different measurements, i.e. $\text{Cov}[\tilde{d}_i, \tilde{d}_j] = 0$.

The uncertainty in $\tilde{d}_k$ can be a statistical uncertainty due to the limited length of the measurement, but it can also contain variations within the road class as, for example, all highways do not have exactly the same properties. The first one can be reduced by more measurements, while the second one will remain and can only be reduced by reconsidering the road classification. The variation within each road class can be evaluated by making measurements on different roads within the road class.

The expectation of the total pseudo damage is then

$$\mathbf{E}[d_{\text{life}}] = \mathbf{E}[\mathbf{E}[d_{\text{life}}|\tilde{d}]] = \mathbf{E}[L(\mu_1\tilde{d}_1 + \mu_2\tilde{d}_2 + \cdots + \mu_n\tilde{d}_n)]$$

$$= L \cdot (\mu_1\nu_1 + \mu_2\nu_2 + \cdots + \mu_n\nu_n) \tag{8.75}$$

where μ_k is the expectation for the weight w_k (the same as before). The corresponding variance can be computed using the law of total variance

$$\text{Var}[d_{\text{life}}] = \mathbf{E}[\text{Var}[d_{\text{life}}|\tilde{d}]] + \text{Var}[\mathbf{E}[d_{\text{life}}|\tilde{d}]]. \tag{8.76}$$

The first part is equal to

$$
\mathbf{E}[\mathrm{Var}[d_{\mathrm{life}}|\tilde{d}]] = \mathbf{E}\left[L^2 \cdot \left(\sum_{k=1}^{n} \sigma_k^2 \tilde{d}_k^2 + 2 \sum_{i<j} \sigma_{i,j} \tilde{d}_i \tilde{d}_j \right) \right]
$$

$$
= L^2 \cdot \left(\sum_{k=1}^{n} \sigma_k^2 \mathbf{E}[\tilde{d}_k^2] + 2 \sum_{i<j} \sigma_{i,j} \mathbf{E}[\tilde{d}_i \tilde{d}_j] \right)
$$

$$
= L^2 \cdot \left[\sum_{k=1}^{n} \sigma_k^2 \left(\mathrm{Var}[\tilde{d}_k] + \mathbf{E}[\tilde{d}_k]^2 \right) + 2 \sum_{i<j} \sigma_{i,j} \mathbf{E}[\tilde{d}_i]\mathbf{E}[\tilde{d}_j] \right]
$$

$$
= L^2 \cdot \left[\sum_{k=1}^{n} \sigma_k^2 \left((\nu_k \rho_k)^2 + \nu_k^2 \right) + 2 \sum_{i<j} \sigma_{i,j} \nu_i \nu_j \right]
$$

$$
= L^2 \cdot \left[\sum_{k=1}^{n} \sigma_k^2 \nu_k^2 \left(\rho_k^2 + 1 \right) + 2 \sum_{i<j} \sigma_{i,j} \nu_i \nu_j \right], \tag{8.77}
$$

where σ_k^2 is the variance of the weight w_k and $\sigma_{i,j}$ the covariance between weights. The second part is

$$
\mathrm{Var}[\mathbf{E}[d_{\mathrm{life}}|\tilde{d}]] = \mathrm{Var}[L \cdot \sum_{k=1}^{n} \mu_k \tilde{d}_k] = L^2 \cdot \sum_{k=1}^{n} \mu_k^2 \mathrm{Var}[\tilde{d}_k] \tag{8.78}
$$

$$
= L^2 \cdot \sum_{k=1}^{n} \mu_k^2 (\nu_k \rho_k)^2.
$$

The variance for the damage intensity is therefore

$$
\mathrm{Var}[d_{\mathrm{life}}] = L^2 \cdot \left(\underbrace{\sum_{k=1}^{n} \sigma_k^2 \nu_k^2 + 2 \sum_{i<j} \sigma_{i,j} \nu_i \nu_j}_{\text{usage scatter}} + \underbrace{\sum_{k=1}^{n} (\nu_k \rho_k)^2 (\mu_k^2 + \sigma_k^2)}_{\text{load environment uncertainty}} \right) \tag{8.79}
$$

where the first part represents the scatter in customer usage (the same as Equation (8.60)), and the second part is due to the uncertainty in $\tilde{d}$, which can contain both statistical uncertainty due to the limited length of the measurement, and scatter due to variation within the road class.

Apart from the load uncertainties discussed so far, also other uncertainties often exist, such as calibration errors of measurement systems, differences between commercial truck and field measurement truck, and variation in driving behaviour. These sources should also be judged in terms of variations in life, and can then be added to the total load uncertainty.

Example 8.16 (Three Road Types, Continued)
The expectation and variance for w_k is the same as in Example 8.15. In Example 8.9, the expectation for the damage intensity is estimated to $v_1 = 0.00261$, $v_2 = 0.00164$, $v_3 = 0.00112$. The coefficient of variation is estimated to $\rho_1 = 0.313$, $\rho_2 = 0.537$, $\rho_3 = 0.250$. The expectation for the damage is

$$
\begin{aligned}
\mathsf{E}[d_{\text{life}}] &= L(\mu_1 v_1 + \mu_2 v_2 + \mu_3 v_3) \\
&= 600 \cdot 10^3 \cdot (0.45 \cdot 0.00261 + 0.30 \cdot 0.00164 + 0.25 \cdot 0.00112) \\
&= 600 \cdot 10^3 \cdot 0.00195 = 1.17 \cdot 10^3.
\end{aligned}
\tag{8.80}
$$

The variance is

$$
\begin{aligned}
\text{Var}[d_{\text{life}}] &= L^2 \cdot \left(\sum_{k=1}^{3} \sigma_k^2 v_k^2 + 2 \sum_{i<j} \sigma_{i,j} v_i v_j + \sum_{k=1}^{3} (v_k \rho_k)^2 (\mu_k^2 + \sigma_k^2) \right) \\
&= (600 \cdot 10^3)^2 \cdot (1.17 \cdot 10^{-7} + 3.12 \cdot 10^{-7}) = (600 \cdot 10^3)^2 \cdot 4.29 \cdot 10^{-7} \\
&= 1.54 \cdot 10^5 = (393)^2.
\end{aligned}
\tag{8.81}
$$

The coefficient of variation has increased to

$$
\mathsf{R}[d_{\text{life}}] = \frac{\sqrt{\text{Var}[d_{\text{life}}]}}{\mathsf{E}[d_{\text{life}}]} = \frac{393}{1.17 \cdot 10^3} = 0.336
\tag{8.82}
$$

compared to 0.176 in the previous example. In terms of equivalent load the result is

$$
\mathsf{E}[A_{\text{eq}}] \approx \left(\frac{\mathsf{E}[d_{\text{life}}]}{N_0} \right)^{1/\beta} = \left(\frac{1.17 \cdot 10^3}{10^6} \right)^{1/5} = 0.259
\tag{8.83}
$$

and the coefficient of variation has increased to

$$
\mathsf{R}[A_{\text{eq}}] \approx \frac{\mathsf{R}[d_{\text{life}}]}{\beta} = \frac{0.336}{5} = 0.0672
\tag{8.84}
$$

compared to 0.0352 in the previous example.   □

8.6   Vehicle-Independent Load Descriptions

To be able to use the same measurements for different vehicles it is necessary that the measurements are vehicle- and driver-independent. It is then important to find quantities which generate loads on trucks but do not depend on a particular truck. What is needed is a description of the load environment. The relevant properties of the load environment of the trucks depend on the roads, the climate, the traffic intensity, and the legislation. These load influence factors are discussed in Section 1.3. In the following we will include climate, traffic intensity and legislation in the road properties.

In that case the customer distribution consists of three components: customer usage, vehicle-independent road properties, and the vehicle model. It is also favourable if there is a global road classification so that the difference between markets can be described by the frequencies of visits to the different classes.

The question then arises of how to define the classes. Which properties of the roads are relevant?

To be somewhat more specific, assume that we are interested in fatigue caused by lateral loads. Then clearly, the frequency of curves and their curvatures are of interest. But to calculate the fatigue damage we also need the centripetal acceleration. This acceleration depends on the speed of the vehicle. Is the speed a road, a driver or a vehicle property?

For the time being, let us assume that the average speed depends on the curvature. In one model of this kind the speed is proportional to a negative power of the curvature, when the curvature is large enough to affect the speed. Then, for simplicity we write the total damage for a road of length l with $N(l)$ curves with maximal curvatures

$$C_i, i = 1, 2, \ldots, N(l),$$

as

$$D(l) = \sum_{i=1}^{N(l)} D(C_i),$$

where $D(C_i)$ is the damage caused by curve number i. This is a somewhat rough approximation but is enough to illustrate our reasoning. We see now that the damage depends on the distribution of the curvatures and the intensity of the curves.

Let's say we want to use the following classification: highway, secondary road, thoroughfare and street. This is a good classification if the damage merely depends on the class and not on the market.

We will now indicate a way to check whether there is a market effect or not. The tool we will use is the standard analysis of variance (ANOVA) model. We have I=4 classes and, say, J markets and measure K roads for each of the IJ combinations of class and market. We calculate the damage of 30 km of each of these IJK roads and denote the result as

$$y_{ijk} = \ln(d_{ijk}) = \mu + \tau_i + \beta_j + (\tau\beta)_{ij} + e_{ijk} \tag{8.85}$$

where μ is the average logarithmic damage, τ_i is the effect of the class number i, β_j is the effect of the market number j, $(\tau\beta)_{ij}$ is the interaction effect of class i and market j, and e_{ijk} denotes the random deviation. The damage d_{ijk} can be calculated from measurements on a map or from actually driving on the roads. In this latter case the centripetal accelerations can be measured and the damage model checked.

In order to test if there is a significant market effect, the sum of squares of the deviations of the logarithmic damage $\ln(d_{ijk})$ from its mean is written as a sum of squares of the estimations of the class effects, market effects, interaction effects and random effects, see Table 8.2.

The presence of an interaction effect or a market effect is checked by comparing the ratios $s_{\tau\beta}^2/s_e^2$ and s_β^2/s_e^2, respectively, with appropriate percentiles of the F distributions. If there is no effect present, the deviation from zero is random and the mean squares are all of the same magnitude and their ratios follow the F distribution. More details about analysis of variance can be found, in, for example, Searle et al. [213], Box et al. [35] and Montgomery [162].

Table 8.2 ANOVA table for class effects, market effects, interaction effects and random effects

Source of variation	Sum of squares	Degrees of freedom	Mean square
class	S_τ	I-1	$s_\tau^2 = S_\tau/(I-1)$
market	S_β	J-1	$s_\beta^2 = S_\beta/(J-1)$
interaction	$S_{\tau\beta}$	(I-1)(J-1)	$s_{\tau\beta}^2 = S_{\tau\beta}/[(I-1)(J-1)]$
random	S_e	IJ(K-1)	$s_e^2 = S_e/[IJ(K-1)]$
Total	S_d	IJK-1	

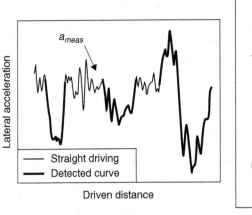

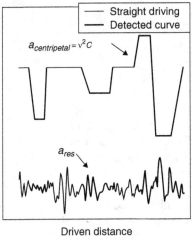

Figure 8.9 The measured lateral acceleration is due to the curvature and the residual (Karlsson [132])

Example 8.17 (Road Classification)

From on-board logging, the curvature of a road (cf. Figure 8.9) is extracted and the centripetal acceleration is calculated. Each curve contributes damage and the amount of damage for each 30 km part of a road is calculated. The results shown in Table 8.3 were obtained for logarithmic-scaled damage, where a constant 19.0 has been added to all numbers.

From the data we calculate the different sum of squares as follows: Let the average of all logarithmic pseudo damage values for road class number i be denoted by

$$\bar{y}_{i..} = \sum_{j,k} y_{ijk}/(JK) = \sum_{j,k} \ln(d_{ijk})/(JK)$$

and let the average of all logarithmic pseudo damage values be

$$\bar{y}_{...} = \sum_{i,j,k} y_{ijk}/(IJK) = \sum_{i,j,k} \ln(d_{ijk})/(IJK).$$

Table 8.3 Damage for each measured 30 km part

Road class Market	Highway	Secondary Road
Brazil	−1.6; 0.2; 0.4; −0.2; −1.1; 2.0	−1.7; 0.2; −0.6; 0.1; 1.0; 0.8
Germany	−0.4; −0.8; −0.6; 0.6; −0.6; −0.9	0.3; 0.5; 1.1; −0.3; 1.0; 0.8
Russia	−0.2; −0.2; 0.7; 0.6; −1.1; −1.5	−0.4; −0.7; 0.8; 0.3; 0.3; −0.7

Road class Market	Thoroughfare	Streets
Brazil	0.0; 1.5; 1.3; 1.3; 0.3; 1.7	1.8; 3.6; 2.2; 2.3; 1.2; −0.2
Germany	2.0; 0.5; 0.6; −0.5; −0.7; 1.0	0.8; 1.6; 0.2; 3.0; 1.2; 2.1
Russia	1.5; 1.1; 0.6; 1.7; −0.2; 1.1	1.4; 1.4; 0.4; 1.4; 0.6; 0.2

Then we obtain for highways, secondary roads, thoroughfares and streets the average logarithmic pseudo damage values

$$\overline{y}_{1..} = \sum_{j,k} y_{1jk}/18 = \sum_{j,k} \ln(d_{1jk})/18$$

$$= (-1.6 + 0.2 + \cdots + 2.0 + (-0.4) + (-0.8) + \cdots + (-0.9)$$

$$+(-0.2) + (-0.2) + \cdots + (-1.5))/18 = -0.26;$$

$\overline{y}_{2..} = 0.16; \overline{y}_{3..} = 0.82;$ and $\overline{y}_{4..} = 1.40$, respectively. The corresponding sum of squares is

$$S_\tau = \sum_{i,j,k} (\overline{y}_{i..} - \overline{y}_{...})^2 = JK \sum_{i} (\overline{y}_{i..} - \overline{y}_{...})^2$$

$$= 18 \cdot ((-0.26 - 0.53)^2 + (0.16 - 0.53)^2$$

$$+ (0.82 - 0.53)^2 + (1.40 - 0.53)^2) = 29.0$$

where the total mean $\overline{y}_{...} = 0.53$. The market sum of squares is calculated by an analogous formula

$$S_\beta = \sum_{ijk} (\overline{y}_{.k.} - \overline{y}_{...})^2.$$

The interaction effect is measured by

$$S_{\tau\beta} = \sum_{i,j,k} (\overline{y}_{ij.} - \overline{y}_{i..} - \overline{y}_{.k.} + \overline{y}_{...})^2$$

and the random effect by

$$S_e = \sum_{i,j,k} (y_{ijk} - \overline{y}_{ij.})^2,$$

where

$$\overline{y}_{ij.} = \sum_{k} \overline{y}_{ijk}/K.$$

We obtain the ANOVA table shown in Table 8.4.

Table 8.4   ANOVA table for Example 8.17

Source of variation	Sum of squares	Degrees of freedom	Mean square
class	$S_\tau = 28.95$	I-1 = 3	$s_\tau^2 = 9.65$
market	$S_\beta = 1.14$	J-1 = 2	$s_\beta^2 = 0.57$
interaction	$S_{\tau\beta} = 4.49$	(I-1)(J-1) = 6	$s_{\tau\beta}^2 = 0.75$
random	$S_e = 45.78$	IJ(K-1) = 60	$s_e^2 = 0.76$
Total	$S_d = 80.37$	IJK-1 = 71	

We conclude that the class effect which gives the ratio $s_\tau^2/s_e^2 = 9.65/0.76 = 12.7$ is extremely significant. The corresponding F percentile is 6.17 at 0.1% level of significance. A market effect is not observed in these data since $s_\beta^2 = 0.57$ and the variance of the random effect, $s_e^2 = 0.76$, are of the same magnitude. The same conclusion is drawn for the interaction effects.

The question which remains to be considered is whether the class effect is not only significant but also important. The sizes of the class effects have then to be studied. The estimate of τ_i is $\tau_i^* = (\overline{y}_{i..} - \overline{y}_{...})$.

We obtain the estimates of the effects of highways, secondary roads, thoroughfares and streets as −0.79, −0.37, 0.29, and 0.87, respectively. This means that the damage due to lateral forces when driving on streets compared to highways will increase by a factor $e^{0.87+0.79} = e^{1.66} = 5.3$, which seems to be an important effect.   □

So far we have discussed lateral forces as an illustration of how vehicle independent load models can be carried out. For more detail on modelling of lateral loads, see Karlsson [132]. When modelling longitudinal forces on brakes and transmissions the topography is relevant and should be included in the road model. For chassis and suspensions, vertical forces are important and there are road models available describing the road surface, see e.g. Bogsjö [30], Bogsjö et al. [33], Johannesson and Rychlik [122] and Öijer and Edlund [175, 176].

8.7   Discussion and Summary

Traditionally, the design of trucks is based on measurements from test tracks. The relation to the loads on trucks in service is not easy to find. But there are several sources which give some indication such as warranty claims and the demand for spare parts. To find the actual service loads there are several approaches. The most direct approach is to measure the trucks in service. In Section 8.2, the strength of probability sampling is illustrated and different ways to carry out an efficient sampling plan are presented.

After choosing the truck loads to be measured, an obvious question, discussed in Section 8.3, is the length of the measurement needed to get a satisfactory accuracy of measurements on the truck in question.

Three strategies for obtaining the distribution of the load severity of customers are introduced. As has already been mentioned the most direct approach is to use the principles of

probability sampling to select customers. The number of customers needed to obtain sufficient precision in the measurements can be reduced if a standard statistical distribution can
be used to describe the load severity. Then only a number of parameters have to be estimated instead of a whole customer distribution. The first approach, presented in Section 8.4,
is thus to estimate the customer load distribution from load measurements on customer.

A second approach, presented in Section 8.5, is to do a road classification and estimate
the severity of driving on the roads in each class. Then the severity of each customer is
not measured directly. The severity of a customer is obtained from the frequency of the
customer's visit to each road class.

A third approach, discussed in Section 8.6, is based on a vehicle-independent load description where the customer severity is obtained by combining models for customer usage,
vehicles and roads. This approach is useful when new trucks are designed for different
markets and also when the importance of education of drivers is to be estimated.

Due to the increasing availability of data during ordinary operation of vehicles (accelerations, pressures etc. used for ABS, ESP, ...) it becomes more and more promising to
use this data for the service load description. Based on simple real-time simulation models
which are running on a specific on-board CPU, this data may be even enhanced to derive
more durability related signals. A simple example is to estimate the axle torsion of a trailer
using the air spring pressures. This offers new possibilities for collecting valuable data in the
field. Some investigations in this direction are described, for instance, in Müller *et al.* [167].

Chapter 8 is devoted to an important discussion about obtaining information from field
measurements. But there is also a number of practical problems which have to be resolved.
A good practice is to do a pilot study before the full-scale campaign is launched. The
pilot study is used to calibrate the equipment and to check that the model assumptions are
reasonable. Also the methods of the analysis are checked. In design of experiments, a rule
of thumb is to use 20% at the most of the available resources for the pilot study.

9

Derivation of Design Loads

9.1 Introduction

This chapter tries to answer the question of what loads should be used in the design process and during subsequent verification in order to meet the reliability targets, given a certain customer population.

The structure to be designed can be a component like a steering rod, a subsystem like a suspension, or the full vehicle.

The design procedures can be based on the load-strength modelling which is described in detail below, design codes, or numerical experiments using vehicle models.

The methods for verification can be driving on the road (most often proving grounds, but in principle also public roads), executing rig tests, or numerical verification based on mathematical models.

In order to achieve this goal, we need to describe the load representation we are going to use and, most importantly, the statistical aspects involved since the inherent scatter in both load and strength necessitates statistical analysis. This requires the population of interest as well as the reliability target to be clearly defined. A possible implementation of the process of deriving test loads in the truck industry with emphasis on the statistical aspects is described in Speckert *et al.* [220]. A couple of specific methods for the generation of accelerated test rig signals are presented in Halfpenny [111]

9.1.1 Scalar Load Representations

For statistical analysis, it is quite common and practical (though not strictly necessary) to work with scalar numbers representing the severity of a load, for example

- *pseudo damage number d*: an abstract number d with a non-physical unit, but carrying the fatigue-relevant weighted sum of load cycles

$$d = \sum_{\text{(all cycles } i)} S_i^{\beta}$$

e.g. induced by customer usage, driving on the test track, or by rig testing,

Guide to Load Analysis for Durability in Vehicle Engineering, First Edition. Edited by P. Johannesson and M. Speckert.
© 2014 Fraunhofer-Chalmers Research Centre for Industrial Mathematics.

- *equivalent amplitude* A_{eq}: a transformation of the pseudo damage number into the scale of the physical unit of the load cycles

$$A_{eq} = \left(K \frac{d}{N_0} \right)^{1/\beta}$$

  for example, in unit kN (note that its interpretation depends on the extrapolation factor K and the chosen number of cycles N_0),
- *equivalent test life*: for example, an equivalent mileage in kilometres, a number of test track laps, or a number of load cycles on a rig.

See Section 3.1.12 for a detailed discussion of these notions. It is important to note that pre-knowledge of the fatigue model (the exponent β) is assumed when calculating these numbers. This introduces uncertainty about the model assumptions which in the load-strength model will be captured by uncertainty parameter τ_β.

The concepts described below are essentially independent of the different representations. Of course, care needs to be taken when interpreting the results. For instance, a (safety) factor of 2 is very large in terms of equivalent amplitudes, but fairly small in terms of pseudo damage or life.

9.1.2   Other Load Representations

In addition to these scalar numbers, one can try to handle more complex data formats like rainflow matrices or PSD functions. These formats do not require any knowledge of the fatigue exponent and contain additional information.

What load properties are important depends on the structure, for example, for a full vehicle the frequency content is clearly of importance, but for a stiff component only the rainflow cycle content is relevant. In Section 4.1 we discussed this when editing load signals. In the present context of defining design loads and strength requirements we again have to decide what properties are important (see Section 4.1.1) and what criteria for equivalence (see Section 4.1.2) we want to apply.

Since the statistically derived design load level does not uniquely define the load signals, these considerations can and should be taken into account in addition.

9.1.3   Statistical Aspects

The reliability target will be defined in statistical terms, for example, an upper bound on the failure rate for the design load representing a certain severe customer with a specified number of kilometres. The estimation of the customer load distribution, the design load, and the strength distribution is based on available information which is never complete (few customer load measurements, few test samples, ...). Consequently, any decision based on these results may be wrong. Nevertheless, a process leading to a decision about the release of a component is needed. This may be implemented by rejecting the hypothesis that the design does not meet the reliability target with reasonable confidence or by other statistical reasoning.

The loads used within that process are discussed in this chapter, and the verification procedures are discussed in Chapter 10.

9.1.4 Structure of the Chapter

The *first focus* of this chapter is on the statistical aspects during the derivation of the load severity in terms of one of the representations mentioned above. For all load representations such as pseudo damage numbers, equivalent amplitudes, or rainflow matrices, the design specifications need to be transformed into load time signals (to be used for the excitation of models or as input on a rig) or into driving schedules for the proving ground.

There are several ways of doing that, which is the *second focus* in this chapter. With the exception of Section 9.8, which deals with optimum track mixing, the technical details of the methods involved in the generation of time signals for the design loads are omitted. Instead we refer to publications or to Chapter 4 as much as possible and concentrate on the overview.

Section 9.8 dealing with optimum track mixing is rather detailed as the method has not been explained so far and we consider it very important in practice. This is the *third focus* within the present chapter.

The detailed structure of the chapter is as follows:

- Section 9.2: Definition and derivation of design loads and strength requirements based on reliability targets. This can be done in terms of pseudo damage numbers or equivalent load-strength measures.
- Section 9.3: Find a synthetic load model (for example, sine waves or several constant load blocks with different amplitudes) and adjust the corresponding parameters to meet the derived severity of the design load. Especially for uniaxial loads (e.g. a single force acting on a steering rod) this might be an interesting option. It will be more difficult in multi-input load situations, since we usually cannot derive a phase relation for the synthetic load channels.
- Section 9.4: Find a random load model and adjust the corresponding parameters to achieve the derived properties of the design load. Here too, we have to be careful when dealing with multi-input loads.
- Section 9.5: Adjust measured load spectra or rainflow matrices to meet the derived design load using the methods described in Section 4.4. Construct a time signal based on these histograms using the methods from Section 4.5.1. Again multi-input loads need special attention.
- Section 9.6: Adjust standardized load spectra such that they describe the derived design load.
- Section 9.7: Use pre-defined driving schedules on the proving ground. Often these test track loads have evolved during many generations of trucks and are based on long-term engineering experience.
- Section 9.8: Apply the methods of optimum track mixing. The result of the corresponding optimization is a certain mixture of the basic tracks, which may be public road measurements or test track events. Then we can go back to the underlying time signals and derive everything else we might need (rainflow matrices, load spectra, PSD functions ...).
- Finally, a summary is given in Section 9.9.

9.2 From Customer Usage Profiles to Design Targets

Before we are able to formulate a process to derive loads for design and verification, we need to clearly define the notions we are going to use.

9.2.1 Customer Load Distribution and Design Load

We assume that we 'somehow' have assessed the distribution of the customer loads for the population of interest (i.e. 'all trucks' or '4x2 tractors sold in Eire'). This means that we are able to estimate the mean load or a certain quantile of this population in terms of pseudo damage values or equivalent load-strength measures as described in Chapter 7. The estimation should be based on a sufficiently large set of measurements in customer service which can also be used to establish an estimate of the variance.

In general it is not easy to say how large a 'sufficiently large set of measurements' is. It depends on the expected variance. If the population is 'all trucks', we need more measurements than if the population is '4x2 tractors sold in Eire'. The question can be attacked, for example, with the statistical concept of confidence intervals or bootstrap methods. For more on the estimation and the judgement of its accuracy, see Chapter 8 and the references therein.

The *design load* as we use it here is a quantity with a well-defined relation to the load distribution. It may be a description of the loads accumulated by a target customer up to a certain number of target kilometres. To be more specific, the design load is typically defined, as a moderately severe customer and can be specified as a certain high load quantile, say, the 90%- or 95%-customer. The only restriction in choosing the design load is that the quantile must not be too extreme because of the uncertainty in the load distribution. As a rule of thumb, we should not exceed the 95%-quantile.

In the following we denote the design load by L_{dl} or by L_{α_L}, if we explicitly want to indicate that it represents the α_L-quantile of the customer load distribution. An illustration can be found in Figure 9.1.

The design load specification may be also given by company-specific rules based on experience such as a certain number of test track laps or a certain number of cycles at a specified amplitude. In those cases, the relation to the customer load distribution might be unknown. In the following, we concentrate on the former definition based on the clear relation to the customer load distribution.

9.2.2 Strength Distribution and Strength Requirement

We assume that we are able to estimate the mean or median or a moderate quantile of the strength distribution, for instance, by rig tests (see Chapter 10). In addition, we have an estimate of the variance which may be based on experience or on rig test results if sufficiently many components can be tested (see Chapter 7). Again, the distribution is given in terms of pseudo damage values or equivalent load/strength measures such that the data is on the same scale as the load data.

The *strength requirement* can be expressed on the basis of the strength distribution by choosing a certain quantile, say, the 50%-quantile (median), the 10%-quantile, or the

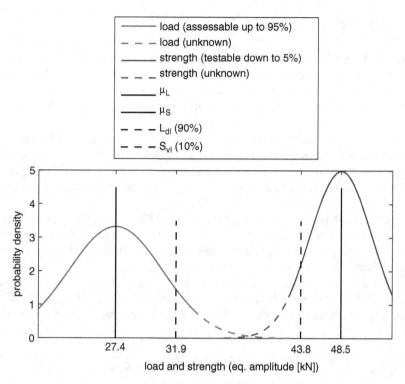

Figure 9.1 Load and strength distributions, design load and strength requirement. The assumption that the load and strength distribution, respectively, is assessable up to 95% and down to 5%, respectively, is an example and can differ in a specific situation

5%-quantile of the strength distribution. One restriction in choosing it is that the quantile must not be too extreme because of the uncertainty in the tails of the strength distribution.

Since the strength requirement needs to be checked based on test rig experiments, another restriction is given by the amount of testing effort we can afford. As we will see in Section 10.3, the 5%-quantile is harder to verify than the 10%-quantile. As a rule of thumb, we should not go below the 5%-quantile.

In the following, we denote the strength requirement by S_{vl} or by S_{α_S}, if we explicitly want to indicate that it represents the α_S-quantile of the strength distribution. We will also use the notation verification load L_{vl}, representing a load having the same severity as the strength requirement S_{vl}. In Figure 9.1 both the design loads and the strength requirements are illustrated. The figure also shows by dotted lines that the region of trust in the distribution is limited.

Once we have derived the design load and the strength requirement in terms of pseudo damage numbers or equivalent amplitudes, we need to translate these into rainflow matrices, time signals, or test track lanes for further processing. Of course, in a multi-input setting we have to estimate the mean, the quantile, and the variance for all important load channels. Sections 9.3–9.8 describe several possibilities to handle that problem.

9.2.3 Defining the Reliability Target

Definition 9.1 (Safety factor.) *The safety factor is the quotient between a strength and a load. If we choose the median load and the median strength, it is called a central safety factor, $\phi = S_{0.50}/L_{0.50}$. If we choose quantiles, for example, the 90%-quantile for the load and the 10%-quantile for the strength, it is called a partial safety factor, $\varphi = S_{0.10}/L_{0.90}$.*

Since we restrict ourselves to moderate quantiles both for the load and the strength distribution, we can use a distribution model determined by the estimated means and variances for the load and the strength and easily transform central safety factors into partial safety factors or vice versa. The simplest of such models which is often used is the log-normal model.

In Figure 9.2 all the definitions given so far are illustrated. The grey solid line (left) shows the customer load distribution as far as we have confidence in it. The grey dotted line indicates the region of high uncertainty, which here is beyond the 95%-quantile. The grey solid line (right) shows the strength distribution as far as we have confidence in it, the grey dotted line the region of high uncertainty, in this case beyond the 5%-quantile. In addition, the median of load and strength, the 90% load quantile, the 10% strength quantile, and the corresponding safety factors are shown. Note that the design loads and strength requirements

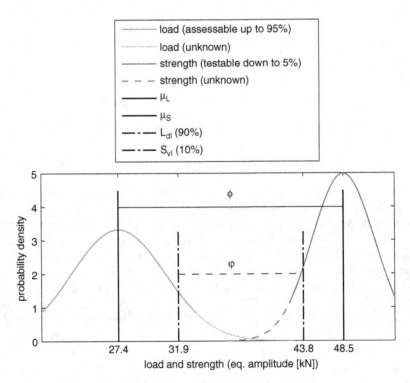

Figure 9.2 Load and strength distributions, design load and strength requirement, central and partial safety factors

have both been chosen to be well within the region of the distributions we have confidence in. As long as we stay below the 95%-load quantile with the design load and above the 5%-quantile with the strength requirement, the dependence on the specific quantile we choose is small.

As we will see, the reliability target can be defined in different ways. It depends on the specific design approach we want to take. In order to describe it in more detail, we here recall the options given in Chapter 7:

1. In the *pure empirical safety factor approach* (the first-moment method in Section 7.4), the reliability target ϕ_{target} is a lower bound for the safety factor. This bound has to be determined by experience. The simplest method is the central safety factor which does not take the uncertainties into account. Therefore, the design targets are often given as a partial safety factor: the quotient between a low strength quantile and a high load quantile. The main drawback is that there is no possibility to refine the bound by, for example, new test results or better knowledge about load or strength variance.
2. In the *pure statistical approach* (full probabilistic model in Section 7.3), the bound is derived from a target value $P_{f,target}$ for the failure probability and explicit parametric models of load and strength distributions including the tails. Using these distributions, the failure probability P_f can be calculated and compared to the target value $P_{f,target}$. The main drawback is that the uncertainties about the load and strength distributions are so large that the computed failure probability is unreliable. Thus, more details of this approach are omitted.
3. In the *reliability index approach* (second-moment method in Section 7.5), the bound is derived from a target value (the reliability index γ_{target}) which needs to be determined by experience. Compared to the partial safety factor, this is a more complete way to take both the means and uncertainties into account. The reliability index is a statistical measure of how far separated the strength is from the load, in terms of the number of standard deviations.
4. In the *mixed approach* (second-moment method with extra safety factor in Section 7.6.9), we combine approach 3 with approach 1 in order to better take advantage of both statistical considerations and engineering experience. Since this is our proposed way of defining the reliability target, we recall the method in the following.

In Section 7.6, the notion of a reliability index based on estimates on the location and the variance of the load and strength distributions was introduced and discussed extensively. The mixed approach was written in the form

$$\overline{\ln(S)} - \overline{\ln(L)} > S_d + 2\sqrt{\tau_S^2 + \tau_L^2 + \tau_\beta^2}, \tag{9.1}$$

where $\overline{\ln(S)}$ and τ_S are the location (mean) and uncertainty measures of strength, $\overline{\ln(L)}$ and τ_L are the location (mean) and uncertainty measures of the load, and τ_β is the uncertainty measure due to the damage exponent (Wöhler slope). The extra safety distance S_d is defined based on engineering experience. Together with the statistical safety distance $2\sqrt{\tau_S^2 + \tau_L^2 + \tau_\beta^2}$, it is used to make sure that the load and strength distributions are sufficiently separated. Consequently, for this approach, the reliability target is defined in the form of an empirical safety factor $\phi_{target} = \exp(S_d)$.

9.2.4 Partial Safety Factor for Load-Strength Modelling

If we do not want to work with median load and median strength, but rather use moderate quantiles L_{α_L} and S_{α_S} for load and strength, it is helpful to translate Equation (9.1) into an equivalent formula based on the quantiles. This can be done easily, for example, under the assumption that a log-normal distribution approximation is reasonable within the central part of the distributions (for a short discussion on this assumption, see the corresponding remark in the summary below). The central part here means the region between the mean and the moderate quantiles L_{α_L} and S_{α_S}. Then we have $\ln(S_{\alpha_S}) \approx \overline{\ln(S)} + \Phi^{-1}(\alpha_S) \cdot \tau_S$ and $\ln(L_{\alpha_L}) \approx \overline{\ln(L)} + \Phi^{-1}(\alpha_L) \cdot \tau_L$, where Φ^{-1} denotes the inverse of the standard normal cumulative distribution function. Inserting these expressions into Equation (9.1) we get

$$\ln(S_{\alpha_S}) - \ln(L_{\alpha_L}) > \ln(\phi_{target}) + 2\sqrt{\tau_S^2 + \tau_L^2 + \tau_\beta^2} + \Phi^{-1}(\alpha_S) \cdot \tau_S - \Phi^{-1}(\alpha_L) \cdot \tau_L, \quad (9.2)$$

where we have written $S_d = \ln(\phi_{target})$ due to the current terminology. Since the partial safety factor is defined as $\varphi = S_{\alpha_S}/L_{\alpha_L}$, we can rewrite the equation in the form

$$\ln(\varphi) > \ln(\phi_{target}) + 2\sqrt{\tau_S^2 + \tau_L^2 + \tau_\beta^2} + \Phi^{-1}(\alpha_S) \cdot \tau_S - \Phi^{-1}(\alpha_L) \cdot \tau_L. \quad (9.3)$$

A property of this approach is that the reliability target does not depend on the probabilities α_L and α_S of the quantiles we are using. This is due to the presence of the two rightmost terms in the formula, which transform the quantiles to the corresponding mean values. The interpretation of this formula is twofold:

1. All quantities with the exception of ϕ_{target} are estimated by evaluation of measurement campaigns, test rig results, and experience. The reliability target ϕ_{target} itself is pre-defined (by experience) and we have to check whether formula (9.3) is fulfilled or not.
2. Again the reliability target ϕ_{target} is pre-defined (by experience). All quantities with the exception of the strength requirement S_{α_S} are estimated by evaluation of measurement campaigns, test rig results, and experience. Then formula (9.2) is used to calculate the strength requirement in the form $\ln(S_{\alpha_S}) = \ln(L_{\alpha_L}) + \ln(\phi_{target}) + 2\sqrt{\tau_S^2 + \tau_L^2 + \tau_\beta^2} + \Phi^{-1}(\alpha_S) \cdot \tau_S - \Phi^{-1}(\alpha_L) \cdot \tau_L$. Finally, it has to be verified, using the methods described in Chapter 10, that the true α_S-quantile of the strength is not smaller.

Example 9.1 (Derivation of a Strength Requirement)
We return to the example of Section 8.4. The mean of the logarithmic load (equivalent amplitude of vertical wheel forces, calculated with $\beta = 5$, $N_0 = 10^6$) was estimated to $\mu_L = 3.31$, corresponding to an equivalent amplitude of 27.5 kN. The estimated standard deviation of the measurements (36 drivers) was $s = 0.09$, which is rather small. To add uncertainties not captured within the data, we agreed to work with the uncertainty parameter $\tau_L = 0.12$ instead.

Assume further that within the company, one has agreed to the 95%-customer as the design load. Then, we get

$$L_{\alpha_L} = L_{0.95} = \exp(\mu_L + \Phi^{-1}(0.95) \cdot \tau_L) = \exp(3.31 + 1.28 \cdot 0.12) \text{ kN} = 33.5 \text{ kN}.$$

The variance of the strength of the component under consideration is assumed to be bounded above by $\tau_S = 0.08$ and the uncertainty due to the Wöhler exponent is $\tau_\beta = 0.10$. The reliability target is $\phi_{target} = 1.25$, the probability of the strength quantile is $\alpha_S = 0.1$.

Plugging everything into the formula

$$\ln(S_{\alpha_S}) > \ln(L_{\alpha_L}) + \ln(\phi_{target}) + 2\sqrt{\tau_S^2 + \tau_L^2 + \tau_\beta^2}$$

$$+\Phi^{-1}(\alpha_S) \cdot \tau_S - \Phi^{-1}(\alpha_L) \cdot \tau_L,$$

we obtain

$$\ln(S_{0.1}) > \ln(33.5) + \ln(1.25) + 2\sqrt{0.08^2 + 0.12^2 + 0.1^2}$$

$$+\Phi^{-1}(0.1) \cdot 0.08 - \Phi^{-1}(0.95) \cdot 0.12$$

$$= 3.78,$$

thus $S_{0.1} > 44.0\,\text{kN}$. The final step is then to set up rig tests to check that there are less than 10% failures at $44.0\,\text{kN}$ and $N_0 = 10^6$ cycles. □

Remark 9.1 *The magnitude of the estimated uncertainty parameters $\tau_L, \tau_S, \tau_\beta$ and the target safety factor ϕ_{target} depend on the type of data we are working with. Assume that we have the estimations $\tau_L^d = 0.6, \tau_S^d = 0.4, \tau_\beta^d = 0.5, \phi_{target}^d = 3$ in terms of pseudo damage (or life), indicated by the superscript d. Assume further that we work with the Wöhler exponent $\beta = 5$. If we want to transform the uncertainty estimations to the equivalent amplitude scale we obtain the following results: $\tau_L^A = \frac{\tau_L^d}{\beta} = 0.125, \tau_S^A = \frac{\tau_S^d}{\beta} = 0.08, \tau_\beta^A = \frac{\tau_\beta^d}{\beta} = 0.1, \phi_{target}^A = (\phi_{target}^d)^{\frac{1}{\beta}} = 1.25$.*

9.2.5 Safety Factors for Design Loads

Verifying the reliability targets would require reproducing the uncertainties in the population of customer loads as well as in the strength. In practice this is not possible or feasible to do in, for example, numerical calculations, numerical simulations, or physical tests. However, what is often possible to verify is that the design load (or strength requirement) is sufficiently separated from the nominal strength. Therefore, we here propose to transform the reliability target to a central safety factor with respect to the design loads (or the verification load representing the strength requirements). These safety factors can then be verified. Next, we have a more detailed discussion of the transformation of reliability targets to safety factors for design loads.

9.2.5.1 Reliability Targets for Design Loads

The approach here is to use the *central safety factor*, ϕ, the quotient between the median strength and the median load. The first step is to interpret the reliability target in terms of a safety factor. In the following discussion we will use the equivalent load-strength measures that are used in the load-strength model, Section 7.6. Accordingly, we will assume that both the load and the strength are log-normally distributed,

$$\ln \ S \sim N(\mu_S, \tau_S) \quad \text{and} \quad \ln \ L \sim N(\mu_L, \tau_L)$$

respectively. The central safety factor is then

$$\phi = \frac{S_{0.50}}{L_{0.50}} = \frac{\exp(\mu_S)}{\exp(\mu_L)} = \exp(\mu_S - \mu_L).$$

9.2.5.2   Transforming the Reliability Target to a Safety Factor

The reliability target may be given as a safety factor based on engineering experience, for example by using in-house standards at the company. In this guide we suggest deriving the safety factor by also including statistical modelling in order to take care of the uncertainties in load and strength.

The *partial safety factor* is a first-moment method, see Section 7.4, and in its simplest form it compares a severe customer with a weak component. An often used choice for fatigue assessments is comparing the "+2std-customer" with the "−3std-strength", which leads to the partial safety factor

$$\varphi = \frac{S_{0.001}}{L_{0.975}} = \frac{\exp(\mu_S - 3\tau_S)}{\exp(\mu_L + 2\tau_L)} = \frac{\exp(\mu_S) \cdot \exp(-3\tau_S)}{\exp(\mu_L) \cdot \exp(+2\tau_L)} = \phi \cdot \exp(-3\tau_S - 2\tau_L)$$

where we have used the log-normal distribution assumption above, and hence

$$\phi = \varphi \cdot \exp(3\tau_S + 2\tau_L).$$

Here we set the reliability target in terms of the partial safety factor, φ_{target}, which can then be transformed into the central safety factor, ϕ_{target}, by having knowledge about τ_S and τ_L, the uncertainties in strength and load, respectively.

Our recommendation in Chapter 7 is to use a second-moment reliability method, and we propose the load-strength model implemented for fatigue, presented in Section 7.6. The reliability target can then be formulated in terms of a reliability index

$$\gamma = \frac{\mathbf{E}[\ln \ S - \ln \ L]}{\sqrt{\mathrm{Var}[\ln \ S - \ln \ L]}} = \frac{\mu_S - \mu_L}{\sqrt{\tau_S^2 + \tau_L^2 + \tau_\beta^2}} = \frac{\delta}{\tau}$$

which is easily recalculated in terms of a safety factor

$$\phi_{target} = \exp(\delta) = \exp(\gamma_{target} \cdot \tau)$$

where γ_{target} is the reliability target defined in terms of a safety index, see Table 7.1.

In Section 7.6.9 a compromise between statistical modelling and engineering experience is proposed by combining a statistically determined safety factor with an empirical deterministic safety factor, namely

$$\phi_{target} = \phi_d \cdot \phi_s \quad \text{with} \quad \phi_d = e^{S_d} \quad \text{and} \quad \phi_s = e^{2\tau}$$

where the reliability target is represented by the extra safety distance S_d.

9.2.5.3 Defining Safety Factors for Design Loads

The idea is now to transform the safety factor calculated from the reliability target into safety factors with respect to the design load. Therefore, we need to relate the design load to the customer load distribution, and change the safety factor accordingly. The (central) *safety factor with respect to the design load*, ϕ_{dl} can be expressed in terms of the target safety factor ϕ_{target}

$$\phi_{dl} = \frac{S_{0.50}}{L_{dl}} = \frac{S_{0.50}}{L_{0.50}} \cdot \frac{L_{0.50}}{L_{dl}} = \frac{\phi_{target}}{L_{dl}/L_{0.50}},$$

where the factor $L_{dl}/L_{0.50}$ is a measure of how much more severe the design load is compared to the median load. Thus, in durability assessments using the design load, we have to make sure that the safety factor ϕ_{dl} is fulfilled. For example, if we make a numerical fatigue evaluation of a component, we have to make sure that the quotient between the calculated nominal strength and the severity of the design load is at least ϕ_{dl}.

Example 9.2 (Safety Factor for the Design Load)
For a certain component the target safety factor is $\phi_{target} = 3.5$, and the design load is defined as the 95%-customer

$$L_{dl} = L_{0.95} = \exp(\mu_L + 1.64\tau_L) = L_{0.50} \cdot \exp(1.64\tau_L).$$

The safety factor for the design load becomes

$$\phi_{dl} = \frac{\phi_{target}}{L_{dl}/L_{0.50}} = \frac{\phi_{target}}{\exp(1.64\tau_L)} = \frac{3.5}{\exp(1.64 \cdot 0.25)} = \frac{3.5}{1.51} = 2.32$$

with load uncertainty $\tau_L = 25\%$. The design load is 51% more severe than the median customer load, and the safety factor has decreased from $\phi_{target} = 3.5$ to $\phi_{dl} = 2.32$ for the design load. Therefore, in the subsequent durability evaluation of the component subjected to the design load, we need to ensure that the calculated nominal strength is $\phi_{dl} = 2.32$ times higher than the severity of the design load. □

The safety factor for the verification load representing the strength requirement is defined in the same way as for the design load, see Section 10.2.1.

9.2.6 Summary and Remarks

The aim of this section was to set a framework for the "customer correlated calibration" of loads for testing. To this end, we have introduced the notions of design load, strength requirement, reliability target, and safety factors. In addition some different possibilities to work with these quantities have been shown. An application of such ideas in the industry can be found in Dressler *et al.* [84] or Speckert *et al.* [219]. In the following list, the underlying ideas are summarized and some additional remarks are given.

1. **The common scale** of the load and strength variables we are using is either a pseudo damage number or an equivalent amplitude. This was introduced in Chapter 3 and explained and used in more detail in Chapter 7 as well as in the introduction of this

chapter. We often work with the logarithm of these quantities, especially when talking about distribution assumptions or variances. In that case, the relation between design loads and strength requirements can be expressed as a distance. However, it is common to speak of safety factors (as we did in previous subsections), in which case this describes a relation between the original load and strength variables.

2. **The design load** is a quantity with a well-defined relation to the customer load distribution. Due to the uncertainties within the distribution assessment, we work either with the mean or median value or with a moderate quantile L_{α_L}, preferably not higher than the 95%-quantile.

3. **The strength requirement** is a quantity with a well-defined relation to the strength distribution. Due to the uncertainties within the distribution assessment, we work either with the mean or median value or with a moderate quantile S_{α_S}, preferably not lower than the 5%-quantile.

4. **The measures of uncertainty** τ_L, τ_S used to characterize the variance of the load and strength distributions as well as the model uncertainty τ_β have been introduced and discussed extensively in Chapter 7.

5. **The safety factors are quotients** $S_{\alpha_S}/L_{\alpha_L}$. When working with the mean or median values we speak of central safety factors, otherwise of partial safety factors. Sometimes we also speak of a safety factor for the mean load or a safety factor for the design load.

6. **Uncertainty of safety factors:** Since the safety factors are derived from the corresponding load and strength variables which in turn are estimated from a more or less limited database, they inherit the uncertainty we have in the design loads and the strength requirements. In addition, the model uncertainty, for example, the uncertainty in the Wöhler slope β, is also present in the safety factors.

7. **Unification in theory:** In the sub-sections above, we introduced several types of safety factors and formulas to relate them to the design loads and the strength requirements. This may appear somewhat confusing. However, in a specific company or department it is very much advisable to unify the approach taken for the derivation of design loads and strength requirements. Since there are no statistical arguments in favour of partial or central safety factors, it should be possible to settle on one of the methods.

8. **Unification in practice:** Despite the theoretical equivalence, there are some practical arguments for preferring, for instance, the 90% load quantile as a design load instead of the median load. First, the term "design load" in itself suggests a high load rather than a mean load. A stronger argument is that the representation of a high load quantile in terms of time signals or rainflow matrices usually contains higher amplitudes, more extreme events, etc. This is often appreciated with regard to accelerated testing.

9. **Transformations between the different safety factors** as described above rely on distribution assumptions (in our case the log-normal model). This is reasonable for the strength side, but more questionable on the load side. However, this should not be a major problem in practice as the following arguments show:

   (a) If the knowledge about the customer load distribution allows for an estimation of the mean, say, $L_{50\%}$, as well as a certain quantile, for example $L_{90\%}$, then for a safety factor ϕ referring to the mean load and for another safety factor φ referring to the 90% load quantile we have the relation $\phi = L_{90\%}/L_{50\%} \cdot \varphi$. Thus, we do not need a certain assumption about the distribution of the customer load in that case.

(b) If the customer population can be modelled as a combination of different sub-populations as described in Section 8.5, we may argue that a log-normal model is reasonable for each sub-population, leading to a mixed log-normal model for the entire population. In that case we can also calculate quantiles of the entire population using the means and variances of the sub-populations. The transformation between central and partial safety factors becomes more involved but it is still possible.

(c) In a specific company or in a department, it is nevertheless advisable to unify the approach taken for the derivation of design loads and strength requirements. In that case, the need for the transformation is reduced or no longer present.

10. **The reliability targets** depend on the type of approach we are taking (pure empirical, full probabilistic, reliability index, or mixed approach). In the case of the mixed approach, the reliability target is, according to Equation (9.1) and Equation (9.2), given as an empirically defined lower bound for the value S_d or equivalently for $\Phi_{target} = \exp(S_d)$.

9.3 Synthetic Load Models

It is quite easy to construct a time signal meeting a specified pseudo damage number D_{Target}. Just take *any* non-constant signal, calculate its pseudo damage D_S (given a corresponding model, e.g. the Basquin equation $\frac{1}{D} = N = \alpha \cdot S^{-\beta}$ (see Equation (3.23)) and the Palmgren-Miner rule) and repeat it $n = \frac{D_{Target}}{D_S}$ times (assuming $D_{Target} > D_S$). What 'taking *any* signal' means needs to be made more specific. A few possibilities are listed below:

1. Deterministic constant amplitude signals such as sine waves, rectangular signals, and ramp signals.
2. Signals composed of several blocks of different constant amplitude signals.
3. Sine sweep signals with constant amplitude but varying frequency.
4. Irregular signals defined by a certain stochastic process model (see Chapter 6).
5. A superposition of several such models (for instance, sine waves + noise to model unevenness in curves).
6. A certain measurement on a public road or a test track.

There are obviously many options to choose the 'structure' or 'process class' for such signals. In practice, this choice is often done based on technical convenience with regard to test capabilities or based on the assumption that loads (e.g. road profiles) have certain characteristics (e.g. stationary Gauss process). The model error introduced by representing the load by its pseudo damage number will be reduced by choosing a load type close to the service situation.

Instead of simply repeating the signal as often as required to get the target damage, we can also scale it by a factor γ_A. Using the simple damage accumulation rule without endurance limit, the pseudo damage of the signal has to be multiplied by the factor γ_A^β. Thus, for the required repetition number we have $n \to n/\gamma_A^\beta$. Of course, the failure mode must not be changed by that type of acceleration, that is, γ_A must not be too large.

When using random loads, we exploit the fact that we usually get an 'irregular' signal with no specific sequence effects. In addition, we could use the underlying stochastic process to create an arbitrarily long signal. In that case we automatically have an extrapolation of the signal to the target duration including, for example, the extrapolation of maximum peaks, maximum hysteresis cycles, etc. (see also Section 4.4). However, when using the signal on a test rig, we often have to restrict the length of the signal to a few minutes since we have to iteratively determine the drive signal from the test load.

The freedom in selecting a signal can and should be used to fulfil additional requirements not captured in the pure pseudo damage number. Some of these requirements are as follows:

1. Choose the **amplitude** in a suitable range, such that the failure type provoked by that load is the one you expect in customer usage (see the corresponding remarks in Chapter 10). While respecting this condition and possibly the limitations of the test rig, choose the amplitude as large as possible to get the shortest possible test.
2. Having fixed the amplitude of the signal, choose the **frequency** of the signal as high as possible. Thereby restrictions are again limitations of the test rig and, even more important, considerations from the component or system being tested. You need to check carefully which eigenfrequencies are excited by the signal and which are excited during customer loading. As it is difficult to compare the damage potential of two signals with different frequency spectra acting on a vibrating component, the response of the component at interesting spots rather than the input load should be analysed in such cases.
3. Check whether **mean load effects** play a role. If they do, estimate the mean load properties of the customer loading and try to stay close to the mean loads. This is achieved if, for example, the rainflow matrices of the customer loading and the test load are close to each other.
4. Check whether **sequence effects** play a role. If they do, do not use constant amplitude loading but try to stay close to the customer loading properties (if known). Sequence effects are not captured by rainflow matrices. See Haibach [110] for details about sequence effects and corresponding damage models.
5. **Multiple input loads**: So far we have treated single signals only. However, in many applications it will be necessary to deal with multiple input load signals. If this is the case, we have to treat the phase relation between the signals in the sense of Section 3.3. The first problem to solve is to estimate the target pseudo damage numbers for additional load combinations. In principle, one can define a certain set of combinations (either a representative set of combinations or a set tailored to a component of interest) and, during the entire design load derivation process, treat these in the same way as the measured channels. Proceeding like this, we obtain target values for the pseudo damage of these combinations too. Then we can make sure that the synthetic signals and their corresponding combinations all match the targets. The general approach and an example have been described in Section 3.3.

Example 9.3 (A Single Sinusoidal Signal)

As a simple example for the generation of a single sinusoidal load, consider a sine wave $L_0(t) = A_0 \cdot \sin(2\pi f_0 t)$ of length $T_0 = 60\,\mathrm{s}$, amplitude $A_0 = 1\,\mathrm{kN}$, and frequency $f_0 = 5\,\mathrm{Hz}$. Consider further a Wöhler slope of $\beta = 5$ in the damage model. Then the pseudo damage $D_0 = T_0 \cdot f_0 \cdot A_0^\beta = 60 \cdot 5 \cdot 1^5 = 300$. Assume further that the target value for the pseudo damage is $D_{Target} = 6 \cdot 10^6$. Then the number n_0 of repetitions of $L_0(t)$ needed to reach the target value is $n_0 = \frac{D_{Target}}{D_0} = \frac{6 \cdot 10^6}{300} = 20000$. The time needed to test the component is thus $T_{test} = n_0 \cdot T_0 = 20000 \cdot 60\,s \approx 333\,\mathrm{h}$. If the amplitude is increased by a factor of

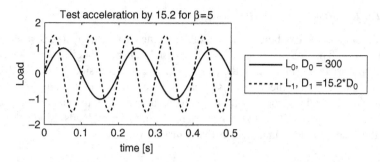

Figure 9.3 Two load sine waves. The amplitude and the frequency of the second one are increased to accelerate the test. The measure of acceleration is given by the ratio of the pseudo damage numbers with a slope of $\beta = 5$

$\gamma_A = 1.5$ (which might be too large if A_0 is a typical amplitude in customer loading) and the frequency is increased by a factor of $\gamma_f = 2$, then we get $D = T_0 \cdot \gamma_f f_0 \cdot (\gamma_A A_0)^\beta = 60 \cdot 2 \cdot 5 \cdot 1.5^5 \approx 4556$ and $n = \frac{D_{Target}}{D} \approx \frac{6 \cdot 10^6}{4556} \approx 1317$. Thus, the testing time now becomes $T_{test} = n \cdot T_0 \approx 1317 \cdot 60\,\text{s} \approx 23\,\text{h}$, which corresponds to an acceleration by a factor 15 (see Figure 9.3). $\qquad\qquad\qquad\qquad\qquad\qquad\qquad\qquad\qquad\qquad\qquad\qquad\qquad\qquad\qquad\quad\square$

9.4 Random Load Descriptions

There is a variety of situations where random loads can be useful in design and verification.

9.4.1 Models for External Load Environment

The external input loads depend on many factors such as road quality, driver behaviour, and traffic intensity. The loads coming from the load environment vary in a random manner during operation and thus stochastic models are useful.

Road roughness – Vertical input. The main source of the vertical load excitation is the uneven road profile the vehicle is travelling over. The task is therefore to model the road roughness. However, also the speed of the vehicle is an important parameter. One such model, "road with pot holes", is presented in Bogsjö [30], and is described in Chapter 6, see Example 6.6 on page 215. Here the road profile is modelled as a stationary Gaussian process with added irregularities at random locations.

Curviness – Transversal input. The main lateral inputs are induced by driving through curves. The intensity of curves, the curvature of the curves, and the speed are important parameters for describing the lateral load excitation. A statistical model for this purpose is presented in Karlsson [132], see also Section 8.6.

Braking – Longitudinal input. For longitudinal loads, braking (and acceleration) events are obvious sources. There are many factors influencing the number of brakings per kilometre and their power, for example, the traffic intensity, mission type, road type, and driver behaviour.

9.4.2 Load Descriptions in Design

Simplified load models can be useful, especially in the early design stages when no physical prototypes exist. The frequency content of a signal can be modelled by a Gaussian load, which is appropriate for modelling the input load to a system, since Gaussian loads exhibit several good properties. If the system is linear and the input is a Gaussian process, then also the output is a Gaussian process but with changed PSD. Another advantage of Gaussian loads is that they are well suited for modelling multi-input loads, since the correlation between the load signals is included in the Gaussian model.

9.4.3 Load Description for Testing

The generation of load signals for testing can be based on random load models. If the frequency content of the input load is important we can use

Gaussian loads that are defined through the PSD. The Gaussian loads are also well suited for modelling multi-input loads, as discussed above.

If only the cycle content is relevant, we can concentrate on modelling the turning points:

Markov loads are defined through the Markov transition matrix. This is a flexible class of random loads that can model the sequence of turning points, see Section 6.8. It is easy to simulate Markov loads, also in the case of on-line simulation.
Level crossing intensity together with additional assumptions like the irregularity factor can define a random load. One such simulation approach is presented in Svensson [226], see also Section 4.5.2 on page 163.
Rainflow matrix or range-pair distribution together with additional assumptions can define a random load. This is discussed in Section 4.5.1.

Since the frequency content is not modelled, the methods are mainly applicable for component loads. Moreover, as has been discussed before, turning point models are not well suited for the multi-input case. Methods for simulating load signals from descriptions of random loads are found in Section 4.5, see also Section 9.5.

9.5 Applying Reconstruction Methods

9.5.1 Rainflow Reconstruction

Sometimes the design target is given by a load histogram instead of a pseudo damage value, for instance, by a rainflow matrix which has been derived using the extrapolation methods described in Section 4.4. Then the derivation of a time signal can be done by means of reconstruction methods described in Section 4.5. As already mentioned there, the reconstructed time signal will match the rainflow matrix perfectly, but there is no direct control over other properties, for example, the spectral properties. Thus, a further step is usually needed.

Here, the test pre-processing procedure (see Section 4.3) can be used. After assigning a sampling rate to the sequence of points from the reconstruction, the method checks the slopes and the frequency of the signal and modifies the signal such that bounds on the maximum slope and the maximum frequency are respected. This is done in such a way that the resulting signal has minimum length and thus minimum duration.

None of the operations, namely the reconstruction of the sequence of turning points from the rainflow matrix and the subsequent test pre-processing, changes the pseudo damage. Thus, we have generated a time signal with the required pseudo damage number, the required shape of the rainflow matrix, and in addition, we have considered some restrictions from the test rig.

Example 9.4 (A 1-DOF Test For a Control Arm)

A rig test for a control arm has to be designed. Due to the geometry of the suspension, the loading of the component is dominated by a force acting in a fixed direction. Thus, it is decided to use a single actuator rig. From a measurement campaign there are strain sensor signals of the control arm under various conditions (different roads, manoeuvres, payload...) and a rainflow matrix has been derived which is expected to cover the 90%-quantile of 10.000 km customer loading. Since the target is 1.000.000 km, the matrix is extrapolated by a factor 100 using the methods described in Section 4.4. In order to speed up the test, small cycles are omitted without affecting the pseudo damage. In Figure 9.4 the matrix before and after the extrapolation and omission is shown. The extension of the largest amplitudes by the extrapolation procedure can be seen clearly.

A time signal has been reconstructed from that matrix, which is shown in Figure 9.5. The signal is very homogeneous. Thus, a representative segment of 1% length can easily be chosen without changing the shape of the load spectrum and which still covers the largest cycles. This is illustrated in Figure 9.6.

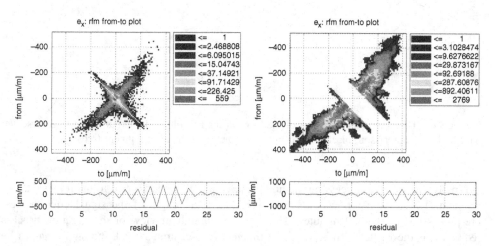

Figure 9.4 The rainflow matrix before (left) and after extrapolation to the target mileage and omission of small cycles

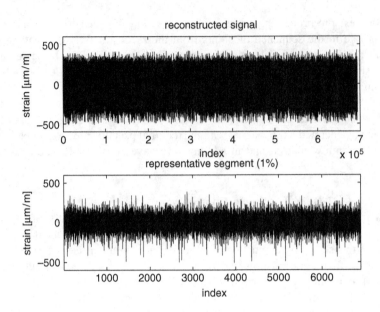

Figure 9.5 The reconstructed signal (top) and a representative segment (bottom)

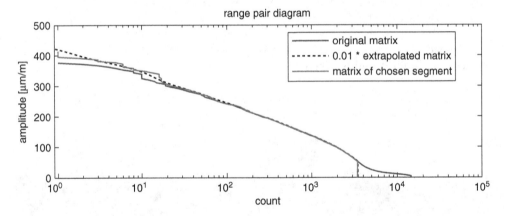

Figure 9.6 The range-pair diagrams of the original matrix (blue), the extrapolated matrix (scaled with 0.01 for comparison, black dashed line), and the representative segment (red)

Finally, the test pre-processing method (see Section 4.3.1) is applied to the representative segment. The result is shown in Figure 9.7. The total length of the signal after the test pre-processing is about 9 minutes. This is short enough in view of the iteration process required for the rig test. To reach the estimated 90%-customer load, the signal has to be repeated 100 times on the rig. However, in order to gain statistical significance and to fulfil

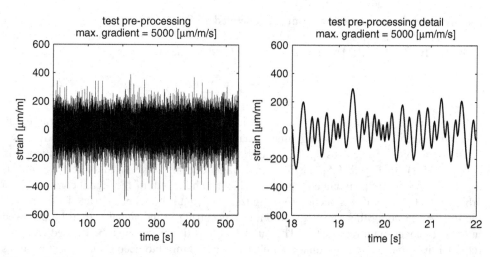

Figure 9.7   The result after application of the test pre-processing method (the entire signal to the left and a detail to the right)

the given reliability target, a safety factor has to be applied on top of that (see Section 9.2 above).   □

9.5.2   1D and Markov Reconstruction

If we start the reconstruction from the range-pair count, we again meet the pseudo damage target exactly since the simple pseudo damage calculation described above does not take mean load effects into account.

But if we start from the level crossing count or the Markov count, we cannot explicitly control the pseudo damage number (see also the corresponding remarks in Section 4.5.1). Instead, we have to compute this number after the reconstruction and, if we want to meet the target for the pseudo damage, calculate a required repetition factor for the signal to meet the target. Thus we have to make a compromise between fitting the histogram we started with and fitting a target pseudo damage number.

9.5.3   Spectral Reconstruction

We can also start with a spectral representation (e.g. a PSD) and construct a corresponding signal from that. In Section 4.5.2 we saw some of these algorithms. Probably the most important case is the reconstruction from a PSD. Applications can be found, for example, in the area of testing electronic devices or parts attached to the engine. Since the PSD is a density, the length of the time signal is not explicitly given. This is reflected if the signal is constructed from an ARMA model, which in turn has been adapted to the PSD. In that case we have to decide about the length of the signal and the repetition factor using, for example, a target value for the pseudo damage if available.

Example 9.5 (Reconstruction from a Predefined PSD)
A component is known to be excited by accelerations with resonance frequencies at about
5 Hz and 16 Hz. An ARMA process (filter) defined by the equations

$$x_k + \sum_{j=1}^{p} a_j x_{k-j} = \sum_{j=0}^{q} b_j \epsilon_{k-j}$$

is fitted to these properties. Once the coefficients a_j, b_j are found, the equation can easily
be used to create a corresponding time signal given a Gaussian white noise ϵ_k. Here we
fitted an 8th-order process using the Yule-Walker method (see the literature mentioned in
Section 4.5.2). In Figure 9.8, the resulting magnitude response of the filter, as well as
the first 5 seconds of a simulated signal are shown. The length of the signal can be chosen
arbitrarily. However, if it is needed for a rig test, the length should be restricted for practical
reasons. When doing this, the large amplitudes occurring in the signal should be observed
in order to cover them reasonably. The number of repetitions can then be defined from the
required duration or pseudo damage number and by taking into account the considerations
regarding safety factors in Section 9.2 above.   □

9.5.4 Multi-input Loads

Up to now we have not considered multi-input loads. As already explained in Section 4.5.1,
there is no combinatorial rainflow reconstruction method respecting the RP rainflow count

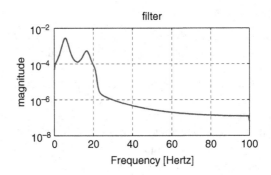

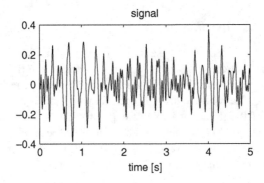

Figure 9.8 The magnitude response of the filter (left) and the beginning of a signal constructed
from the corresponding ARMA process

as in the 1D case. Nevertheless, the existing RP reconstruction approach can be applied and the deviation from the desired RP count can be checked in order to decide if the result is reasonable. Then we can proceed as described before using the test pre-processing method, which takes multi-input loads into account.

Multi-input loads can also be generated from stochastic models like Gaussian vector processes. The Gaussian model is well suited for that purpose. The complete correlation structure of a Gaussian vector process is described by its power spectral density matrix (PSD matrix). Methods for generation are given in, for example, Smallwood [215], Deodatis [70], or Pitoiset *et al.* [187].

9.6 Standardized Load Spectra

A completely different approach to the derivation of design load targets is to use standardized load sequences. The idea is that for certain applications or components like a car trailer coupling, it should be possible to define a load sequence typical of the overall usage.

In Heuler and Klätschke [113], an overview of such standardized spectra is given. There are many examples within aerospace applications, but also other industries like off-shore platforms or wind turbines are covered. For automotive applications, the CARLOS series (car loading standard, again see Heuler and Klätschke [113]) has been developed including sequences for car front suspension parts, car power train, and car trailer couplings. Some are uniaxial, some are multiaxial sequences.

Loading measurements under various operational conditions have been conducted and a statistical analysis has been performed including extrapolation techniques such as the ones explained in Chapter 4. This analysis is combined with assumptions about the customer usage profile in order to obtain a standardized load signal trying to cover the typical customer usage.

Example 9.6 (Trailer Couplings with Bike Carriers)
As an example we briefly mention the derivation of a standard for trailer couplings with bike carriers, which is described in Weiland [247]. By means of an extensive poll (≈ 1200 users), details about the use of the component have been derived and measurements with 3 different cars and trailers have been conducted accordingly. In order to get comparable load data for the different configurations, the forces at the coupler (calculated from measurement strains) had to be normalized. The corresponding factors have been derived from a careful analysis of the data ($f_T = \frac{m_V \cdot m_T}{m_V + m_T}$ for trailer usage and $f_B = m_B \cdot d$ for bike carrier usage, where m_V, m_T, m_B denote the masses of the vehicle, the trailer, and the bikes, respectively, and d denotes a lever arm). From the normalized data and the results from the poll, a neutral load standard could be derived, which can be re-adapted to a specific configuration and used for corresponding tests or numerical verification. In the thesis cited, the results have been exemplarily used for the optimization of the geometry of the coupler.   □

The strength of such an approach lies in the comparability and the fact that the sequences are available at every stage of the development process. Especially in the early stages, they might be used to assess various design alternatives.

While this approach tries to cover the shape of loading spectra, it still needs to be calibrated with respect to a certain desired customer usage quantile when used for verification purposes. As only a few standardized sequences for special components exist, the applicability is limited. Details of the approach and application examples can be found in Grubisic [106], Bruder *et al.* [44], Heuler and Klätschke [113], Sonsino [218] and Weiland [247].

9.7   Proving Ground Loads

The design loads are often defined through a test track. Existing driving schedules for the proving ground may be based on empirical knowledge about the relation between the loads on the test track and the expected service loads. A typical rule might read as follows:

> *A unit of the proving ground load is defined as a certain set of lanes and manoeuvres with a total length of 5 km. It has to be repeated 1000 times. The corresponding load is supposed to cover 1,000,000 km of Western Europe customer usage.*

The rule might depend on the target market as well as on the type of truck. Here are two ways of defining a schedule:

1. By long-term experience. In that case it is only known from previous results that the proving ground load has been demanding enough to cover the customer usage in the past. The relation between the customer load and the design load is not given explicitly. However, in this case it can be back-calculated from the customer load distribution and measured test track loads.
2. A proving ground schedule can also be derived from the description of the design load in terms of pseudo damage values or rainflow matrices for all channels under consideration. The target (i.e. the design load) needs to be derived from load data in customer service as explained in Section 9.2. Often, the target load represents a certain quantile of the customer loading. Since the strength requirement essentially is the design load times the safety factor, we can equally well use the strength requirement as the target for the optimization.

In the latter case, the mapping of the proving ground to the load targets (design loads or strength requirements) can be done using optimization techniques, which take into account all the interesting spots (channels or sensors). The mathematical formulation and the details are given below in Section 9.8.

9.8   Optimized Combination of Test Track Events

In this section we assume that a target load (for example, a certain customer quantile, i.e. the design load) has been derived. Depending on the derivation process, the target load is often characterized by rainflow matrices, range-pair histograms, PSD functions, pseudo damage numbers, or another kind of representation. These targets exist for a typically large number of different spots of the truck (channels), for instance, wheel forces, cabin accelerations, spring displacements, or strains. Let m denote the number of channels and $b_i, i = 1, \ldots m$ denote the representation of channel i (i.e. a rainflow matrix, a PSD function, a pseudo damage number ...).

If the truck drives on the test track, the same channels are measured and the corresponding data formats are derived from the time signals. This is done for all lanes on the proving ground at different speeds or with different payloads leading to a certain number of so-called basic tracks or segments. We arrange the resulting data in a matrix $A_{i,j}, i = 1, \ldots, m, j = 1, \ldots, n$, where n is the number of basic tracks and each cell of the matrix contains data for one channel and one basic track.

The task now is to find a combination $w_j, j = 1, \ldots, n$ of the basic tracks which comes as close as possible to the target loads in the sense of the representations under consideration. Here $w_j \in \{0, 1, 2, \ldots\}$ is the number of repetitions of basic track j. In order to put this task into a mathematical setting, it is necessary to calculate the data representation $m_i(w)$ of the combination given the individual data $A_{i,j}$ and a vector w of repetition factors w_j. This is written for each channel in the form

$$m_i(w) = \sum_{j=1}^{n} A_{i,j} \cdot w_j, i = 1, \ldots m. \tag{9.4}$$

For most of the counting results like rainflow, range-pair, or pseudo damage numbers, the operation in Equation (9.4) is called superposition and is essentially a summation (see Section 4.4 for details). For PSD functions, it is an averaging procedure.

The next basic step is to define a difference measure between the targets b_i and the mixed track m_i. This, of course, depends on the data representation. A distance measure for rainflow matrices or a measure for PSD functions will surely differ and needs to be defined carefully such that the important differences are covered and less important differences are neglected. We will come back to this point later.

9.8.1 Optimizing with Respect to Damage per Channel

9.8.1.1 Distance Measures

For an explanation of the basic optimization process we assume that the b_i, $A_{i,j}$ and consequently also m_i are simple pseudo damage numbers. Then, the superposition process in Equation (9.4) is an ordinary addition and the most obvious distance measure is given by the magnitude of the residuals

$$r_i = b_i - \sum_{j=1}^{n} A_{i,j} \cdot w_j = b_i - m_i, i = 1, \ldots m. \tag{9.5}$$

The pseudo damage numbers depend on the parameter β of the underlying Wöhler or SN curve (see Section 3.1.12). The strength parameter α in the SN curve is of no importance, since damage is used for comparison of different tracks only, and we can get rid of it by simply dividing the target and basic track data by the target pseudo damages for each channel. An additional advantage of that is that the resulting value m_i for channel i of the mixed track can be interpreted as a factor of how much more or less damage we have in the mixed track compared to the target. If $m_i = 2$, we have twice the damage, if $m_i = 0.5$ we have half the damage. The most important advantage, however, is that the residuals r_i become comparable for different channels. In the following we assume that this normalization has been performed.

Since pseudo damage numbers are typically interpreted on a logarithmic scale, we introduce an additional distance measure in the form

$$f_i = \frac{b_i}{m_i} + \frac{m_i}{b_i}, i = 1, \ldots m.$$ (9.6)

This measure has the following properties:

1. Since the ratio of the damage numbers to be compared is used, it is based on the factors between target and current value and can be interpreted very conveniently.
2. Since the definition is based on the sum of both ratios $\frac{b_i}{m_i}$ and $\frac{m_i}{b_i}$, it is symmetric with respect to over- and underestimation. If $m_i = 2 \cdot b_i$, then $f_i = 2.5$, if $m_i = 0.5 \cdot b_i$, then again $f_i = 2.5$. The smallest possible value is 2, which is obtained if the mixed track damage equals the target damage.

9.8.1.2 Weighting Channels and Norms

Having defined distance measures for the channels we need to deal with the large number of channels which are to be controlled. In general, we cannot expect to find a mixed track such that the residuals are zero for all channels. This is due to the large number of channels and the restriction $w_j \in \{0, 1, 2, \ldots\}$. Thus, the best possible mixed track is a compromise between all channels. Mathematically, this is expressed using so-called norms on the vector of residuals. Three of the most important ones are

$$\|r\|_1 = \sum_{i=1}^{m} \gamma_i |r_i|,$$

$$\|r\|_2 = \sqrt{\sum_{i=1}^{m} \gamma_i^2 \cdot r_i^2},$$ (9.7)

$$\|r\|_\infty = \text{Max}\{\gamma_i |r_i|, i = 1, \ldots m\},$$

where γ_i denotes a channel-dependent weighting coefficient, which can be used to reflect the importance of different channels (priority).

The first two of these norms are calculating an average residual which might be small if most of the channels have small residuals and only a few have large residuals. Its final value depends on the residuals of all channels.

The second of these norms is used in the well-known least squares procedures. Again, it calculates an average value and thus depends on all channels. It is more sensitive to outliers than the first norm.

In contrast to that, the third norm depends only on the largest residual. Thus, it models a worst-case interpretation of the compromise to be found. Using this norm means that the maximum residual is minimized. Once the priority of the channels is defined by the engineer using the coefficients γ_i, it is rather natural to proceed with this norm, since it fits best into the framework of multi-objective optimization. Some more remarks on that will be given below in the course of the example.

In Formula (9.7) we can use either the linear residuals r_i as indicated or the logarithmic residuals f_i defined in Equation (9.6). As mentioned above, the latter residuals are somehow closer to the durability application due to its symmetry with respect to over- and underestimation and its interpretation as a factor to the target.

9.8.1.3 Constraints

Instead of trying to minimize the deviations between target and mixed track for all channels, one can also define bounds on the residuals for some channels and minimize the others. If, for example, it is sufficient that the factor between target and mixed track for some channels $i \in C$ is not greater than 3, this can be expressed by the inequalities $m_i \leq 3b_i$, $3m_i \geq b_i$ or equivalently $f_i \leq 3 + \frac{1}{3}$, $i \in C$. Since m_i is a function of the repetition factors w_j, such constraints can be written in the form $g_i(w) \leq 0$, where g_i is an appropriate function of the repetition factors w_j. The channels corresponding to these constraints are eliminated from the vector of all channels. The resulting set of channels to be minimized is denoted by $M = \{1, \ldots, m\} \backslash C$.

Additional constraints not replacing the residuals in the vector to be minimized can be considered too. Examples are simple bounds on the repetition factors like $l_j \leq w_j \leq u_j$, bounds on the total number of kilometres of the mixed track like $\sum_{j=1}^{n} l_j \cdot w_j \leq L_{\max}$, where l_j is the length (or duration) of track j, or relations on the factors for different tracks like $w_{j_1} = w_{j_2}$. All of these constraints can be written in the form $h_k(w) \leq 0$ for appropriate functions h_k, $k \in K$, where K denotes the set of additional constraints.

9.8.1.4 The General Optimization Problem

Summarizing the considerations above we can formulate the task of optimal track mixing in the form

$$\text{Minimize} \| e_M(w) \|, \text{ subject to}$$

$$g_i(w) \leq 0, i \in C$$

$$h_k(w) \leq 0, k \in K$$

$$w_j \in \{0, 1, 2, \ldots\}, j = 1, \ldots, n, \tag{9.8}$$

where $e_M(w)$ is either $r_M(w)$ or $f_M(w)$, M is the set of channels to be minimized, C is the set of channels with bounds on the residuals, $\| \cdot \|$ is any norm, $r_M(w)$ and $f_M(w)$ denote the vectors with components $r_i(w), i \in M$ and $f_i(w), i \in M$, respectively.

Depending on the type of functions involved in Equation (9.8), this is a constrained linear or nonlinear integer optimization problem which can be solved by means of various dedicated optimization algorithms.

Since the target usually represents the load for a whole vehicle life and the basic tracks are short segments, a single repetition of a basic track does not contribute very much to the mixed track damage. Therefore, the repetition factors w_j are often large and the requirement that the factors w_j are integer numbers can be dropped. Solving the optimization problem for real values $w_j > 0$ is considerably simpler. In this case, the non-integer values w_j need to be rounded after the optimization. Although the solution calculated like that is not optimal in a strict sense, it is often good enough to justify the simplification.

If the number m of channels to be optimized is small compared to the number n of basic tracks, then one can usually come very close to the target values. However, in a typical application, m is between 100 and 200 and n is between 50 to 100. Thus there is typically no perfect solution to the problem and a compromise between the channels needs to be found. Mathematically, we speak of a multi-criteria or multi-objective optimization problem. We will not go into details of ordinary or multi-criteria optimization. Instead we refer to Gill *et al.* [102] for general optimization algorithms and Miettinen [160] or Steuer [224] for multi-criteria optimization.

9.8.2 An Instructive Example

9.8.2.1 The Data

We use an example from a construction machine to illustrate the general considerations above. There are $m = 28$ channels and $n = 16$ basic tracks. The channels contain forces, displacements, strains, and accelerations. The target has been derived from an extended measurement campaign with different customers and working conditions. The basic tracks are taken from a few short typical applications of the construction machine. In Figure 9.9 the basic data (normalized as described above) is shown.

9.8.2.2 Overall Optimization with Respect to Different Residuals

To begin with the analysis of this problem, we minimize the residuals $e_i(w) = |r_i(w)|$ or $e_i(w) = f_i(w)$ using the definitions given in Equation (9.5) and in Equation (9.6). We use

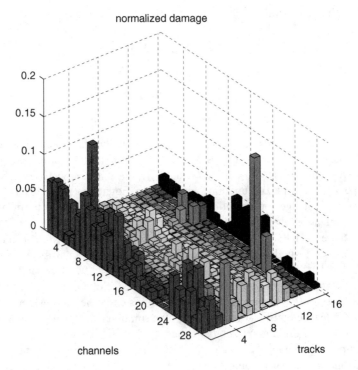

Figure 9.9 The matrix of normalized pseudo damage of the basic tracks. The corresponding target value is one for all channels

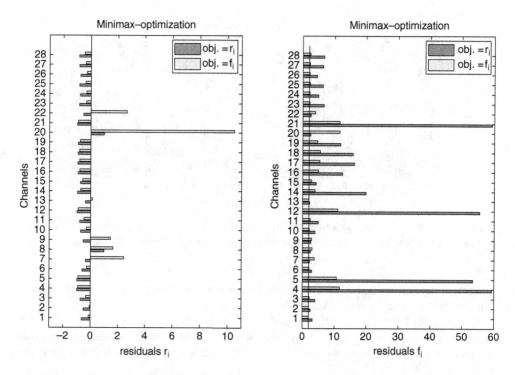

Figure 9.10 Plots of the residuals of all channels when optimizing either $\|r(w)\|_\infty$ or $\|f(w)\|_\infty$. Minimax optimization means that the maximum residual is minimized. In the left figure, the residuals r_i are shown. In the right figure, the residuals f_i are shown

the maximum norm $\|e(w)\|_\infty = Max\{\gamma_i e_i(w), i = 1, \ldots m\}$ as defined in Equation (9.7), where we set all priority coefficients γ_i to one. No additional constraints are considered.

In Figure 9.10 the basic results are shown. As can be seen in the left plot, the minimization of $Max|r_i|, i = 1, \ldots, m$ leads to residuals between -0.98 and 0.98. Since the target value is one, the negative residual -0.98 corresponds to a value of $1 + 0.98 \approx 2$ for the mixed track, that is, an overestimation by a factor 2. However, the positive residual 0.98 corresponds to a value of $1 - 0.98 = 0.02$ for the mixed track, that is, an underestimation by a factor 50. In contrast to that, the minimization of $Max f_i, i = 1, \ldots, m$ leads to a maximum value of 10, which corresponds to a factor of approximately 10 between mixed track and target. Thus, the optimization of the residuals f_i gives a much better result in terms of factors between target and mixed track.

9.8.2.3 Analysis of Channels and Groups

The largest residual of the results above is so large that we want to go deeper into the analysis. This is a typical situation in multi-criteria optimization. In order to get more insight, it is helpful to identify the *active* channels. These are the channels with the largest residuals. Another possibility is to check the improvement one obtains for the maximum residual, if one channel is eliminated from the set of channels to be optimized (leave one out).

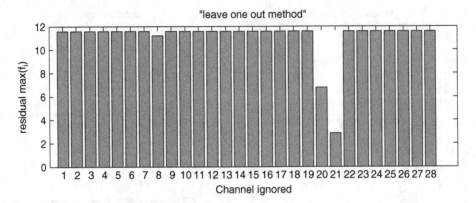

Figure 9.11 Result of the leave-one-out method. The label on the x-axis denotes the deleted channel. The bars show the total residual after elimination of the channel indicated on the x-axis

A strategy to explore this is to loop over all channels and optimize the maximum residual without that specific channel. The result with the smallest total residual is stored.

In Figure 9.11 the result of that method using the residuals f_i is shown. It can easily be seen that channel 21 is the critical one, followed by channel 20. If channel 21 is ignored, then the total residual becomes $\max(f_i(w), i \neq 21) \approx 2.8$, which corresponds to a maximum factor of ~ 2.4.

After elimination of the most critical channel, the procedure can be repeated with the reduced set and the next critical channel is searched. This can be iterated until only one channel is left or the remaining residual is small enough. In Figure 9.12 the result of this iteration is shown. As can be seen in Figures 9.11 and 9.12, eliminating the first critical channel results in a major improvement. The next steps lead to small improvements only. The iterative deletion process was stopped as soon as the remaining residual was below 2.05, which corresponds to a factor of 1.25 between target and mixed track. At that point,

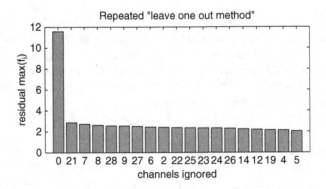

Figure 9.12 Result of the repeated leave-one-out method. The label on the x-axis denotes the deleted channels. The bars show the total residual after elimination of all channels starting from the left up to that point

there are still 10 channels left, meaning that the mixed track meets the target values for these channels with an error smaller than 25%.

The question now is what to do with channel 21. If we optimize for this channel only, we can of course get a 'perfect' solution, but we must not ignore all the remaining channels.

In this situation, it is often helpful to find groups of channels which behave similar to each other during the optimization. This means that if a modification of the track combination increases or decreases the damage of a channel, we also observe an increase or decrease of the damage in the other channels of that group. This behaviour is described by the correlation between the channels, which is given by the formula

$$\text{corr}(i, k) = \frac{\sum_{j=1}^{n} (A_{i,j} - \bar{A}_{i,.}) \cdot (A_{k,j} - \bar{A}_{k,.})}{\sqrt{\sum_{j=1}^{n} (A_{i,j} - \bar{A}_{i,.})^2 \cdot \sum_{j=1}^{n} (A_{k,j} - \bar{A}_{k,.})^2}}, \tag{9.9}$$

where $\bar{A}_{i,.} = \frac{1}{n} \sum_{j=1}^{n} A_{i,j}$.

In Figure 9.13 the correlation matrix of the channels is shown. There are high and low correlation values and after re-ordering the sequence of channels (in the right plot), groups of channels with high correlation can be identified quite easily. Group 1 contains the 13 channels belonging to the upper left block in the right correlation matrix, group 2 the next 6 channels, group 3 the following 3 channels, group 4 the next 4 channels and group 5 the remaining 2 channels.

Usually, the construction of groups is not a straightforward task. It is important that the channels in one group do not affect each other in the optimization. Using the correlation

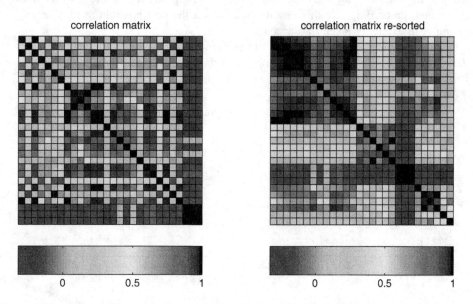

Figure 9.13 Plots of the correlation matrix for all channels. On the left, the correlation between the channels is shown in the given order. On the right, the same correlation matrix after re-ordering the sequence of channels is shown

matrix for building groups is just one possible way to go. In some cases, it might be advisable to take the physical properties of the channels into account (forces, strains, . . .) or to consider the subsystem or component the channels belong to (cabin, frame, . . .).

9.8.2.4 Optimization Based on Groups

Channel 21 belongs to the third group of three channels, which shows a very high correlation within the group and nearly no correlation to the others. These channels are accelerations.

If we now try to optimize the accelerations only, we find that even then, there is no good solution. The factor between mixed track and target is about 6 in that case. However, the remaining channels have unacceptable values. The set of basic tracks is not rich enough to fit the specific relation between the target damage values of the accelerations. Some specific tests or additional basic tracks have to be found to compensate for that.

We conclude that it is better to accept that gap for now and go on with the optimization of the remaining channels only. This leads to a maximum residual $\max(f_i) \approx 2.7$ which corresponds to a maximum factor between mixed track and target of 2.3.

In the present example the latter result might already be acceptable. But we may also want to have better control over the residuals of the individual groups or channels. One way to handle this is to use the priority factors γ_i introduced in Equation (9.7). Doing this is a trial and error approach in the sense that it is not exactly predictable how an increase of the priority for a certain channel will influence the result.

Another approach is to begin the optimization with the group of the highest priority, optimize for these channels only and see how close we can come to the target. Then we set bounds on the residuals of that group and optimize the next group. If the result for the next group is too bad, we can relax the bounds of the first group to achieve a better result for the second. Proceeding like this ensures that we get the best possible result, the priorities having been explicitly taken into account.

In our example we ignore group 3 and perform a group-wise optimization as sketched above in the order 5, 1, 4, 2. Thus, group 5 gets the highest priority and group 2 the lowest. The result is shown in Figure 9.14.

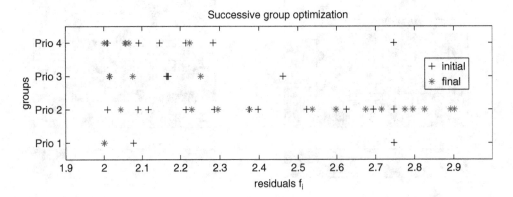

Figure 9.14 Optimization results without group 3. The residuals f_i are shown. In blue (+) the results of an initial optimization for all channels (without group 3) and in red (*) the result of the group-wise optimization in the group order 5, 1, 4, 2

As can be seen, the results of the initial global optimizations are slightly better for the second priority group, but especially for the highest priority group the group-wise optimization gives much better results. It is now a matter of engineering judgement to decide which of these solutions should be chosen.

9.8.3 Extensions*

The only requirements needed for the optimum track mixing approach are to be able to superimpose counting results, derive a measure of distance between the counting results, and have an optimization procedure to minimize the distance. Thus, the possibilities of extending the method are manifold. In the following, we briefly mention some of them.

9.8.3.1 Taking into Account the Shape of the Rainflow Matrices

We have explained in detail how the approach works for pseudo damage numbers. It is clear that in this case, a good agreement between target and mixed track might be found by replacing a large number of small cycles in the target rainflow matrix by a few larger cycles in the mixed track matrix giving the same total damage. This may happen especially when a small number of channels are to be optimized. If this is to be avoided, we have to find a more complex distance measure than the difference between pseudo damage numbers.

An effective extension is to divide the matrix into a set of r clusters or regions (overlap is allowed), calculate a pseudo damage number for the cycles in each of the clusters, and compare the set of cluster damage values. For each of the m channels in the optimization, we get r deviations to be minimized, thus a total number of $m \cdot r$ *virtual channels*. From that point on, we can proceed as before. Of course, due to the larger number of virtual channels, building groups becomes even more important and more involved.

The clusters can be constructed in many ways. A natural thing to do is to build clusters according to the amplitude of the cycles, for example, a group of small cycles, one with medium-sized cycles, and one with large cycles. In this case only the amplitude of the cycles matters. If the mean load of the cycles is important, another approach is to move a rectangular window over the matrix and build a cluster for each location of the window. Both approaches are illustrated in Figure 9.15.

The cluster approach puts a penalty on the replacement of cycles from one cluster by cycles from another cluster and will lead to a better shape fit. Another nice feature of the

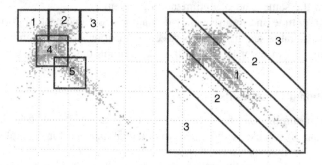

Figure 9.15   Possible clusters of a rainflow matrix

cluster approach is that the dependence on the slope of the Wöhler curve used for the damage calculation decreases. However, the number of objectives increases and the price is a poorer fit for the total damage of each channel. Thus, a compromise between the number of clusters and the accuracy in the total damage needs to be found.

9.8.3.2   Handling an Unknown Wöhler Curve Slope

A consequence of the approach so far is that the results depend on the slope of the Wöhler curve used for the damage accumulation. As mentioned above, the cluster approach reduces that dependence. Another possibility to reduce it is to use multiple Wöhler curves with different slopes. Again, we get a higher number of virtual channels $m \cdot s$, where s is the number of Wöhler curves.

When using this method, a large enough spread in the different slopes should be chosen, e.g. $\beta \in \{3, 6, 9\}$, in order to get an effect. Here too, there will be a tendency to respect the shape of the load spectrum. One advantage of this approach is that it is rather simple to calculate the corresponding damage values, while building clusters as described above either needs specialized software tools or considerably more 'manual' work by the engineer.

9.8.3.3   Taking into Account Correlations

If we want to explicitly take into account the correlation of different channels, we can do so by using the RP approach described in Section 3.3. Instead of optimizing the pseudo damage numbers for the channels only, we can increase the number of channels by a set of rainflow matrices and corresponding pseudo damage numbers belonging to certain combinations of the channels. Again, the number of objectives in the optimization increases.

9.8.3.4   PSD

So far, the comparison between the target histogram and the mixed track histogram has been based on the amplitude domain, namely rainflow matrices, load spectra or pseudo damage numbers. However, it is a straightforward matter to apply the same ideas to the frequency domain. Instead of pseudo damage we can use spectral energy (the integral of the PSD function), instead of cluster damage we can use spectral energy within frequency bands. Then the distance measures are based on the difference in the spectral energy values and we end up with the same kind of optimization problem.

If both the amplitude and the frequency domain are to be controlled during optimization, one has to find a suitable scaling factor between the residuals of the pseudo damage numbers and spectral energies in order to treat both in a common optimization.

9.8.3.5   Additional Constraints

Often, the pure minimization of the distance measures as described so far leads to solutions which have some drawbacks with respect to practical aspects, for example

1. The total length of the optimum mixed track is too long. Can we reduce that without making the fit in the distance measures considerably worse?

2. The number of repetitions of some basic tracks is too high. Is it possible to replace some repetitions with other basic tracks?
3. It is not possible to implement the schedule of the mixed track due to the topology of the proving ground, for instance, basic track j immediately follows basic track k such that we have to have the same number of repetitions for both tracks. Is it possible to include this in the optimization?
4. The number of basic tracks used is high. The total schedule is simpler to handle if only a smaller number of basic tracks were used. Can we achieve this without making the fit in the distance measures considerably worse?

There might be more restrictions of that type. In most cases, they can be formulated quite easily in mathematical terms, for example

1. Bounds on the total length: $\sum_{j=1}^{n} l_j \cdot w_j \leq L_{\max}$, where l_j is the length or duration of basic track j.
2. Bounds on the maximum number of repetitions: $w_j \leq w_{\max}$, $j = 1, \ldots, n$.
3. Relations between different basic tracks: $w_j = w_k$.

As has already been pointed out above, these restrictions can easily be integrated in the general form of the optimization problem described in Equation (9.8). It is harder to control the number $n_{eff} = \sum_{j=1}^{n} \text{sign}(w_j)$ of different basic tracks in the optimized schedule, since $\text{sign}(w_j)$ is not a continuous function. Of course, one can formally set a corresponding bound like $n_{eff} < b_{\max}$, but most of the standard optimization algorithms cannot deal with that. A detailed treatment of this kind of optimization task is beyond the scope of this guide. A workaround is to start with a full optimization and try to remove more and more basic tracks while checking the reduction of quality of the solution.

9.8.4 Hints and Practical Aspects

The most attractive property of the optimum track mixing approach besides its generality is that the result is a combination of data which is 'well known' to the test engineer and without any unrealistic or non-physical behaviour. Thus it is usually better accepted than more synthetically derived load data (see Section 9.3 or Section 9.5). All properties of these loads (phase relation, spectral properties, ...) are realistic, no matter whether they have been controlled explicitly or not. More application examples can be found in Gründer *et al.* [107], Dressler *et al.* [83] or Weigel *et al.* [245].

The following list gives some hints that have been proven to be helpful in many practical applications since the introduction of the methodology in the 1990s.

- Ignore additional restrictions in the first run in order to explore the best fit you can get. Otherwise you do not know whether a certain deviation between target and mixed track is due to lacking basic tracks or due to the additional restrictions.
- Start with a single pseudo damage number for each channel. Build clusters only in those channels where the difference in the shapes of target and mixed track is too large.
- If the first attempts do not lead to a satisfactory result for all channels, remove the less-priority channels to see how good the results can become without them. Optimize the

high priority channels first and then optimize the remaining channels while using bounds on the former ones to maintain the quality you have achieved for them.

- As an alternative to setting bounds on the total length or duration of the mixed track as described above, bounds can also be put on the deviation between target and mixed track, and length or duration can be minimized (runtime optimization). Suitable bounds can be calculated by a preceding histogram optimization.
- Do not use pseudo damage numbers and spectral energies together right from the start. Instead, first concentrate on the amplitude domain and then bring in the frequency domain using increasing scaling factors for the spectral energies. Stop that procedure as soon as the deviations in the pseudo damage numbers become unacceptable due to the increasing importance of the spectral energies.
- When using bounds on the residuals of some channels to perform either histogram optimization for the other channels or runtime optimization, the constraint optimization procedure delivers Lagrange multipliers for the active constraints (constraints which are at their bounds). The larger the Lagrange multiplier is, the greater the gain in the objective function will be if we relax that constraint. Thus, we get hints on how to improve the value of the objective function if some bounds can be weakened somewhat.

9.9　Discussion and Summary

In this chapter, we have investigated the task of deriving load specifications for the design situation and for verification tests. This section summarizes the most important topics and gives some additional remarks.

1. **Design loads, safety factors, and strength requirements:** The first step is to analyse the customer load distribution and derive estimations for quantiles and variances. As has been pointed out in Section 9.2, the quantile alone is not sufficient, since the tail of the distribution beyond the quantile influences the failure probability within the entire population. To this end, the considerations in Chapter 7 are used to estimate a safety factor, which has to be applied to the quantiles to obtain a strength requirement. The result of that step is a design load in terms of pseudo damage numbers and a safety factor for all channels under consideration.

   Whether the design load or the strength requirement is used for the transition to time signals is a matter of taste from a theoretical point of view. For practical reasons it is often advantageous to use the (smaller) design load and apply the factor in terms of required repetitions of the signal on the rig.

2. **From pseudo damage to time signals:** Once a design load target has been derived in terms of pseudo damage numbers, a corresponding set of (multi-input) time signals has to be found in order to drive either a rig test or a numerical verification. There are many ways of doing that (synthetic signals, random loads, standardized loads, optimum track mixing). However, with the exception of the optimum track mixing, all of these methods concentrate on a certain aspect (e.g. pseudo damage) and need special attention in a multi-input setting.

3. **Range of validity of design loads:** The results of most of the methods are valid only for the vehicle which has been used to perform the underlying measurements. The transfer to other vehicles or variants of the vehicle needs careful investigation. Here simulation

techniques can help in order to decide whether a transmission is valid or not. See also the considerations about invariant loading in Section 5.4.

4. **Proving ground schedules have a rather wide range of validity:** The result of the optimum track mixing is potentially more general. The proving ground schedule derived for one vehicle can be applied to another one, giving at least realistic loads for that vehicle. Of course, if the schedule represents a 90%-quantile for the first vehicle, the same schedule might represent a 80%-quantile for the second, since the condensation factor of the proving ground may depend on the vehicle.

5. **Proving ground schedules indirectly respect all load properties:** The result of the optimum track mixing is attractive not only because of the better portability to other vehicles or variants, but also because it automatically gives realistic load properties in the frequency and correlation domain, even if only the amplitude domain (pseudo damage numbers) has been optimized explicitly.

6. **Proving ground schedules can be used for numerical verification:** Since the proving ground is often digitized, the schedule derived for one truck can be transferred to a pure numerical model (e.g. a new design without a prototype). Thus, this type of load description can be used in an early phase of the development.

7. **Speeding up rig tests:** The representation of a design load target by a proving ground schedule automatically contains a certain condensation with respect to time or length compared to the customer usage. However, synthetic loads or specialized random load models are potentially more condensing, speeding up the rig test even more. For example, this can be utilized if a single load for a stiff component needs to be derived. As we have seen in the former sections, things become more complex if multiple loads are considered or frequency behaviour is important.

8. **Numerical derivation of load channels from measured quantities:** If the load channels needed for a verification test are not included in the list of measured quantities, then the derived design load targets do not contain them either. In that case, they might be derived by the simulation of a numerical model of the vehicle or a part of the vehicle (see Section 5.2.4, where the calculation of section forces based on multibody models has been described). This works if the excitation of that model can be done based on the derived design load targets, that is, if we have time signals representing the target. In that situation it is important that the design load signals are realistic with respect to the amplitude domain, the frequency domain, and the correlation domain, since typically the response of the model depends on all these properties. For example, we cannot use synthetic time signals constructed from a pseudo damage number or a rainflow target. Here again, the design load target formulated as a proving ground schedule is advantageous.

10

Verification of Systems and Components

10.1 Introduction

10.1.1 Principles of Verification

In this chapter, the word *verification* means ensuring that "you built the right product". This means that we are interested in finding out if a certain **design** of a component or system complies with its specified characteristics.

The idea of verification is in some respect impossible, since "ensuring" that you built the right product would need testing all products against all usages. This apparent fact has been formulated by the philosopher Karl Popper who claims that all knowledge is necessarily tentative,

> We cannot reasonably aim at certainty. Once we realize that human knowledge is fallible, we realize also that we can *never* be *completely certain* that we have not made a mistake ... Since we can never know anything for sure, it is simply not worth searching for certainty; but it is well worth searching for truth; and we do this chiefly by searching for mistakes, so that we can correct them.
>
> – Popper [190].

In our actual case the problem at hand may not primarily be mistakes due to design or unknown physical behaviour. Rather, it is the unknown details of the strengths and loads that govern our uncertainty and the following necessity of using simple empirical models for the life assessments. The fact, discussed in Chapter 7, that both strength and load are influenced by large uncertainties highly disturbs the verification task. When performing a verification test the very design cannot be put on test, but a random choice from the uncertain population of test objects will be subjected to specific choices of load. Our lack of certain knowledge about the representativeness of the performed test will always make the verification uncertain and "ensuring that you built the right product" must be interpreted as a statement that the product complies with its intended characteristics with sufficiently high probability.

Guide to Load Analysis for Durability in Vehicle Engineering, First Edition. Edited by P. Johannesson and M. Speckert.
© 2014 Fraunhofer-Chalmers Research Centre for Industrial Mathematics.

10.1.2 Tests for Continuous Improvements vs. Tests for Release

One question about the concept of verification is: what is a successful verification test? If the product tested survives the test, then we may conclude that the product is at least as strong as we demand, but we do not know how much stronger it is. If the product fails, then we gain much more information about the strength, at least for the specific tested product. In addition, failures may give rise to design improvements. In conclusion, survival may not always be a success. Perhaps a slightly more severe test would have given more knowledge both with respect to reliability and with respect to the reason for weaknesses in the design. The author Patrick O'Connor puts it as follows:

> ..., tests to provide assurance of strength, reliability or durability of a design must be planned to generate failures, or at least evidence that failure will not occur, or will be very unlikely to occur, within the product's expected conditions of use and lifetime. The uncertainty and variability inherent in forecasts of such properties mean that testing to generate failures outside the operating regime is usually the only practicable way of providing assurance that failures will not occur inside. Therefore the test programme must distinguish between the tests that should demonstrate success, and those that are planned to generate failures.
>
> – O'Connor [173].

O'Connor recommends a method called *Highly Accelerated Life Testing* (HALT). This method does not regard survival of the test object as a success, but designs tests with the purpose of failure, regardless of what severities are needed, even if they are far beyond specifications. The resulting failures are analysed and the following questions are asked:

1. Could this failure occur in service?
2. If so, could we prevent it from happening?

The method is inexpensive as it forces failures to happen in a short time and gives rise to *continuous improvements* of the design. In contrast, tests at expected severity, "verifying" that the construction survives, are expensive and time-consuming and are not likely to discover weak spots that may cause failure at extreme events in service.

A limitation of the HALT method is that a highly accelerated life test may cause failures by other mechanisms than the one that is of interest and that it lacks a decision rule in cases when the product cannot be tested until failure. Particularly with regard to safety critical components, there is a strong need for a decision rule answering the question: when can the product be released? In spite of the large uncertainties in probabilistic statements about rare events, the probabilistic approach may be a good tool for the construction of such decision rules. However, the probabilities at hand should not be interpreted as true measures of failure frequency, but rather be seen as notional values. The authors Der Kiureghian and Ditlevsen discuss the problem in a recent publication:

> The arbitrariness in the choice of the distribution model and the 'tail-sensitivity' of small probabilities has lead to the recommendation that probabilistic structural design codes standardize probability distributions for load and resistance

quantities. One point of view is that in such a construct the computed probabilities should be considered as notional values and that caution should be exercised in using them in an absolute sense, e.g., for computing the expected costs of rare events. However, this view assumes that absolute probability exists as a physical entity outside the mathematical model by which it is computed.

– Der Kiureghian and Ditlevsen [71].

In Section 10.3, probabilistic methods are described that could be the basis for decision rules for product release.

All difficulties in finding the full truth by verification tests emphasize the importance of previous experience of design. It is not possible to construct tests by mathematical methods that overcome these difficulties, but previous engineering experience makes it possible to

1. **construct tests that have high probability to discover weaknesses,**
2. **define test procedures, based on probabilistic schemes that are judged to provide safe components**.

10.1.3   Specific Problems in Verification of Durability

The primary subject of this *Guide* is load analysis with respect to durability of vehicles. This means that it may not be right to use the practical way of verification according to O'Connor, testing outside the operating regime. It is important to distinguish fatigue failures from other failure mechanisms, in particular since they are well represented in service failures. Testing at too high loads will stress other failure mechanisms and thereby hide the fatigue properties that may be dominant at more moderate load levels. **Verification with respect to fatigue strength must generate failures caused by the same mechanism as expected field failures**, making the second verification approach based on probabilistic schemes important for durability verification.

The fatigue phenomenon has the special characteristic that failures almost always occur as a result of defects. These can be weld pores or discontinuities because of misalignment introduced in the welding process. They can be scratches or marks from the assemblage procedure or pits due to corrosion. On a smaller scale, inclusions and pores are always present in the material. Graphite formations in castings and microstructural anomalies in the phase and grain structure can also be regarded as defects. The existence and variation of the defect content are the source of the large scatter in fatigue performance and must be taken into consideration. None of the defects is given at the drawing of the product and cannot be included in the stress and strain analysis. Therefore, this analysis is preceded by calculations based on rough empirical models which introduce model errors. Unknown model errors add uncertainty to the problem of verification; both scatter and uncertainty must be considered.

- The large scatter and uncertainty demand solutions to the problem of choosing objects and loads for verification tests that represent the service usage. How do we choose the test objects? How do we define the test load?
- The vehicle is constructed for a long life, but testing time is limited, both due to short lead times for designs, and due to costs. Accelerating tests is therefore an important issue.

What methods can be used for accelerating tests? How do we make sure that we test for a relevant failure mechanism?

- The test result interpretation must consider the randomness of the observations. How should tests be designed to give proper answers to the question: Does the product comply with the requirements with sufficiently high probability?

10.1.4 Characterizing or Verification Tests

It may be convenient to distinguish between two types of test procedures, namely, on the one hand, tests for characterizing strengths and loads focusing on their expected values and spread, and, on the other hand, verification tests for checking assembled structures with respect to unexpected behaviours or for checking them against defined safety requirements. The first test type is a part of a *verification procedure* rather than a verification test.

10.1.4.1 Characterizing Tests

For the second-moment load-strength method, evaluated in Chapter 7, the statistical part of the safety distance is based on means, variances, and identified uncertainties. The basis for this part is characterizing material or component strengths and customer loads. Such characterizations are done for different purposes:

1. for verification of the design, both at the design stage and at later stages of development, and
2. to compare materials, components, or customers.

For these characterizing purposes, it is both possible and feasible to use loads that are close to reality in order to resemble the true failure mechanisms as far as possible. Achievements in this area update uncertainties and thereby refine the statistical part of the required safety distance.

Characterizing tests should be performed with loads close to normal service situations to resemble the main failure mechanism and not exceed the "design load", perhaps defined as the "95%-customer", too much. In particular, the lack of resemblance should be moderate enough to be possible to be quantified as a model uncertainty in the reliability evaluation.

Load time histories for laboratory tests can be constructed by methods described in Chapter 4 and customer behaviour can be characterized by means of load analysis tools described in Chapters 8 and 9.

10.1.4.2 Verification Tests

Again referring to Chapter 7, the deterministic part of the safety distance represents the unknown tails of the load and strength distributions. This tail can usually not be judged based on laboratory tests or field measurements, since it is a combination of very rare load events and unexpected structure weakness. Verification tests are preferably designed primarily to discover such unexpected structure weaknesses and are accomplished by subjecting the structure to severe loads, representing rare events that may occur in service.

The difficulty of judging the statistical properties of the load and strength tails makes probability assessment doubtful. This fact presents two possibilities:

1. avoid probabilistic statements and instead create rational systems of continuous improvements, and keep track of quantification and updating engineering judgements, or
2. use the probabilistic tool, but regard probabilities as notional values without reference to real failure frequency.

In both cases a verification test consists of 1) choosing a severe load (verification load) and 2), if possible, choosing a weak object in order to verify the total safety factor used in the design. The verification load may be chosen without quantitative reference to the customer usage, but the aim is rather achieving maximal load without violating the failure mechanism. The weak object may be chosen among different configurations or by weakening component objects on purpose. Since the aim of the verification test is to discover very rare failures, it is usually necessary to violate aims at correct failure mechanisms, but, of course, to make the test relevant, the failures must be reasonably close to the ones expected in service.

The true severity of a verification test can never be quantified since it is performed on only one or a few samples, that is, these few samples may happen to be from the weak or the strong part of the component distribution. Therefore, the choice of verification load should be based on relevance by means of failure mechanisms rather than judgement of quantified severity in relation to customer loads.

The types of load-time histories for verification tests are more limited than in the case of characterizing tests, since large severities can only be achieved by large acceleration factors in rig tests or by tests on roughly constructed test tracks.

10.1.5   Verification on Different Levels

The target life of a vehicle is by means of driving distance, millions of kilometres and, with respect to time, decades. Thus, verification tests against durability must be accelerated in some way. We can distinguish between two different types of acceleration, namely 1) increasing load levels in order to shorten time to failure, and 2) decreasing time by eliminating non-damaging time. The possibilities of using these two types of acceleration are different depending on the failure mechanisms and depending on the level of complexity. On the highest level, when testing the whole vehicle, both the load levels and the frequency content are limited by the possible forces that can be applied on the outer structure. On the other hand, on the lowest level, for material testing, the load levels and frequency are almost only limited by the demands of the right failure mechanism. On all levels below the whole vehicle level, some calculations must be used for the verification of the vehicle performance. Therefore, model errors will be introduced.

10.1.5.1   Vehicle Level

On the vehicle level, verification is expensive and can only be performed for a few vehicle configurations and load spectra. This puts very high demands on the choice of the items to be tested and on the choice of the applied load. At this point, the idea of the HALT approach should be considered.

Time acceleration can be done primarily by excluding time intervals with non-damaging loads, but the possibility of raising the frequency of damaging cycles is highly limited. Therefore, fatigue mechanisms are difficult to stress to failure within reasonable time and the tests primarily verify other failure mechanisms. The idea of high severity outside the normal operating region in accordance with the O'Connor recommendation may be the dominating verification strategy, with the aim of finding the weakest points in the structure. However, weak points by means of resistance against high cycle fatigue are often not found, but may have to be subjected to tests at lower levels of complexity.

10.1.5.2 Subsystem Level

On the subsystem level (for instance, axles) the possibilities of verification increases with respect to load application, but on the other hand, additional uncertainties are introduced by cutting the subsystem from the whole vehicle; forces that are introduced at the cutting edges will not be fully equivalent to the service case.

10.1.5.3 Component Level

On a less complex system level, the component level, it is often possible to make fatigue tests to failure, without violating the demands on correct failure mechanism since test simplicity makes it possible to accelerate mainly by increasing the frequency of load cycles. The main features that are responsible for failure, surfaces and simpler assemblages, can be tested as they will appear in service. Of course, the problem of applying forces equivalent to field forces gets worse.

On this level, and below, most tests are of the characterizing type. Cases when verification is still the primarily interest in a component test are for instance:

- when the component is designed for other criteria than endurance and need to be verified with respect to new failure mechanisms.
- when small changes have been made in the component, by design, by new suppliers, or by new batches.
- tests on safety critical components, where test procedures are performed according to safety verification rules.

In the safety critical component case, verification using tests based on probabilistic schemes may be useful.

10.1.5.4 Material Level

The lowest level of complexity in vehicle engineering is the material level. Here, the limitations on the acceleration level can usually be judged by pure failure mechanism considerations. However, in many cases the material fatigue strength is difficult to relate to the strength of the components, subsystems, and the vehicle.

For truck durability, the material properties are indeed important. However, for the durability of a component, the geometry, surface roughness, and production scatter are in most cases even more important. Therefore, testing on the material level is generally not recommended. Durability tests should instead be performed on the component. However, components in the engine and powertrain that have smooth surfaces in the critical areas can often be quantitatively related to material resistance.

10.1.6 Physical vs. Numerical Evaluation

All numerical models of physical structures are approximate. Modern software and computer development have eliminated great many of problems regarding the calculation of stresses and strains in homogeneous bodies, but modelling assemblages raise ambiguities caused by phenomena such as friction, statically undetermined states, and inhomogeneous welds. In addition, behaviour regarding fatigue is primarily governed by defects that are not present at the drawing stage and is hard to predict in advance. Therefore, numerically determined designs based on characterizing tests on material level contains more uncertainties than designs based on characterizing tests on components. On the other hand, component tests are more expensive and a suitable trade-off must be found between the characterizing efforts and confidence in the design. The final uncertainty in the numerical strength prediction must therefore be verified.

Verification tests on large structures can only be performed on one or perhaps a few samples. The aim of these tests is primarily to find unexpected characteristics of interactions and assemblies that have not been identified at the design stage. On the basis of the few verification tests, it is, however, possible to verify other configurations and load combinations by using solely numerical tools. By modelling the configuration that has been tested by an MBS/FEM model, local stress and strains measured during the test may be used for adjusting the computer model by tuning. The updated model can then be used to study small modifications in the input, in the configuration, and in geometry and material properties for verification of other cases.

10.1.7 Summary

Verification is, in principle, an impossible task, since knowledge about rare events, human mistakes, and extreme defects is limited. This fact gives rise to the conclusion that quality improvements are best accomplished by forcing failures in early development, i.e. to design severe tests beyond expected severity. There is, however, still a safety requirement, which should be "guaranteed" by clearly defined testing schemes defining decision rules for release. These may be

- Survival of severe tests that are judged to be far beyond possible cases in service (HALT).
- Fulfilment of requested results from probabilistic defined verification test schedules (Section 10.3).
- Fulfilment of safety factors based on characterizing tests (Chapter 7).

10.2 Generating Loads for Testing

Again referring to the O'Connor philosophy of verification:

> It cannot be emphasized too strongly: testing at 'representative' stresses, in the hope that failures will not occur, is very expensive in time and money and is mostly a waste of resources. It is unfortunate that nearly all standardized

approaches to stress testing demand the use of typical or maximum specified stresses. This approach is widely applied in industry, and it is common to observe prototypes on long-duration tests with 'simulated' stresses applied. For example, engines are run on test beds for hundreds of hours, cars are run for thousands of miles around test tracks, and electronic systems are run for thousands of hours in environmental test chambers. Tests in which the prototype does not fail are considered to be 'successes'. However, despite the long durations and high costs involved, relatively few opportunities for improvement are identified, and failures occur in service that were not observed during testing.

– O'Connor [173].

10.2.1   Reliability Targets and Verification Loads

In design, the goal is to separate the customer load distribution and the strength distribution as much as is needed in order to get sufficiently high reliability. For verification, the situation is rather the opposite; the aim is to move the test load and the strength as close to each other as possible in order to get failures in a reasonably short time, without altering the failure mechanism. However, in order to verify the design goals it is important to relate the test scenario to the design scenario. This can be done by establishing how much more severe the verification test is compared to the design situation.

The goal here is to define verification loads that can be used in the verification process to ensure that the structure meets the design targets. Therefore, it is important to transform the original reliability target in terms of targets for the verification load, which is discussed in Section 9.2.

In Section 10.3, different approaches for planning and evaluation of verification tests are discussed. These require different kinds of verification loads.

1. *Characterization of the strength distribution*, Section 10.3.2. In order to be able to estimate the mean and the variance, the verification load needs to be severe enough to generate failures, at least in some of the tests.
2. *Verification of a central safety factor*, Section 10.3.3. This corresponds to verifying the median strength, and thus the severity of the verification load needs to be at the median of the required strength or above. Safety factors for design and verification are discussed in Section 9.2.5.
3. *Verification of a low strength quantile*, Section 10.3.4. The verification load is here defined in such a way that its severity corresponds to a given quantile in the required strength distribution. This approach is discussed in Section 9.2.4.

10.2.2   Generation of Time Signals based on Load Specifications

When performing tests it is necessary to have a load signal representing the specifications of the verification load. Different ways of deriving design and verification loads is the topic of Chapter 9. Often the verification load is given by test track measurements. However, it can also be given as a measured rainflow matrix or an observed PSD, from which we need to reconstruct a load signal, see Section 9.5 and Section 4.5. If signals from a previous design are used, they need to be modified, which is discussed in Section 5.4. Further, when testing sub-systems or components, the loads can be computed from the system loads, see Chapter 5 for a review. Sometimes the verification load is specified in terms of a random load model, see Section 9.4 and Chapter 6.

10.2.3 Acceleration of Tests

Since we are dealing with accumulated damage, the most straightforward approach to increase the load severity is to repeat the load; two repetitions implies twice the damage. This procedure is commonly used in testing, that is, repeating a given load signal until failure (or run-out level). An alternative is to generate a long load signal, for example, using a model for a random load.

However, there is also a need for test acceleration so that the test can be performed in a shorter time and/or at a reduced cost. The acceleration can be achieved either by making the load more severe or making the test specimens weaker. We will here concentrate on load acceleration and review different methods.

Time acceleration can be performed by applying the load at a higher frequency. In that case it is important that the amplitude response of the structure does not change due to the change of input frequencies. Therefore, it is mostly applicable in component testing. Another kind of time acceleration is to remove segments of the signal that induce negligible damage, see Section 4.3.3. This technique is often used when preparing signals for test rigs, and is suitable for the vehicle level as well as for the component level, since it does not change the frequency of the remaining signal.

Cycle acceleration by omitting small cycles that cause negligible damage can be performed by using the rainflow filter, see Section 4.3.1. This is an efficient way of reducing the length of the signal, without significantly affecting its damage content. Since it acts on the turning points of the signal, the frequency information of the signal is lost, and the method is therefore usually only applicable in component testing.

Amplitude acceleration by multiplying the signal by a scale factor is an efficient way of making the load more severe, but still keeping the frequency content. However, it should be used with care since the increased load levels may change the failure mechanism. Further, the degree of acceleration depends on the damage increase due to the increased load levels. Consequently, a damage model need to be used in order to estimate the acceleration factor. This makes the amount of acceleration uncertain. For example, when using the Basquin model, an increase in load by 20% implies a damage increase by a factor 1.7 for $\beta = 3$, while a factor 2.5 for $\beta = 5$.

The above methods are often combined in order to get an efficient acceleration. Moreover, the load signal may also need to be adapted to the test equipment, see Section 4.3.1 on page 119.

10.3 Planning and Evaluation of Tests

In this section, we deal with the question how to evaluate tests for high cycle fatigue. Suppose that we are given either a load distribution for customer usage or a test load which could be derived from such a distribution or from experience. Our task is to verify that the system under investigation is strong enough to endure this load. Given the large amount of uncertainty about loads and strengths, we cannot guarantee survival under all circumstances. Instead, we focus on the more realistic goal of ensuring that the probability of a failure is sufficiently small.

The methods discussed below are particularly applicable when testing components and subsystems, as we presume that there is a single dominant failure mode and that any load signal can be converted to a scalar value using a law for damage accumulation such as the Palmgren-Miner rule, discussed in Appendix A. In this context, the strength of a component is the (pseudo) damage number of the load on the test rig or track that causes a failure.

As fatigue tests are lengthy and expensive, we need to draw conclusions on extreme events (component failure under normal service conditions) from relatively few strength samples. It is thus important that our statistical tools use the available information efficiently. Even then, it is usually impossible to obtain satisfactory results without some additional assumptions about the strength distribution. Anything else either requires an inordinate number of components for testing or leads to extreme overdesign.

There are two conventional approaches to evaluating fatigue tests, both of which are used in practice. The first method estimates the probability distribution of component strength and calculates the failure probability relative to the given customer load distribution or test load. The second method fixes a certain test load and attempts to establish that the failure probability is below a level still considered acceptable. Both rely on a parametric model for the strength distribution, which means that they are almost equivalent in practice. The first approach tends to require longer testing, as it is only applicable if at least some components fail. However, this also ensures that we always gain information about the actual quality of the design. The second approach can handle tests where no failures occur, and it simplifies test planning, as there is an explicit relationship between the number of samples required and the duration of each test. Its drawback is that it becomes more difficult to spot overdesign.

While both methods deal with explicit failure probabilities, these should merely be seen as notional values. The uncertainty introduced by small sample sizes and the arbitrary choice of the distribution model means that one cannot be certain, for instance, that the true probability of failure is 10^{-6}, even if that is specified by the procedure. We merely know that demanding a certain low probability in order to release a component has caused no problems in the past, so the procedure itself is 'verified' by experience. This can be made more explicit if we use the second-moment methods introduced in Chapter 7 and fleshed out in Section 9.2. Below, we propose a variant of the estimation approach, where, instead of fitting a parametric model, we estimate the safety factor that separates the load and strength distributions. The release decision is based on the size of this factor measured on a scale derived from safety considerations and empirical knowledge. This is not fundamentally different from using a notional failure probability, but it highlights the fact that we rely on experience above all.

Some examples for application options of such methods in the vehicle industry may be found in Weigel et al. [244] or Weihe et al. [246].

10.3.1 Choice of Strength Distribution and Variance

The standard methods for evaluating fatigue tests require the choice of a class of probability models for component strength. In case we want to estimate the strength distribution, we need to restrict its shape in order to obtain useful results from a small number of samples. And to ensure a sufficiently low rate of failure, we need a distribution for evaluating tests which use more severe loads than the given test load.

The usual way to prescribe a class of distributions is to specify a parametric model with a small number of parameters. As in the case of loads, popular choices are the log-normal and Weibull distribution. Both have two parameters, so the scale (mean) and shape (variance) of either distribution can be determined independently of each other. In case one believes that the components have a guaranteed minimum strength, one can also use a three-parameter Weibull distribution, although estimating the offset from zero is difficult and requires a large number of samples. Appendix B contains more information about the different stochastic models.

For the remainder of this section, we only deal with the log-normal and Weibull distributions. As there is no reliable method to distinguish between the two using a small number of samples (say, 20 or less), the choice of one of them has to be done on the basis of experience. As both models are simplifying approximations, this is more a question of establishing a consistent process for evaluating tests than one of correctness. Still, it should be noted that the Weibull distribution is more compatible with the usual notion of a lifetime distribution than the log-normal:

- The failure rate of a Weibull distributed random variable is strictly increasing (for shape parameter $c > 1$), which is the correct behaviour for components subject to fatigue. By contrast, the failure rate of the log-normal distribution is only increasing for small quantiles. This is sufficient for the analysis of weak components, but shows that the model is not a particularly good choice for a strength distribution overall.
- The Weibull distribution satisfies a "weakest link" condition, i.e. the minimum of several Weibull random variables (all having the same shape parameter c) is again Weibull distributed. This yields a consistent model for the strength of a system with multiple identical parts that fails as soon as the first part breaks. The log-normal distribution has no such property.

Even if we restrict the choice of model to a class with only two parameters, we are often unable to obtain good estimates for the scatter of the distribution. Estimating σ for a log-normal or c for a Weibull random variable from a small sample is subject to large uncertainties. Figure 10.1 illustrates this for random samples of size 5 from a known distribution: the shapes of the fitted densities are highly variable and there is a considerable spread between the estimated 1%-quantiles. These effects can be taken into account, but conservative variance estimates lead to severe requirements for component strength. The usual solution to this dilemma is to treat the shape parameter as a property of the material (only) and use recommended values found in material science handbooks and industry standards. Table 10.1 lists some examples taken from Haibach [110].

Remark 10.1 (Pseudo damage and load cycles.) *The parameters σ and c in Table 10.1 describe the scatter of the pseudo damage value for loads leading to failure. But for rig tests with constant amplitude loads, damage is often simply measured in load cycles. Fortunately, both the log-normal and Weibull distribution have the property that the same shape parameter applies when evaluating pseudo damage or number of cycles with fixed amplitude – only the scale parameters will be different.*

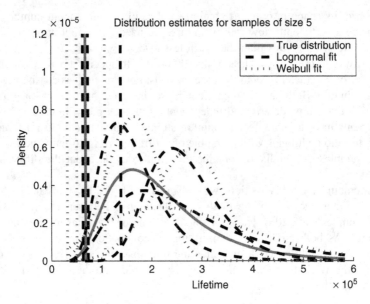

Figure 10.1 Estimated log-normal and Weibull densities for 3 samples of size 5 each. The true distribution is log-normal to base e with parameters $\mu = 12.2$ and $\sigma = 0.46$. Vertical lines show the position of the 1% quantiles.

Table 10.1 Typical scatter parameters for damage distributions

	log-normal[*]		Weibull
	Base e	Base 10	
Material and shape	σ	σ	c
Cut steel, moderate crenation (Wöhler exponent $\beta = 5.0$)	0.454	0.197	2.603
Forged and tempered steel (Wöhler exponent $\beta = 5.0$)	0.587	0.255	2.013
Welded steel (Wöhler exponent $\beta = 3.0$)	0.428	0.186	2.756

[*]Values for σ to base 10 are taken from Haibach [110]. Those to base e are derived by multiplying with $\ln(10) = 2.303$.

After fixing the shape parameter, the distribution is fully determined by either its scale parameter (log-normal μ and Weibull a), or the value of any one quantile. In particular, the distribution-based approach and the quantile-based approach for evaluating fatigue tests lead to equivalent statements.

10.3.2   Parameter Estimation and Censored Data

Given a number of strength samples and a parametric class of distributions, the question is what parameters give the 'best' description of the observations. A common problem is that some results are censored, meaning that the test was stopped before any visible damage occurred. This may happen if the runtime is limited from the beginning, due to unforeseen complications, or because the component broke for different reasons than the failure mode under investigation. In this case, some strengths will be known exactly, while others are given as lower bounds ('strength is greater than ...'). For long tests that are only checked infrequently, censored data can also take the form of an interval ('strength is in between ...').

In the case of the log-normal model with uncensored data, the recommended estimators for μ and σ are the sample mean and sample standard deviation of the logarithmic observations. For the Weibull distribution, or for either model in the case of censored data, the solution is less obvious. Fortunately, there is a consistent statistical tool for estimating parameters in all these situations, the so-called *maximum likelihood method*. It is based on the principle that among all permissible models, we should choose the one giving the highest probability to the values we have actually observed. At first glance, the maximum likelihood method does not appear applicable, as we are working with continuous distributions, where the probability to observe any given value exactly is zero. However, as the probability density at a point x indicates how likely we are to observe a value near x, we can use the latter as our measure instead.

We denote by $f(x|\theta)$ the probability density of the pseudo-damage distribution with unknown parameter(s) θ, and by $F(x|\theta)$ the corresponding distribution function, i.e.

$$F(x|\theta) = \int_0^x f(y|\theta) \, dy \tag{10.1}$$

is the probability that the strength of a component is x or less. Here, θ can be the pair (μ, σ) for the log-normal distribution, (a, c) for the Weibull distribution, just μ or a in case we fix the shape parameter in advance, or in fact the parameter vector for any other parametric model.

Given N independent observations $x_1, \ldots, x_N$, of which the first N_0 are uncensored data points, the next N_1 are lower bounds, and the last ones are intervals $x_n = [a_n, b_n]$, we define the *likelihood function* as

$$L(\theta|x_1, \ldots, x_N) = \prod_{n=1}^{N_0} f(x_n|\theta) \prod_{n=N_0+1}^{N_0+N_1} (1 - F(x_n|\theta))$$

$$\times \prod_{n=N_0+N_1+1}^{N} (F(b_n|\theta) - F(a_n|\theta)) \tag{10.2}$$

Note that in case observations are known exactly, we use the density as has been discussed above. For lower bounds, we use the probability that the strength is greater, and for intervals the probability to be inside.

According to the maximum likelihood principle, we choose that value of θ as our estimator which maximizes L for the given observations. In general, the maximization problem does not have a closed-form solution and needs to be solved numerically, but this is handled easily by most statistics software. The only issue is the existence and uniqueness of a maximizer. For example, the problem has no solution if all observations are lower bounds, as the probability of this event is equal to one in case of infinite lifetime. A sufficient condition for the existence of a unique maximizer is to have at least one uncensored observation for each unknown parameter, but certain combinations of censored data are also feasible. In practice, tests need to be severe enough to ensure that some failures occur. If this is not possible, a distribution estimate is likely to fail and one should employ the alternative method based on statistical tests discussed in Section 10.3.4.

The maximum likelihood method is known to yield good estimators in the sense that they converge to the true parameters of a distribution as the number of samples goes to infinity, and do so with minimal variance. These properties are known as asymptotic consistency and efficiency. They hold under rather general conditions, which are certainly satisfied by the log-normal and Weibull distributions. The method is also consistent with the 'naive' approach, as the maximum likelihood estimator for the parameter μ of a log-normal distribution coincides with the sample mean, and the estimator for σ is equivalent to the sample standard deviation for a large sample size. For more on the maximum likelihood method, see, for example, Montgomery and Runger [163].

Example 10.1 (Distribution Estimates for Censored Data)
Assume that we perform a rig test with a repetitive load that was run until failure, but no longer than 200,000 load cycles per component. The results for 5 samples are:

Component	1	2	3	4	5
Load cycles	135, 030	119, 120	> 200, 000	> 200, 000	> 200, 000

The last three components were still intact after the test, so we only know that their lifetime is greater than 200,000 cycles. If we believe in a log-normal model, we derive the likelihood function according to Equation (10.2) and maximize over μ and σ to obtain the estimates

$$\hat{\mu} = 12.27 \qquad \hat{\sigma} = 0.48. \tag{10.3}$$

For the Weibull distribution, the maximum-likelihood estimators are

$$\hat{a} = 250, 000 \qquad \hat{c} = 2.63. \tag{10.4}$$

Instead of using the maximum-likelihood method, one could consider the naive approach of treating the censored data points as equal to the lower bound. This appears conservative at first glance, as the mean strength is underestimated systematically – but so is the variance! For the example data, we obtain

$$\tilde{\mu} = 12.02 \qquad \qquad \tilde{\sigma} = 0.25$$

$$\tilde{a} = 185, 000 \qquad \qquad \tilde{c} = 6.10 \tag{10.5}$$

As we are interested in small quantiles of the lifetime distribution, these values yield a much too optimistic assessment of component strength. For example, the 1%-quantile for the log-normal distribution is $\hat{q}_{1\%} = 68, 000$ cycles using the maximum likelihood estimators, but $\tilde{q}_{1\%} = 92, 000$ cycles for the naive approach. Using the Weibull model, the respective values are $\hat{q}_{1\%} = 44, 000$ and $\tilde{q}_{1\%} = 87, 000$.

The same effect occurs if we drop all success runs and calculate the estimators only from failed samples. In contrast, the maximum likelihood method uses all available information and is known to have good statistical properties, so it is the tool of choice for evaluating censored data.   □

10.3.3 Verification of Safety Factors

Instead of working directly with the strength distribution, we may use second-moment reliability methods to assess how well the strength of a component is separated from the expected service loads. In that case, the reliability target can be set in terms the reliability index, that is, using the number of prediction standard deviation as a measure of the separation between load and strength. In order to verify the reliability index by testing, we would need to replicate all uncertainties in both load and strength. In practice this is not possible, and therefore another verification approach is needed. In Section 9.2.5 we proposed one such measure of separation that can be verified by testing, namely the central safety factor ϕ, which can be calculated from the reliability index. As ϕ is the quotient between the median strength and the median load, verifying the central safety factor corresponds to verifying the median of the required strength distribution. This can be done by using the method described in Section 10.3.4, i.e. making a hypothesis test for the 50%-strength quantile. However, we will here instead aim at estimating the median strength, which can be transformed into an observed safety factor. Further, we will construct a confidence interval for the observed safety factor, which can be used to decide whether a component should be released or not. The procedure is illustrated in Example 10.2.

In order to show how the confidence interval can be used, assume that we have a component with a required safety factor of 2, and that we want to verify the demand with 95% confidence, corresponding to a 5% significance level. From the result of the verification test we have calculated the 90% confidence interval for the safety factor to [2.21; 2.59]. In this case the component can be released, as there is only a 5% chance that the actual safety factor is below 2.21, which is higher than the required safety factor of 2. An advantage of the confidence interval compared to the hypothesis testing is that the confidence interval gives a quantitative answer of the strength, in our case an interval for the safety factor, instead of only 'pass' or 'not pass' as in the case of hypothesis testing.

The reliability target safety factor ϕ_{target} can be transformed into a safety factor ϕ_{vl} with respect to the verification load.

$$\phi_{vl} = \frac{S_{0.50}}{L_{vl}} = \frac{S_{0.50}}{L_{0.50}} \cdot \frac{L_{0.50}}{L_{vl}} = \frac{\phi_{target}}{L_{vl}/L_{0.50}}$$

where the factor $L_{vl}/L_{0.50}$ is a measure of how much more severe the verification load is compared to the median customer load. The more severe we make the verification load the more likely it is that the test will result in a failure. If the safety factor for the verification load is 1, there should be a 50% chance of failure. If we use the principles of HALT, i.e. we construct a verification load that is so severe that it most likely provokes a failure, then we should aim for a safety factor of less than 1 for the verification load (run-out level for the test).

Example 10.2 (Verification of a Safety Factor)
We have a safety critical component with a required (central) safety factor of $\phi_{target} = 3.5$ in terms of equivalent load-strength measures with $\beta = 5$. The goal is to verify with 95% confidence that the requirement is fulfilled.

We have defined the design load to be 10 times more severe than the median customer, in terms of pseudo damage. In equivalent load scale the design load is thus $10^{1/5} = 1.58$ times more severe than the median customer. The verification procedure is now to repeat the design load until failure (or run-out level of 200 repetitions). Thus, the verification load is defined as 200 repetitions of the design load. In the equivalent load scale the verification load is $(10 \cdot 200)^{1/5} = 4.57$ times more severe than the median customer, and the safety factor for the verification load becomes $\phi_{vl} = 3.5/4.57 = 0.76$.

Testing $n = 8$ components resulted in the following observed lives

$$\{124, 143, 73, 149, 70, 72, 82, 125\}$$

representing the number of repetitions of the design load until failure. This can be recalculated in terms of an observed safety factor, with respect to the median customer load, for each of the tests as $\phi_i = (10x_i)^{1/5}$, giving

$$\{4.16, 4.28, 3.74, 4.31, 3.71, 3.73, 3.83, 4.16\}.$$

Can we now ensure that the safety factor is greater than $\phi_{target} = 3.5$? To answer this question we construct a 90%-confidence interval of the logarithmic mean

$$\left[\overline{\ln \phi} \pm t_{n-1, 0.95} \cdot \frac{s_{\ln \phi}}{\sqrt{n}} \right] = \left[1.381 \pm 1.895 \cdot \frac{0.066}{\sqrt{8}} \right]$$

$$= [1.381 \pm 1.895 \cdot 0.023] = [1.381 \pm 0.044]$$

which gives a 90% confidence interval for the median safety factor

$$[\exp{(1.381 \pm 0.044)}] = [3.81; 4.16].$$

Hence, we can conclude with 95% confidence that the value is above 3.81, and hence the required safety factor of 3.5 is fulfilled. The result is illustrated in Figure 10.2.   □

Note that the verification procedure above only verifies the target safety factor, and not necessarily the original reliability target, since both the load and the strength uncertainties

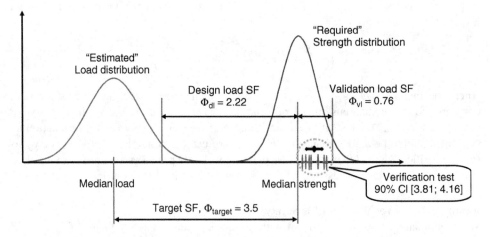

Figure 10.2   Verification of a safety factor; Example 10.2

are different in customer service compared to the test set-up. In order to verify the reliability target we would need to replicate not only the mean values but also the variations in the customer distribution and in the population of components. However, in practice this is not feasible, since it would lead to far too long test times and huge costs. The uncertainties need to be handled by other means, for example, the load-strength method.

10.3.4 Statistical Tests for Quantiles

Instead of estimating the strength distribution of a component, we can ask the question whether the probability that it survives a given test load is sufficiently high. This approach leads to a *statistical test*, which is a decision problem (called hypothesis testing) with two possible outcomes:

Hypothesis (H_0): The component is insufficiently strong.

Alternative (H_1): The component is strong enough.

When we make a decision, one of two possible errors can occur. The *error of the first kind* is to reject the hypothesis although it is true. This would mean that we release an unsafe component, which is the situation we want to avoid if at all possible. The *error of the second kind* is to accept the hypothesis although it is false. In this case, we do not recognize that the component is already strong enough and incur additional development costs.

By convention, statistical tests are designed to minimize the probability of an error of the first kind. This is the reason why we choose H_0 as the undesirable outcome – the error of the first kind is now the worse of the two. Its probability is called the *consumer's risk*, whereas the probability for an error of the second kind is the *producer's risk*.

To evaluate a statistical test, we need to fix the largest probability α for an error of the first kind that we are willing to tolerate. This value is called the *significance level of the test* – common choices are 0.2%, 1%, or 5%. Note that α is not related to the failure probability of the component in question. Instead, it tells us that if we conduct the test properly and choose outcome H_1, the probability to be wrong and release an unsafe component is at most α. On the other hand, if we decide on H_0, we do not know the probability that we made a mistake. As the principle is to err on the side of caution, a doubtful outcome of the test means that we must accept the hypothesis and treat the components as being insufficiently strong. It is thus impossible to control both errors at the same time.

The statistical test we use to evaluate fatigue tests is based on the number of successful trials. If we conduct N independent tests with results

$$y_n = \begin{cases} 0 & \text{test } n \text{ failed (component broken)} \\ 1 & \text{test } n \text{ succeeded (component intact)} \end{cases} \tag{10.6}$$

the number of successes

$$S = \sum_{n=1}^{N} y_n \tag{10.7}$$

is said to have a *binomial distribution* with parameters N (number of trials) and p (success probability per trial), in short $S \sim B(N, p)$. As we want to show that the probability of failure for the test load is less than some value p_0, the hypothesis and alternative are equivalent to

$$H_0 : p \leq 1 - p_0 \qquad H_1 : p > 1 - p_0 \qquad (10.8)$$

We are only willing to reject H_0 if the observed number of successes is large enough. For a more precise statement, we need to take into account the desired significance level and the number of tests we want to perform.

Note that, so far, we have made no assumption regarding a probability distribution for strength. In fact, the binomial test does not even require that we can convert the test load into a scalar value using a damage accumulation law. Unfortunately, the following example shows that this approach is usually not feasible in practice.

Example 10.3 (Pure Binomial Test)

Suppose that we need to show that all but $p_0 = 1\%$ of the components under investigation survive the test load. The significance level is $\alpha = 5\%$, i.e. we are willing to accept that we release an insufficient component 5% of the time. In the context of fatigue testing, both requirements are fairly mild.

How many samples do we need to guarantee level α? As the test rejects H_0 if sufficiently many components survive, it needs to be able to do this at least if all tests are successful. The probability for this to happen is

$$P(S = N) = p^N. \qquad (10.9)$$

The consumer's risk is maximal if the probability of a failure is exactly 1%, i.e. if the component is almost strong enough and $p = 1 - p_0$. In other words, we need to ensure that

$$P(S = N \text{ and } H_0 \text{ is true}) \leq (1 - p_0)^N \leq \alpha$$

$$\Rightarrow N \geq \frac{\ln(\alpha)}{\ln(1 - p_0)} = 298.1. \qquad (10.10)$$

This means we need to test at least 299 components to obtain the desired significance level, which is far too many for most applications. The problem is the small failure probability p_0. For example, we only need 5 successful tests to ensure at the 5% level that the median strength ($p_0 = 50\%$) is at least equal to the given load. $\qquad \square$

The example shows that binomial tests with a small number of samples can only provide reliable information about moderate or high failure probabilities. To obtain useful results in the context of fatigue testing, we need additional assumptions. These usually take the form of a damage law and a stochastic model for (pseudo) damage of the kind discussed in Section 10.3.1. As we do not want to estimate the variance parameter, this has to be specified as well.

Example 10.4 (Binomial Test with Distribution)

As in Example 10.3, we want to test at the level $\alpha = 5\%$ whether the failure probability of a component given some test load is less than $p_0 = 1\%$. But this time, we believe that the strength distribution is log-normal with a standard deviation of $\sigma = 0.46$.

The log-normal distribution is fully determined by its standard deviation and the value of q_0 of its 1% quantile, which is simply the (pseudo) damage number associated with the test load. We can derive the distribution parameter μ from the relationship between the quantiles and parameters of a normal distribution:

$$\ln(q_0) = \mu + \sigma \Phi^{-1}(p_0) \iff \mu = \ln(q_0) - \sigma \Phi^{-1}(p_0). \tag{10.11}$$

The letter Φ represents the distribution function of the standard normal distribution. We can now translate the original test for the failure probability p_0 and load q_0 into an equivalent test with a higher load. For instance, we might want to run the test $L = 3$ times as long, which leads to a load of $q_L = Lq_0$. Using the expression we have derived for μ, we can calculate the failure probability as

$$p_L = \Phi\left(\frac{\ln(q_L) - \mu}{\sigma}\right) = \Phi\left(\frac{\ln(L)}{\sigma} + \Phi^{-1}(p_0)\right) = 52\%. \tag{10.12}$$

The minimal number of tests that can guarantee the desired significance level $\alpha = 5\%$ for a failure probability of $p_L = 52\%$ is $N = 5$, which is usually feasible. □

The binomial test is well established in practice for evaluating fatigue tests. However, it does not use the available information to full effect in case failures occur, as no distinction is made between parts that fail early and parts that almost reach the target life. An improved test using fractional counts is developed in Feth [93]. It is based on the idea that a component that fails still contributes some value less than 1 to the number of successful trials S, depending on how long it survived.

10.3.4.1 Test Planning

Using a distribution model with fixed scatter parameter, we can derive valid test schedules for different numbers of components and run lengths. All guarantee the desired significance level α, but the cost may vary in terms of money or time invested. This allows us to choose the most efficient test with regard to practical restrictions.

Table 10.2 shows some feasible configurations for the test discussed in Example 10.4, and for a test under the same conditions for a failure probability of $p_0 = 0.1\%$. The value L is the run length expressed as a multiple of the quantile q_0, m is the maximal number of failures for which the hypothesis can still be rejected, and N is the number of components

Table 10.2 Sample test schedules

	$\sigma = 0.46$ $p_0 = 10^{-2}, \alpha = 5\%$				$\sigma = 0.46$ $p_0 = 10^{-3}, \alpha = 5\%$		
	$L = 3$	5	10		$L = 3$	5	10
$m = 0$	$N = 5$	2	1	$m = 0$	$N = 11$	3	1
1	7	3	2	1	18	6	3
2	10	5	3	2	24	8	4
3	13	6	4	3	30	9	5

that need to be tested in order to guarantee the significance level $\alpha = 5\%$. The tables were generated by fixing L and m to some value and then taking as N the minimal number of components satisfying

$$\sum_{k=0}^{m} \binom{N}{k} p_L^k (1 - p_L)^{N-k} \leq \alpha. \qquad (10.13)$$

The term on the left-hand side is the consumer's risk in the case that the component is marginally too weak, i.e. the probability to reject the hypothesis although the failure probability is exactly p_0. The probability p_L is the corresponding failure probability for the increased test load q_L, which can be calculated as

$$p_L = \Phi\left(\frac{\ln(L)}{\sigma} + \Phi^{-1}(p_0)\right), \qquad \text{(log-normal)}$$

$$p_L = 1 - (1 - p_0)^{L^c}. \qquad \text{(Weibull)} \qquad (10.14)$$

While this approach to test planning is very flexible, we need to be cautious in one respect. All of the tests designed in this manner have a significance level of α, so they are equivalent with respect to the error of the first kind. However, they can differ considerably in their ability to reject the hypothesis if the alternative is true and the components are in fact sufficiently strong.

The probability of a test to reject the hypothesis if it is false is called its *power*. For the binomial test, it is a function of the true value of the strength distribution quantile q_0. If the component is barely strong enough, the power of any test schedule will be low. But we can compare different schedules based on their ability to detect components that are markedly better than the minimum requirement:

Example 10.5 (Power for Different Schedules)
Consider again the test of Example 10.4. We denote by R the ratio of improvement over the target quantile q_0, i.e. the true quantile of the strength distribution is R times greater than the minimum requirement. In the log-normal model, this translates to a true mean of

$$\mu_R = \ln(q_0) + \ln(R) - \sigma \Phi^{-1}(p_0). \qquad (10.15)$$

The probability to observe a failure if we test a single component for $q_L = L q_0$ cycles is

$$p_R = \Phi\left(\frac{\ln(L) - \ln(R)}{\sigma} + \Phi^{-1}(p_0)\right). \qquad (10.16)$$

If $R > 1$, the hypothesis is false and we want to reject it if possible. The probability for this to happen is the power $\beta(R)$ of the test. For a success run with N components, this is equal to

$$\beta(R) = (1 - p_R)^N. \qquad (10.17)$$

Table 10.2 offers three schedules for a success run in case $p_0 = 1\%$. The power for different values of R is shown in Figure 10.3. For example, the schedule with 5 components and $L = 3$ has a 50% chance of releasing components that are 1.8 times stronger than required. The extreme test with a single component and $L = 10$ has the same power only for $R = 3.5$.

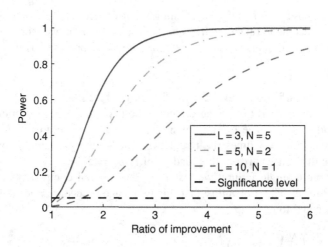

Figure 10.3 Power for different test schedules ($\sigma = 0.46$, $p_0 = 10^{-2}$, $\alpha = 5\%$)

Note also that the probability of rejecting the hypothesis if $R = 1$ (components are marginally too weak) is in all cases below the significance level $\alpha = 5\%$. This is due to the fact that we have to round up the number of components N when solving equations like Equation (10.10). In particular, the test schedule we obtain is usually more strict than required. This 'gap' between the actual and desired significance level can be closed by employing the fractional counting method developed in Feth [93].   □

Under the assumption of a log-normal distribution, tests with many components and short runtime have a higher power than those with few components and long runtime. In the extreme case, a single part tested for a very long time has a low probability of survival regardless of quality. Conversely, a test with many components may even be able to reject the hypothesis if some failures occur – and these are much less likely due to the shorter runtime. If one has the choice between several schedules, it is thus recommended to use the test with the most components.

10.3.4.2   Sequential Testing and Prior Information

Schedules like those given in Table 10.2 suggest the possibility of 'saving' a test that was unable to reject the hypothesis, but only barely failed to do so. Assume that we tested $N = 5$ components according to the plan for $p_0 = 10^{-2}$ and $L = 3$, but $m = 1$ of them broke. We cannot reject the hypothesis, as no failures are permissible. However, we note that for $N = 7$, a single failure is acceptable. The naive approach now would be to test two additional components and reject H_0 if none of them fails. The question is whether this still yields a valid statistical test.

Unfortunately, the answer is 'no' – unless we are willing to increase the significance level. Distortions due to rounding notwithstanding (compare Example 10.5), the binomial test is constructed in such a manner that the probability for an error of the first kind is equal to α in the worst case $p = 1 - p_0$. This assumes that we accept H_0 if we observe more than m failures. If we decide to reject the hypothesis despite this rule, we increase the probability of releasing a weak component, so the extended decision process has a significance level greater than α.

Example 10.6 (Significance Level for an Extended Test Plan)

Assume that we performed the test described at the beginning of this section with $\sigma = 0.2$, $p_0 = 10^{-2}$, and $\alpha = 5\%$. The original plan called for $m = 0$ failures in $N = 5$ trials with run length equal to $L = 3$ times the target life. The failure probability that separates the hypothesis from the alternative at this load level is $p_L = 52\%$.

As we observe a single failure, we test $\tilde{N} = 2$ additional components and reject H_0 if none of them breaks. The probability that this two-stage procedure rejects the hypothesis if $p = 1 - p_0$ is equal to

$$\sum_{k=0}^{m} \binom{N}{k} p_L^k (1 - p_L)^{N-k} + \binom{N}{m+1} p_L^{m+1} (1 - p_L)^{N-(m+1)+\tilde{N}}$$

$$= \binom{5}{0} 0.52^0 0.48^5 + \binom{5}{1} 0.52^1 0.48^6 = 0.08. \qquad (10.18)$$

By extending the test, we are implicitly using a significance level of 8%. Note how the value of the first sum is already as close to α as possible due to the choice of m in Equation (10.13), so we are bound to exceed the initial level. □

The ad hoc extension of a test plan is problematic for the same reason it is not recommended to fix the significance level of a test after making observations. As the experimenter is interested in a positive outcome, he will be inclined to adjust the rule for rejecting H_0 in case of borderline results. The actual significance of such a decision procedure is dubious. For instance, the extended schedule discussed in Example 10.6 has a significance level of 8% based on the assumption that extra trials are only attempted in case of a single failure. If retries are permitted also with two failures, α becomes even larger. In order to get an honest assessment of the reliability of our decision rule, we need to decide in advance when we are to stop. Remember that the significance level is defined as the maximum consumer risk we are willing to tolerate, which is not an appraisal we should revise on the fly.

To obtain a consistent process, we need to delineate the entire schedule in advance. This could mean conducting the initial test at a lower significance level than required by our specifications and permitting a retrial only in case of exactly one failure too many. The necessary number of additional trials can then be calculated by solving an equation similar to Equation (10.18) for fixed α. Schedules of this kind are known as sequential tests. A discussion is found, for example, in Govindarajulu [104], while Martz and Waller [155] treat them in conjunction with Bayesian methods.

The desire to conduct extra trials usually stems from additional information that cannot be represented in the binomial test. For instance, we might observe that the components which did fail almost reached the target life. In this case, the test statistic developed in Feth [93] based on fractional counts may still permit rejecting the hypothesis at the specified level. In other situations, we might be reasonably certain about the quality of the components and want to incorporate our assessment into the mathematical model. This issue is more common when dealing with quality control during production, but the same methods can be used during the design stage if the implications are understood. The tool of choice in this case are Bayesian tests, where beliefs about the unknown parameter can be specified in terms of a prior distribution. Bayesian methods in general are discussed in Berger [21], Bernardo and Smith [24], or Box and Tiao [34]. Their application in the context of reliability analysis is the subject of Martz and Waller [155].

10.4   Discussion and Summary

This chapter is largely devoted to the practical evaluation of *probability-based formal procedures*, with test plans based on formal consistent rules that, by experience, give safe designs. However, also two other procedures for verification are put forward, namely

1. *Highly Accelerated Life Testing*, HALT, based on the idea that failures give more information than non-failures and give rise to improvements regardless of the severities that exceed what is expected.
2. *Load-Strength analysis based on characterizing tests.* Strength and load properties are investigated by characterizing experiments. Scatter and uncertainties are analysed within a statistical framework to verify the design against reliability targets by means of established safety factors.

Table 10.3 summarizes some of the features of these three procedures.

The main purpose of the severe tests is to form a basis for continuous improvements by discovering weak spots in complex structures. The second moment load-strength analysis also gives opportunities for continuous improvements by identifying the areas of weakest knowledge and can also be the basis for constructing formal decision rules for release. Probability-based formal procedures are primarily constructed for consistent decision rules with no immediate reference to the specific component or structure in question.

The load severity is far above target in the case of the severe tests, since the very purpose is to stress the object to failure. In the case of characterizing tests it is important that the object is characterized with respect to the failure mode under study and any severity above target need to be judged by means of the possible error that it introduces. For the probability-based formal procedures minor violation of failure mode may be accepted, but consistency is of the utmost importance.

The different procedures have their main applications at different levels of complexity. Characterizing tests are often too time-consuming to be applied to complex structures. HALT tests are not applicable to most fatigue failure modes, since these modes would be hidden by immediate failures if the severity is far above target. However, formal test procedures based on consistent decision rules could be applied at all levels of complexity.

Acceleration of tests should, as in the case of severity, be minor for characterizing tests with respect to amplitude acceleration, but can often be achieved safely by different types of time or cycle acceleration. For HALT tests on complex structures, eigenfrequencies in the structure may limit time and cycle acceleration, and amplitude acceleration is dominating. Just as in our considerations about severity, the probability-based formal procedures may be accelerated by all types of acceleration, based on engineering judgment, but again, consistency must not be violated.

Table 10.3 Summary of features of the three verification procedures

Verification by …	Main purpose	Load severity	Level of complexity	Acceleration	Necessary engineering judgement	Rule for release
1. Severe tests, HALT	• find weak spots • continuous improvements	Far above target	Full vehicle, subsystem	Large	Load severity	Could the observed failure happen in service?
2. Characterizing tests, load/strength	• find weak areas of knowledge • decision rule	At target	Components, material	Minor	Extra safety factor	Fulfilment of target safety factor
3. Formal test procedures	• consistent decision rules	Above target	Component, subsystem, full vehicle	Well defined	Distribution type, fixed parameters	Fulfilment of target prob. of failure

Since knowledge is limited, each verification method must contain a certain amount of engineering judgment. In the case of Highly Accelerated Life Testing we must judge an appropriate test severity and judge if an observed failure in a reinforced test could occur in service. In the case of load-strength analysis we judge an extra safety factor and different targets. In the case of probability-based formal procedures, the release criteria are based on engineering judgments.

Appendix A

Fatigue Models and Life Prediction

A.1 Short, Long or Infinite Life

Fatigue is the failure mechanism that is caused by repeated load cycles with amplitudes well below the ultimate static material strength. Physically, the damage caused at the microstructural level by each load cycle is plastic, that is, irreversible strain. However, this local damaging strain can usually not be measured nor be calculated, and therefore the fatigue strength is often related to stress measures whose local values can be derived from known outer loads by elastic theory. For further studies on fatigue, see e.g. Bannantine *et al.* [12], Berger *et al.* [20], Chaboche and Lemaitre [53], Collins [58], Dahlberg and Ekberg [62], Dowling [77], Haibach [110], Schijve [209], Stephens *et al.* [223] and Suresh [225].

A.1.1 Low Cycle Fatigue

When plastic strains are measurable also on a macroscopic level, the fatigue failure occurs after a quite limited number of cycles and one refers to the term *low cycle fatigue* (LCF). Since the plastic strain occurs in large volumes of the material, the fatigue strength is often related to the material's bulk properties that can be obtained by laboratory experiments. The design concepts regarding low cycle fatigue are often applied in cases where less than, say, ten thousand slow large cycles appear in the structure's life. This is typically the case for thermal cycles connected to start/stop cycles for air engines and other turbines or mechanical cycles for aircraft start, stop and mission events.

A.1.2 High Cycle Fatigue

When only elastic strains are possible to detect and used for predictions one refers to *high cycle fatigue* (HCF). This concept deals with lives from tenth of thousands of cycles to a few millions and the damage evolution is localized to plastic strain at a crack tip, causing crack growth. Cracks develop early in life from defects and may grow in a stable

Guide to Load Analysis for Durability in Vehicle Engineering, First Edition. Edited by P. Johannesson and M. Speckert.
© 2014 Fraunhofer-Chalmers Research Centre for Industrial Mathematics.

manner from the size of some hundred microns to a critical size of several millimetres. The localization of the damaging plastic strain to one or a few cracks often makes material tests irrelevant to fatigue life; sharp notches or defects by means of porosity, inclusions, surface irregularities or welds are of significant importance. Thus, reference tests should be performed on components containing similar defects as the final product.

A.1.3   Fatigue Limit

For components that are subjected to more load cycles than can be controlled in the fatigue life concepts above, one must design against the *fatigue limit*. The concept of fatigue limit assumes that there is a certain stress level for a component which will not cause any damage at all, and cycling at stress levels below this limit will give infinite life. In practice infinity may be regarded as the largest number of cycles that will be applied due to other limitations of the product life. Experience has shown that if the fatigue limit is determined by experimental tests up to about ten million of cycles, it is also relevant for a vehicle life.

The fatigue limit is, like the high cycle fatigue properties, governed by the defects in the material in question. Namely, there is a relationship between the fatigue limit and the threshold for crack growth, where the critical crack length is related to the largest defect size in the stressed material.

A.2   Cumulative Fatigue

The three different approaches for fatigue strength: low cycle, high cycle, and fatigue limit are all empirical relationships traditionally based on experimental tests in laboratory. These tests were made with an outer load, strain or stress controlled, with constant amplitude. Based on such experiments it was possible to establish a material or component strength by means of a strain/life diagram, a stress/life diagram or an endurance limit for a specified life in cycles, interpreted as the fatigue limit.

These strength properties for the material or the component are intended to be used for design, that is, for life prediction. But the service loads are usually not time histories of constant amplitude. Instead they can be regarded as random processes and the constant amplitude properties need to be translated to variable amplitude situations. In most applications this is done by using the Palmgren-Miner cumulative damage rule, [183, 161], which usually is written

$$D = \sum_{j=1}^{M} \frac{n_j}{N_j}. \tag{A.1}$$

Here D is the accumulated damage measure, equal to one at failure, M is the number of load levels, n_j is the number of load cycles for the j:th level and N_j is the life predicted from the Wöhler curve, $N_j = \alpha \cdot \Delta S_j^{-\beta}$ with stress range ΔS_j.

A.2.1   Arguments for the Palmgren-Miner Rule

In the case of crack growth, it is possible to derive the Palmgren-Miner cumulative damage rule, see for example, Svensson and de Maré [228]. In that case the damage is crack length and failure is related to the situation when the crack length is equal to a critical length which causes immediate failure. This can be seen in the following derivation.

A.2.1.1 The Crack Growth Equation

Stable crack growth in metallic materials is often described by the differential equation called the Paris' law:

$$da/dN = C' \cdot \Delta K^m = C \cdot \Delta S^m \cdot a^{m/2} \tag{A.2}$$

where ΔK is the stress intensity factor range, m is the material dependent growth exponent, C' and C are factors depending on geometry and material, ΔS is the far field stress range, and a is the crack length. If this equation is assumed to be independently valid for each load cycle we obtain for the ith cycle the crack increment:

$$\Delta a_i = C \cdot \Delta S_i^m \cdot a_i^{m/2}. \tag{A.3}$$

Taking the sum of all increments from the initial crack length a_0 to the critical length a_c and rearranging gives

$$\sum_{i=1}^{N} \frac{\Delta a_i}{C \cdot a_i^{m/2}} = \sum_{i=1}^{N} \Delta S_i^m. \tag{A.4}$$

The left-hand side of this equation can be approximated by the corresponding integral,

$$\sum_{i=1}^{N} \frac{\Delta a_i}{C \cdot a_i^{m/2}} \approx \int_{a_0}^{a_c} \frac{da}{C \cdot a^{m/2}} = \frac{a_0^{m/2-1} - a_c^{m/2-1}}{C(m/2 - 1)} = \alpha. \tag{A.5}$$

This property is constant through the whole damage process and is here denoted α to correspond to the Wöhler curve representation. Replacing the sum on the right-hand side of Equation (A.4) we have

$$\alpha = \sum_{i=1}^{N} \Delta S_i^m \tag{A.6}$$

$$1 = \frac{1}{\alpha} \sum_{i=1}^{N} \Delta S_i^m = \sum_{i=1}^{N} \frac{1}{\alpha \cdot \Delta S_i^{-m}} = \sum_{i=1}^{N} \frac{1}{N_i} \tag{A.7}$$

which corresponds to the usual Palmgren-Miner rule (A.1).

The derivation of the cumulative rule in the crack growth case was done using two critical assumptions. Firstly, in Equation (A.3) we assume that the individual load contributions are independent. This assumption is not always valid, since material memory effects exist. Secondly, the rearrangement of Equation (A.4) demands that the expression containing the crack length a_i must be separated from the function containing ΔS_i. This is apparently the case for this ordinary Paris law and is also usually the case for the existing different modifications of the crack growth equation, except for cases when a stress intensity threshold is included.

A.2.1.2 The Plastic Strain Motivation

Jono and co-workers have done a lot of experiments showing that the cumulative damage rule is a good model if the load variable in the summation is based on measured *plastic strain range*; Jono [129]. These results imply that observed deviations from the damage accumulation model should be attributed to other errors, for instance, non-linear relationships between the true plastic strain ranges and stress-based approximations.

A.2.2   *When is the Palmgren-Miner Rule Useful?*

As seen from the derivation of the rule based on crack growth the assumptions that must be fulfilled are 1) the load and crack functions must be separable, and 2) the order of load application must not be influential.

Appendix B

Statistics and Probability

The main purpose of this appendix is to give references to further reading in different topics in statistics and probability, but we also present some useful statistical distributions.

B.1 Further Reading

For applied statistics and probability we recommend Montgomery and Runger [163], and for statistical estimation Casella and Berger [51], Freedman *et al.* [94] and Pawitan [184]. Design of experiment can be studied in Box *et al.* [35] and Montgomery [162]. For survey sampling we recommend Särndal *et al.* [207], and Lohr [148]. For general information about the bootstrap technique, see Davison and Hinkley [65], Efron and Tibshirani [88] and Hjorth [114]. Stochastic process can be read about in Brillinger [39], Brockwell and Davis [40], Grimmett and Stirzaker [105] and Priestley [192]. For general information about reliability, see O'Connor [172], Rausand and Høyland [193], Ditlevsen and Madsen [74], Madsen *et al.* [151] and Melchers [159]. Extreme value theory can be studied in Castillo [52], Coles [57] and Leadbetter *et al.* [142].

B.2 Some Common Distributions

We here give the probability density function (PDF), cumulative distribution function (CDF), expectation, and variance for some common statistical distributions.

B.2.1 Normal Distribution

$$f(x|\mu, \sigma) = \frac{1}{\sqrt{2\pi}\sigma}\, e^{-\frac{(x-\mu)^2}{2\sigma^2}}, \quad F(x) = \Phi\left(\frac{x-\mu}{\sigma}\right), \quad -\infty < x < \infty \qquad (B.1)$$

$$\mathbf{E}[X] = \mu, \quad \mathrm{Var}[X] = \sigma^2 \qquad (B.2)$$

Guide to Load Analysis for Durability in Vehicle Engineering, First Edition. Edited by P. Johannesson and M. Speckert.
© 2014 Fraunhofer-Chalmers Research Centre for Industrial Mathematics.

B.2.2 Log-Normal Distribution

$$f(x|\mu, \sigma) = \frac{1}{\sqrt{2\pi}\sigma x}\, e^{-\frac{(\log(x)-\mu)^2}{2\sigma^2}}, \quad x > 0 \tag{B.3}$$

$$\mathbf{E}[X] = e^{\mu+\sigma^2/2}, \quad \text{Var}[X] = (e^{\sigma^2} - 1)e^{2\mu+\sigma^2} \tag{B.4}$$

B.2.3 Weibull Distribution

$$f(x|a, c) = ca^{-c}x^{c-1}e^{-(\frac{x}{a})^c}, \quad F(x) = 1 - e^{-(\frac{x}{a})^c}, \quad x \geq 0 \tag{B.5}$$

$$\mathbf{E}[X] = a\Gamma\left(1 + \frac{1}{c}\right), \quad \text{Var}[X] = a^2\Gamma\left(1 + \frac{2}{c}\right) - \left(a\Gamma\left(1 + \frac{1}{c}\right)\right)^2 \tag{B.6}$$

B.2.4 Rayleigh Distribution

$$f(x|\sigma) = \frac{x}{\sigma^2}e^{-\left(\frac{x^2}{2\sigma^2}\right)}, \quad F(x) = 1 - e^{-\frac{x^2}{2\sigma^2}}, \quad x \geq 0 \tag{B.7}$$

$$\mathbf{E}[X] = \sigma\sqrt{\frac{\pi}{2}}, \quad \text{Var}[X] = 2\sigma^2\left(1 - \frac{\pi}{4}\right) \tag{B.8}$$

B.2.5 Exponential Distribution

$$f(x|m) = \frac{1}{m}e^{-\frac{x}{m}}, \quad F(x) = 1 - e^{-\frac{x}{m}} \quad x \geq 0 \tag{B.9}$$

$$\mathbf{E}[X] = m, \quad \text{Var}[X] = m^2 \tag{B.10}$$

B.2.6 Generalized Pareto Distribution

$$f(x|\xi, \mu, \sigma) = \frac{1}{\sigma}\left(1 + \frac{\xi(x - \mu)}{\sigma}\right)^{(-\frac{1}{\xi}-1)}, \tag{B.11}$$

$$F(x) = 1 - \left(1 + \frac{\xi(x - \mu)}{\sigma}\right)^{-1/\xi}, \tag{B.12}$$

$$\mathbf{E}[X] = \mu + \frac{\sigma}{1 - \xi}, \xi < 1, \quad \text{Var}[X] = \frac{\sigma^2}{(1 - \xi)^2(1 - 2\xi)}, \xi < 0.5 \tag{B.13}$$

B.3 Extreme Value Distributions

The extreme value distributions are models for the maximum of many random variables.

B.3.1 Peak over Threshold Analysis

The theoretical base for the method is the so-called Peak Over Threshold (POT) technique in statistical extreme value theory, see e.g. Davison and Smith [66] for an overview. Only the extreme excesses over a high level u are modelled, that is, the height of the excursions above u. If we fix a high load level and study the excesses over this level, then under certain conditions these excesses approximately follow an exponential distribution for high enough threshold levels u. We then have the approximation for the excess $Z = Max - u \in Exp(m)$, with cumulative distribution function

$$F(z) = 1 - \exp(-z/m), \qquad m = \text{"mean excesses over } u\text{"}. \qquad (B.14)$$

The estimation of the parameter in the exponential distribution is the sample mean of the excesses $z_1, \ldots, z_N$

$$\hat{m} = \frac{1}{N} \sum_{i=1}^{N} z_i. \qquad (B.15)$$

Appendix C

Fourier Analysis

C.1 Fourier Transformation

The aim of this appendix is to supply the most important notions and formulas of Fourier analysis. For those familiar with the subject it might serve as a short reminder. For beginners it should serve as a guideline for a more comprehensive introduction to the cited text books. Among the large number of available books we mention only Oppenheim and Willsky [180], Marple [152], Brigola [38], Press *et al.* [191], and Oppenheim and Schafer [179].

For reasons of simplicity we use the complex notation based on the harmonic functions $e^{i \cdot x}$ instead of the real sine and cosine functions. The relation between both is given by

$$e^{i \cdot x} = \cos(x) + i \cdot \sin(x), \tag{C.1}$$

where $i = \sqrt{(-1)}$ is the imaginary unit. Using Equation (C.1) all formulas below can be transformed into their real counterparts.

For square integrable functions $x \in L_2$ such that

$$\int_{-\infty}^{\infty} |x(t)|^2 dt < \infty \tag{C.2}$$

there is the well-known Fourier transform

$$\hat{x}(f) = \int_{-\infty}^{\infty} e^{-i2\pi ft} x(t) dt \tag{C.3}$$

which maps each function $x \in L_2$ onto another function $\hat{x} \in L_2$. This mapping is one to one, that means that there is a uniquely defined inverse transformation given by

$$x(t) = \int_{-\infty}^{\infty} e^{i2\pi ft} \hat{x}(f) df. \tag{C.4}$$

The transformation can be interpreted as a decomposition of the function x into harmonic functions $e^{i2\pi ft}$ with weight $\hat{x}(f)$, where the frequencies f vary from $-\infty$ to ∞. The value $|\hat{x}(f)|$ of the transformed function is the amplitude of the corresponding harmonic function

Guide to Load Analysis for Durability in Vehicle Engineering, First Edition. Edited by P. Johannesson and M. Speckert.
© 2014 Fraunhofer-Chalmers Research Centre for Industrial Mathematics.

$e^{i2\pi ft}$. The integral over all squared amplitudes is a measure of the energy E of the signal. This energy can be calculated using either the original representation x or the transform $\hat{x}$.

$$E = \int_{-\infty}^{\infty} |x(t)|^2 dt = \int_{-\infty}^{\infty} |\hat{x}(f)|^2 df. \tag{C.5}$$

The Fourier transform can also be written in terms of sine and cosine functions in the space of real numbers but the complex formulation used here is more elegant and often preferred. For real x we have

$$\hat{x}(-f) = \hat{x}^*(f). \tag{C.6}$$

where x^* denotes the complex conjugate of x. There are some other important properties of the Fourier transform concerning the behaviour with respect to differentiation and convolution, which are briefly mentioned:

$$\frac{d}{dt} \xrightarrow{FT} i2\pi \cdot f \Rightarrow \hat{\dot{x}}(f) = i \cdot 2\pi f \cdot \hat{x}(f) \tag{C.7}$$

$$z(t) = (x * y)(t) = \int_{-\infty}^{\infty} x(t - \tau) \cdot y(\tau) d\tau \Rightarrow \hat{z} = \hat{x} \cdot \hat{y} \tag{C.8}$$

The representation $\hat{x}$ of x in the frequency domain is an alternative without any loss of information. The condition $x \in L_2$ in Equation (C.2) excludes, for example, constant or even harmonic functions. Although the theory can be extended to such cases, we will not go into these details.

There is a similar representation for periodic functions that is explained in the following section.

C.2 Fourier Series

For T-periodic functions x $(x(t) = x(t + T)$ for all $t)$ we have the representation

$$x(t) = \sum_{k=-\infty}^{\infty} c_k e^{i2\pi kt/T} \tag{C.9}$$

where

$$c_k = \frac{1}{T} \cdot \int_0^T e^{-i2\pi kt/T} x(t) dt. \tag{C.10}$$

As in Section C.1 we use the complex notation for simplicity. The existence of the sum in formula (C.9) requires the coefficients c_k to decay to 0. The order of convergence depends on the smoothness of the function x. The smoother the function x, the less coefficients c_k are required for a reasonable approximation of x by its truncated Fourier series

$$x(t) \approx \sum_{k=-M}^{M} c_k e^{i2\pi kt/T}. \tag{C.11}$$

Similar to the Fourier transform we can introduce the energy E of the function x and calculate it either using the original function x or the coefficients c_k as follows:

$$E = \int_0^T |x(t)|^2 dt = T \cdot \sum_{k=-\infty}^{\infty} |c_k|^2. \tag{C.12}$$

C.3   Sampling and the Nyquist-Shannon Theorem

In terms of signal processing, the load acting on a component is an analog signal, meaning that it can be represented by a function $x(t)$ in continuous time with a continuous range of values. On the other hand, measured loads are digital signals, consisting of floating-point values on an equidistant time grid. The error incurred by rounding an observation to its closest floating-point representation is usually negligible, as it is smaller than the accuracy of the measurements. The real problem is that we do not know the state of the process in between sampling times. For the purpose of mathematical analysis, we thus treat load data as a discrete time signal with a continuous range of values.

The main result in this context is the Nyquist-Shannon Sampling Theorem, which gives a sufficient condition for when it is possible to reconstruct a continuous time signal uniquely from a discrete time sample:

Theorem C.1 (Nyquist-Shannon Sampling Theorem) *Let $x(t)$ be a signal with bandwidth $f_b > 0$, meaning that $\hat{x}(f) = 0$ for all frequencies f with $|f| > f_b$. Then, the signal $x(t)$ is uniquely determined by the discrete sample $x_k = x(t_k)$, where*

$$t_k = k \cdot \Delta t \tag{C.13}$$

for $k \in \mathbb{Z}$ and some time step $\Delta t > 0$, provided that the sampling frequency $f_s = \frac{1}{\Delta t}$ satisfies

$$f_s > 2 f_b \tag{C.14}$$

Half the actual sampling rate is also referred to as the *Nyquist-frequency*, which we denote by

$$f_N = \frac{f_s}{2} \tag{C.15}$$

Provided the conditions of the Theorem C.1 are satisfied, the unique reconstruction of the continuous time signal from discrete samples is given by

$$x(t) = \begin{cases} \sum_{k=-\infty}^{\infty} x_k \frac{\sin(\pi(t-t_k)/\Delta t)}{\pi(t-t_k)/\Delta t} & \text{if } t \neq t_k \text{ for all } k \\ x_k & \text{if } t = t_k \text{ for some } k \end{cases} \tag{C.16}$$

This function is continuous at all grid points $x(t_k)$, as

$$\sin(\pi(t_j - t_k)/\Delta t) = \sin(\pi(j-k)) = 0 \qquad \lim_{s \neq 0, s \to 0} \frac{\sin(s)}{s} = 1 \tag{C.17}$$

where $j, k \in \mathbb{Z}$ are arbitrary. In practice, we will of course only have a finite number of sample points and assume that all other x_k are equal to 0.

C.4   DFT/FFT (Discrete Fourier Transformation)

The main tool of frequency analysis is the discrete Fourier transformation (DFT). It transforms a sampled time signal $x(t_k)$ into the frequency domain $\hat{x}(f_k)$. The inverse DFT (IDFT) transforms a frequency domain signal to the time domain.

$$x_k = x(t_k) \xrightarrow{\text{DFT}} \hat{x}(f_k) = \hat{x}_k, \qquad (C.18)$$

$$\hat{x}_k = \hat{x}(f_k) \xrightarrow{\text{IDFT}} x(t_k) = x_k. \qquad (C.19)$$

The method is defined in the following formulae:

$$\hat{x}_j = \sum_{k=0}^{N-1} x_k e^{i\frac{2\pi}{N}jk} \ \text{ for } \ f_j = \frac{j}{N\Delta_t}, \qquad (C.20)$$

$$x_k = \frac{1}{N} \sum_{j=0}^{N-1} \hat{x}_j e^{-i\frac{2\pi}{N}jk}, \ \ \hat{x}_{N-j} = \hat{x}_j^* \ \text{ for real } x_k. \qquad (C.21)$$

Most engineering software supports DFT/IDFT. In practice the very efficient fast Fourier algorithm (FFT/IFFT), see Cooley and Tukey [59], is used (most efficient if the number of samples is a power of 2). Its performance is almost linear in the number of sampling points.

If the sampled data comes from a signal containing higher frequencies than the Nyquist frequency, the Fourier transform becomes distorted (aliasing effect). This can be seen from the formula

$$\frac{1}{N}\hat{x}_j = c_j + \sum_{k=1}^{\infty} (c_{j+kN} + c_{j-kN}), \qquad (C.22)$$

which gives the relation between the exact Fourier coefficients c_j and the coefficients $\hat{x}_j$ calculated by the DFT/FFT algorithm. To reduce the aliasing effect, the signal needs to be low pass filtered before sampling (Shannon's theorem) as has been explained above.

Appendix D

Finite Element Analysis

This appendix introduces some basic notions about continuum mechanics and finite element approximations without going into details. The goal is to present the basic concepts and notions needed for understanding the sections about the capabilities and limits of finite element modelling in the context of load analysis as described in Chapter 5. A rigorous derivation as well as a complete presentation of the corresponding theory are beyond the scope of this guide. Instead, we refer to some of the existing textbooks about this subject, see, for example, Hughes [116] or Bathe [14].

D.1   Kinematics of Flexible Bodies

Fundamental to the description of a flexible (deformable) body is its kinematics. The latter describes the motion of a flexible body per se, irrespective of the loads acting upon it. The deformation resulting from the loading is, however, the primary unknown quantity of kinematics – hence, the so-called displacement vector, typically denoted by $\mathbf{u}$, is introduced. $\mathbf{u} = (u_1, u_2, u_3)$ is a three-dimensional vector with components u_i, $i = 1, \ldots, 3$. If we let $\mathcal{R}$ be the referential configuration of the material body under consideration (i.e. the configuration it assumes at some initial time t_0), and $\mathcal{D}$ a deformed configuration which the body has attained after a time t has elapsed, cf. Figure D.1, the displacement vector is given by

$$\mathbf{u}(x, t) := \mathbf{x}_{\mathcal{D}} - \mathbf{x}_{\mathcal{R}}$$

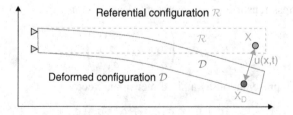

Figure D.1   Referential and actual configuration of the material body under consideration

Guide to Load Analysis for Durability in Vehicle Engineering, First Edition. Edited by P. Johannesson and M. Speckert.
© 2014 Fraunhofer-Chalmers Research Centre for Industrial Mathematics.

where $\mathbf{x}_D$ and $\mathbf{x}_R$ denote the position of a particle in the deformed and the referential configuration respectively, and t denotes time. The position vector $\mathbf{x}$ has components $\mathbf{x} = (x_1, x_2, x_3)$. Note that, if we let a superposed dot denote the time derivative of a quantity, $\dot{\mathbf{u}} = \dot{\mathbf{x}}_D = \mathbf{v}$ is the velocity of a particle located at position $\mathbf{x}_D$. To describe the strains induced by the deformation, the engineering strain tensor,[1] often interpreted as a 3×3 matrix with components

$$\varepsilon_{ij} = \frac{1}{2}\left(\frac{\partial u_i}{\partial x_j} + \frac{\partial u_j}{\partial x_i}\right) , \quad i, j = 1, 2, 3 \tag{D.1}$$

is introduced. Particularizing Equation (D.1) to particular components leads e.g. to $\varepsilon_{11} = \partial u_1/\partial x_1$ and $2\varepsilon_{12} = \partial u_1/\partial x_2 + \partial u_2/\partial x_1$ etc. Employing the engineering strain tensor ε as defined in Equation (D.1) implies that considerations are restricted to what is known as geometric linearity of a problem – in other words, only small deformations and small strains can be adequately captured by Equation (D.1).

In most numerical applications, however, yet another representation of the engineering strain tensor is used: as only 6 of its 9 components are independent (ε is a *symmetric* tensor), ε is frequently represented by a six-dimensional vector so that

$$\varepsilon \hat{=} (\varepsilon_{11}, \varepsilon_{22}, \varepsilon_{33}, 2\varepsilon_{12}, 2\varepsilon_{13}, 2\varepsilon_{23}) . \tag{D.2}$$

D.2   Equations of Equilibrium

Restricting attention to mechanical problems only (e.g. neglecting the coupling of the mechanical response of the system with e.g. thermal effects) implies that the fundamental equation on which all subsequent analysis is based, is the equation of mechanical equilibrium or, equivalently, the principle of virtual work.

The latter can be expressed in the form

$$\int_D \delta\varepsilon^T \sigma \, dv = \int_D \delta u^T f^b dv + \int_{\partial D} \delta u^{S^T} t \, da + \sum_i \delta u^{i^T} f^i , \tag{D.3}$$

where δu are virtual displacements, $\delta\varepsilon$ are the corresponding virtual strains, σ are the stresses, f^b are the body forces (gravity), t are prescribed surface stresses (traction vectors), and f^i are point forces acting on the body. The left side of this equations is the inner virtual work and the right-hand side denotes the outer virtual work. Equation (D.3) needs to be fulfilled for arbitrary virtual displacements, compatible with the boundary conditions.

The Cauchy stress tensor is often represented by a symmetric 3×3 matrix,

$$\sigma \hat{=} \begin{pmatrix} \sigma_{11} & \sigma_{12} & \sigma_{13} \\ \sigma_{12} & \sigma_{22} & \sigma_{23} \\ \sigma_{13} & \sigma_{23} & \sigma_{33} , \end{pmatrix} \tag{D.4}$$

alternatively, the representation as a six-dimensional vector

$$\sigma \hat{=} (\sigma_{11}, \sigma_{22}, \sigma_{33}, \sigma_{12}, \sigma_{13}, \sigma_{23}) \tag{D.5}$$

can be employed.

---

[1] In general, symbolic notation of the engineering strain tensor is given by the symmetric part of the spatial displacement gradient, of $\varepsilon := \text{sym}\nabla\mathbf{u}$.

D.3 Linear Elastic Material Behaviour

A solution of Equation (D.3) requires the prescription of what is known as a material (or constitutive) law, typically relating stress to strain. The simplest of such a material law is relating stress and strain in a linear fashion, where the "factor of proportionality" is the elasticity tensor $\mathbf{C}$:

$$\sigma = \mathbf{C}\varepsilon. \tag{D.6}$$

In the one-dimensional case, Equation (D.6) reduces to Hooke's law $\sigma = E\varepsilon$, in which E denotes the modulus of elasticity (also called Young's modulus). The elasticity tensor $\mathbf{C}$ is determined through elastic constants and it is emphasized that among the five elastic constants generally available (Young's modulus E, Poisson's ratio v, bulk modulus κ, shear modulus G (or μ) and Lamé constant λ) only two are independent – the remaining three can always be expressed as combinations of the two independent constants via e.g.

$$G = \mu = \frac{E}{2(1+v)}, \quad \kappa = \frac{E}{3(1-2v)}, \quad v = \frac{E}{2G} - 1, \quad \lambda = \frac{vE}{(1+v)(1-2v)}. \tag{D.7}$$

A typical representation of Equation (D.6) reads

$$\begin{pmatrix} \sigma_{11} \\ \sigma_{22} \\ \sigma_{33} \\ \sigma_{12} \\ \sigma_{13} \\ \sigma_{23} \end{pmatrix} = \begin{pmatrix} 2\mu+\lambda & \lambda & \lambda & 0 & 0 & 0 \\ \lambda & 2\mu+\lambda & \lambda & 0 & 0 & 0 \\ \lambda & \lambda & 2\mu+\lambda & 0 & 0 & 0 \\ 0 & 0 & 0 & \mu & 0 & 0 \\ 0 & 0 & 0 & 0 & \mu & 0 \\ 0 & 0 & 0 & 0 & 0 & \mu \end{pmatrix} \begin{pmatrix} \varepsilon_{11} \\ \varepsilon_{22} \\ \varepsilon_{33} \\ 2\varepsilon_{12} \\ 2\varepsilon_{13} \\ 2\varepsilon_{23} \end{pmatrix} \tag{D.8}$$

and describes, due to the fact that the entries of $\mathbf{C}$ are chosen as constants, linear elastic material behaviour. Non-linearities in the material behaviour (as occurring e.g. for materials the response of which is non-linear elastic, visco-elastic, visco-plastic, purely plastic, etc.) require more advanced constitutive relations which will, however, not be discussed here. Other types of non-linearities include, for example, geometric-structural non-linearities (occurring if large deformations, buckling phenomena, or the coupling of deformation and load direction have to be accounted for), non-linearities arising from the presence of contact boundary conditions between components, and non-linearities arising if thermo-mechanically coupled problems have to be solved.

The solution of Equation (D.3) directly yields the displacement vector $\mathbf{u}$, the resulting stresses $\sigma = \sigma(\mathbf{u}(\mathbf{x}))$ have to be computed in an *a posteriori* step based on the chosen material law. Typically, Equation (D.8) is employed, in which the components ε_{ij} have to be replaced in terms of the displacement-components u_i according to Equation (D.1).

D.4 Some Basics on Discretization Methods

The numerical solution of Equation (D.3) requires a discretization of the geometry the continuous material component is occupying as well as a discretization of the continuous vector-fields of primary interest, for example, the displacement-field $\mathbf{u}(\mathbf{x})$.

Typical examples of such finite elements are intervals (for one-dimensional problems), triangles or rectangles (for 2D problems) and tetrahedrons, prisms, pyramids or hexahedra

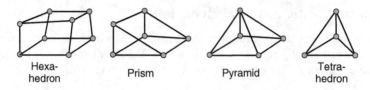

Hexa-   Prism   Pyramid   Tetra-
hedron   hedron

Figure D.2 Typical examples of (3D) finite elements: tetrahedrons, prisms, pyramids, or hexahedra

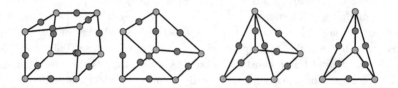

Figure D.3 Higher order finite elements, characterized by additional nodes

for 3D problems, to name but a few, cf. Figure D.2. In the simplest case, the elements are defined by nodes which are located at the corners of an element (cf. Figure D.2), however, more advanced elements are characterized by additional nodes, marked red in Figure D.3. Note that we have mentioned only solid elements. Other types of elements, e.g. shell-elements, do also exist but will not be discussed here.

The discretization of the displacement field is achieved by replacing the continuous function $\mathbf{u}(\mathbf{x}, t)$ by a finite-dimensional vector, containing as components the values of $\mathbf{u}$ evaluated at the nodes of the finite elements. Hence, the resulting vector is often simply called the *vector of nodal values* or *node vector*. The approximated behaviour of $\mathbf{u}$ between those nodes depends on the choice of what is known as shape functions: shape functions of minimal regularity (e.g. piecewise linear polynomials) yield a continuous displacement $\mathbf{u}$ across the element boundaries, cf. Figure D.4. Introducing the discretized geometry, discretized displacement, and an appropriate element formulation into the equation of mechanical equilibrium transform the latter into a system of linear equations, viz.

$$\mathbf{Kv} = \mathbf{f} \tag{D.9}$$

where $\mathbf{K}$, $\mathbf{v}$ and $\mathbf{f}$ are known as the stiffness matrix, the vector of nodal degrees of freedom, and the load vector, respectively. Note that Equation (D.9) describes a static system. The stiffness matrix is of dimension $n \times n$, where n is the number of degrees of freedom of the system, Moreover, its entries depend on the shape functions and material properties such as

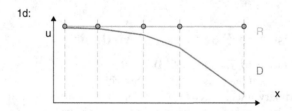

Figure D.4 Approximation of the continuous displacement by piecewise linear functions

e.g. E and ν. The vectors **v** and **f** are of dimension $n \times 1$, respectively. Note that in setting up the system of linear equations, Equation (D.9), one has to account for the specification of appropriate boundary conditions as well as loads acting on the component.

Remark D.1 *To obtain (exact), continuous solutions* **u** *to the equation of mechanical equilibrium, we have to go back to its corresponding variational form given in Equation (D.3). If an exact solution* **u** *exists, it is the physically correct displacement that minimizes the energy of the structure (provided that it cannot perform any rigid body motions). The convergence of the approximate numerical solution (often denoted by* $\mathbf{u}_h$*) to* **u** *can then either be achieved by increasing the refinement of the discretization of the geometry (h-FEM; software automatized: h-adaptivity, mesh-adaptivity) or by increasing the polynomial degree of the shape functions used in the discretization of* **u** *(p-FEM; software automatized: p-adaptivity, advantageous because no new meshes are needed), or by a combination of both (hp-FEM).*

D.5 Dynamic Equations

With the help of D'Alembert's principle we can introduce the inertia forces $\int_{\mathcal{D}} \rho \cdot \delta u^T \ddot{u} dv$ into the equation of the virtual work. Applying the discretization of the displacement field in the same way as above leads to the equation

$$\mathbf{M}\ddot{\mathbf{v}} + \mathbf{K}\mathbf{v} = \mathbf{f}, \tag{D.10}$$

where M denotes the mass matrix of the discretized component. In contrast to the loss of energy observed in vibrational structures, the solution of this equation does not exhibit damping. This is usually achieved by assuming velocity dependent damping forces in the form of $\int_{\mathcal{D}} \kappa \cdot \delta u^T \dot{u} dv$. Applying again the discretization procedure leads to the equation

$$\mathbf{M}\ddot{\mathbf{v}} + \mathbf{D}\dot{\mathbf{v}} + \mathbf{K}\mathbf{v} = \mathbf{f}, \tag{D.11}$$

where D is the damping matrix of the component.

A detailed introduction to continuum mechanics and the finite element method can be found, for example, in Hughes [116] or Bathe [14].

Appendix E

Multibody System Simulation

This appendix introduces some basics about multibody simulation without going into details. The aim is to present the basic concepts and notions needed to understand the sections about the capabilities and limits of multibody modelling in the context of load analysis as described in Chapter 5. A rigorous derivation as well as a complete presentation of the corresponding theory are beyond the scope of this guide. Instead, we refer to some of the existing textbooks about this subject, see, for example, Eberhard and Schiehlen [86], Amirouche [4], Schwertassek and Wallrapp [212] and the references therein.

We start the introduction to MBS with a simple investigation of linear models in Section E.1, which shows some similarities to linear FE models. In Section E.2, the approach to general systems is sketched.

E.1 Linear Models

Setting up the equations of motion for linear dynamic, time-dependent systems can be achieved using an approach, which, in essence, is a generalization of the concepts used in the mechanics of point masses. Consider the system sketched in Figure E.1. Apparently, it has a single degree of freedom, so that Newton's law specializes to

$$m\ddot{u}(t) + ku(t) = p(t), \tag{E.1}$$

where $p(t)$ is the load function. If friction is accounted for by introducing e.g. an idealized, linear viscous damper which develops a force proportional to the velocity, cf. Figure E.2, Newton's equation of motion reads

$$m\ddot{u}(t) + c\dot{u}(t) + ku(t) = p(t). \tag{E.2}$$

Generalizing Equation (E.2) to the case of multiple degrees of freedom renders in a straightforward manner the relation

$$M\ddot{u}(t) + D\dot{u}(t) + Ku(t) = p(t) \tag{E.3}$$

Guide to Load Analysis for Durability in Vehicle Engineering, First Edition. Edited by P. Johannesson and M. Speckert.
© 2014 Fraunhofer-Chalmers Research Centre for Industrial Mathematics.

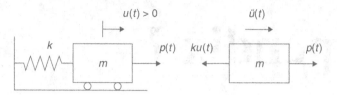

Figure E.1   Sketch of a single-degree of freedom system based on a generalization of the concepts of point mechanics

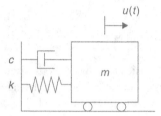

Figure E.2   Accounting for friction in a single-degree of freedom structure by introducing e.g. an idealized, linear viscous damper

in which M, D and K are the mass matrix, the damping matrix and the stiffness matrix, respectively. Note that in Equation (E.3), we have replaced the friction parameter c as occurring in Equation (E.2) by D instead of using C (the latter could be mistaken for the elasticity tensor introduced in Section D.3).

E.2   Mathematical Description of Multibody Systems

Multibody systems are used to model the interaction of several parts linked together by some sort of connections. The parts are called bodies and may be either rigid or flexible. A rigid body has a mass, a centre of mass, and the inertia tensor, which typically is expressed in a body fixed coordinate system (local frame). The possible motion in space is described by 6 coordinates or degrees of freedom (DOF), namely 3 translational and 3 rotational. Besides the centre of mass, there are additional markers (points including a coordinate system or frame) which describe the locations of the connections to other bodies.

If there are no connections between the bodies each body can move freely and the total number of DOF of the mechanism simply is $6 * n$, where n is the number of bodies. The connections can be divided into two types:

- constraints and
- force elements.

The constraints comprise the so-called joints like a revolute joint, a spherical joint, or a translational joint. A joint restricts the relative motion of the two bodies it belongs to. If there are two bodies connected via a spherical joint, then one body is allowed to move freely in space, whereas the second body is only allowed to rotate around the three axis of the first one. The total number of degrees of freedom of this simple mechanism is 6 (body 1) +6 (body 2)−3 (spherical joint) = 9.

A more general constraint is given by an arbitrary relation of the form

$$g(q^{(1)}, q^{(2)}, \dots) = 0 \tag{E.4}$$

where $q^{(1)}, q^{(2)}, \dots$ denote the coordinates of the bodies. Note, that for simplicity, we do not consider time-dependent constraints here. The constraints reduce the total number of DOF as has been explained above for the spherical joint. To satisfy the constraints during the motion of the mechanism, forces are needed which keep the bodies in their defined relative positions. These forces are called reaction forces. They are not explicitly defined during modelling a system, rather they are a consequence of the constraints and a result of the simulation.

Force elements define an interaction between the bodies without reducing the number of DOF. For example, a linear spring between two bodies controls the distance between the bodies. If the bodies come close to each other, the spring tends to separate them, if the distance becomes large, the spring brings them back together. However, the relative motion of one body is not strictly restricted to a certain path as is the case for constraints. Here, the forces between the bodies are explicitly defined during modelling and the relative motion of the bodies is a result of the simulation.

In addition to the bodies and connections, there are (outer) loads acting on the system. Examples of outer loads are the gravitational forces, or the force applied to a specimen on a test rig using a hydraulic cyclinder. The set of bodies, connections, and loads acting on the system completely defines the motion and all reaction forces based on the underlying physical laws which are the well-known Newton-Euler equations.

E.2.1   The Equations of Motion

An equivalent way to formulate these laws is given by the so-called Euler-Lagrange equation, which reads as follows:

$$\frac{d}{dt}\left(\frac{\partial T}{\partial \dot{q}}\right) - \frac{\partial T}{\partial q} = F(q, \dot{q}, t) - \lambda^T G(q) \tag{E.5}$$

$$g(q) = 0, \tag{E.6}$$

where T denotes the kinetic energy, q the vector of all coordinates, $\dot{q}$ the velocities, F the outer forces, g the constraint equations, $G = \frac{\partial g}{\partial q}$ the derivative of g, and λ the Lagrange multipliers associated with the constraints g. The term $\lambda^T G(q)$ is the vector of constraint forces.

Since the kinetic energy $T(q, \dot{q}) = \frac{1}{2}\dot{q}^T M(q)\dot{q}$ is a quadratic form in the velocities, a straightforward calculation leads to the well-known set of equations

$$M(q)\ddot{q} = f(t, q, \dot{q}) - G(q)^T \lambda \tag{E.7}$$

$$g(q) = 0, \tag{E.8}$$

This set of equations is a system of differential algebraic equations (DAE), which can be solved uniquely if appropriate initial value conditions $q(0) = q_0, \dot{q}(0) = v_0$ are supplemented.

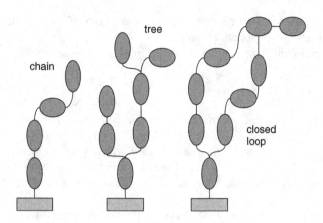

Figure E.3   Topology of multibody systems

In stating the equations of motion of a multibody system we have not precisely defined the coordinates q. In fact there are several different sets of coordinates that can be used. Here, we only address the absolute and relative coordinates. When using absolute coordinates, the position of a body is given by 3 translational coordinates of one point of the body (usually the centre of mass) and the angular orientation of a local frame of the body with respect to a fixed reference frame. In that case, all constraint equations induced by the joints are part of the vector g in Equation (E.5) or Equation (E.7).

When using relative coordinates the notion of kinematic chains and trees becomes important. In Figure E.3 different topologies of multibody systems are shown. Only the constraints between the bodies are displayed, the force elements do not influence the topology. For the chain and the tree type systems, each body has exactly one parent body, such that its position and orientation can be uniquely defined using relative coordinates with respect to the parent body. For a body tied to its parent via a spherical joint only 3 angular coordinates are needed since the position (translational coordinates) is defined by the parent. The mathematical description of such systems using relative coordinates leads to pure differential equations of the form

$$M(q)\ddot{q} = f(t, q, \dot{q}). \tag{E.9}$$

Of course, one can also use absolute coordinates for such systems leading again to a description of the form (E.7).

In the system to the right of Figure E.3 the closed loop structure always leads to equations of the form (E.7) even if relative coordinates are used. However the number of unknowns and equations depends on the type of coordinates used.

E.2.2 Computational Issues

A direct application of well-known numerical techniques to solve ordinary differential equations to Equation (E.7) leads to problems (instability) due to the algebraic part of the equations. By differentiating the constraint equations g we can transform the DAE to one which is closer to a set of ordinary differential equations (ODE). This process is called index reduction, where the index is the number of differentiations needed to transform the equations to an equivalent set of ODEs. It turns out that the underlying system (E.7) is of index 3. A single differentiation leads to the index 2 formulation

$$M(q)\ddot{q} = f(t, q, \dot{q}) - G(q)^T \lambda \tag{E.10}$$

$$G(q)\dot{q} = 0. \tag{E.11}$$

Another differentiation leads to the index 1 formulation

$$M(q)\ddot{q} = f(t, q, \dot{q}) - G(q)^T \lambda \tag{E.12}$$

$$G(q)\ddot{q} + \dot{q}^T g_{qq}(q)\dot{q} = 0. \tag{E.13}$$

Mathematically, these formulations are equivalent. However, the treatment and the behaviour of the numerical solution depend on the formulation. For each of the formulations various algorithms exist. Most of the commercial software tools offer dedicated solutions for the index 1, 2, or 3 formulations. In principle, the user of the tools need not know all details of the algorithms, but at least some basic properties regarding stability and error control should be taken into account. Since this heavily depends on the system and its description (in relative or absolute coordinates) as well as on the formulation and the numerical algorithm, no general recommendation for a specific solver can be given.

In addition to the performance of the algorithm, the accuracy of the solution (on the displacement, velocity, or force and acceleration level) must be taken into account. See, for example, Arnold [5], Ascher and Petzold [6], Brenan *et al.* [37] for more about some general computational aspects and the documentation of the different commercial tools.

The presentation of the material in this section closely follows von Schwerin [235], where more details and references can be found.

Appendix F

Software for Load Analysis

The aim of this appendix is to give a brief review of some selected software packages dealing with fatigue-oriented load analysis methods. Short comments are given on some of the dedicated commercial software packages, others are just listed. Moreover, most of the examples in the *Guide* have been performed using Matlab, often together with the freely available WAFO toolbox, which is specialized on statistical analysis of loads. The review is not complete but should rather be seen as a snapshot of some current software packages that can be used for load analysis in the durability context.

F.1 Some Dedicated Software Packages

Tec Ware

  Company: LMS

  Software: TecWare

  Web-Page: http://www.lmsintl.com

  TecWare is a software package dedicated to load analysis with respect to fatigue applications. Advanced functions (e.g. the RP approach or the optimum track mixing) were developed in the 1990s in close collaboration with the German automotive industry. Besides a broad offer of basic functions for load data analysis and signal processing its strength lies in the 'high-end' features such as the RP approach or extrapolation techniques.

Glyph Works

  Company: HBM nCode

  Software: Glyph Works

  Web-Page: http://www.ncode.com

  The origin of Glyph Works goes back to the 1980s when nCode was established and the method and software development started. Like TecWare it is dedicated to load analysis with respect to fatigue applications and offers a roughly comparable set of features. A strength of recent developments lies in automation features.

Guide to Load Analysis for Durability in Vehicle Engineering, First Edition. Edited by P. Johannesson and M. Speckert.
© 2014 Fraunhofer-Chalmers Research Centre for Industrial Mathematics.

RPC pro
   Company: MTS
   Software: RPC Pro
   Web-Page: http://www.mts.com
   RPC pro is, unlike TecWare and Glyph Works, a package dedicated to data acquisition and
      test rig environment. Besides standard fatigue load analysis methods, specific strengths
      are, for example, signal defect detection and correction and methods for drive file
      iteration or test sequence generation.

F.2   Some Software Packages for Fatigue Analysis

ANSYS Fatigue Module
   Company: ANSYS
   Software: ANSYS Fatigue Module
   Web-Page: http://www.ansys.com
FEMFAT
   Company: Magna International Inc.
   Software: FEMFAT
   Web-Page: http://www.femfat.com
fe-safe
   Company: Safe Technology
   Software: fe-safe
   Web-Page: http://http://www.safetechnology.com/fe-safe.html
LMS Virtual.Lab Durability
   Company: LMS
   Software: LMS Virtual.Lab Durability
   Web-Page: www.lmsintl.com
MSC Fatigue
   Company: MSC Software Corporation
   Software: MSC Fatigue
   Web-Page: http://www.mscsoftware.com
DesignLife
   Company: HBM nCode
   Software: DesignLife
   Web-Page: http://www.ncode.com

F.3   WAFO – a Toolbox for Matlab

WAFO – a Toolbox for Matlab
   Company: Freeware[1] (Mathworks for Matlab)
   Software: WAFO toolbox (requires Matlab)

[1] GNU General Public License.

Web-Page: http://www.maths.lth.se/matstat/wafo/

(Matlab: http://www.mathworks.com)

WAFO (Wave Analysis for Fatigue and Oceanography) is a toolbox of Matlab routines for statistical analysis and simulation of random waves and random loads, see WAFO Group [237, 238], Brodtkorb *et al.* [41]. The main purpose of WAFO is for scientific research, and thus the aim of WAFO is not to contain all the features of the commercial softwares above. Nevertheless, it is used in industry, for example, in automotive and offshore applications. The most important features of WAFO regarding durability analysis is described below, and many of these features are unique to WAFO and cannot be found elsewhere. Many of the calculations and figures in the Guide are done using WAFO, especially the calculations in Chapter 6 on random loads.

WAFO contains basic methods for load analysis, such as rainflow cycle counting, level crossing counting, rainflow filtering, damage calculation, and PSD estimation. The unique part of WAFO is modelling and analysis of random loads, such as Gaussian loads, transformed Gaussian loads, some non-Gaussian loads, Markov loads including switching Markov loads. For the models there are, for example, routines for estimation of the load model parameters, simulation of load signals, theoretical computation of the distribution of rainflow cycles, theoretical computation of the level crossing intensity, and theoretical computation of the expected damage. Further, WAFO contains many statistical tools, including statistical distributions, parameter estimation, kernel density estimation, design of experiments, and statistical extreme value analysis.

Bibliography

[1] S. Åberg, K. Podgórski, and I. Rychlik. Fatigue damage assessment for a spectral model of non-Gaussian random loads. *Probabilistic Engineering Mechanics*, 24: 608–617, 2009.

[2] ADAMS. *Modal Flexibility Method in ADAMS/Flex*. Ann Arbor, MI, USA, 1998.

[3] AFNOR. Fatigue sous sollicitations d'amplitude variables. Report A03-406, 1993. *In French*.

[4] F. Amirouche. *Fundamentals of Multibody Dynamics*. Birkhäuser, Boston, 2006.

[5] M. Arnold. Simulation algorithms in vehicle system dynamics. Technical report, Martin-Luther-Universität Halle-Wittenberg, FB Mathematik und Informatik, 2004.

[6] U. Ascher and L. Petzold. *Computer Methods for Ordinary Differential Equations and Differential–Algebraic Equations*. SIAM, Philadelphia, PA, 1998.

[7] ASTM. Standard practices for cycle counting in fatigue analysis, ASTM E 1049-85. *Annual Book of ASTM Standards*, Vol. 03.01, pp. 710–718, 1999.

[8] M. Bäcker, T. Langthaler, and H. O. und M. Olbrich. The hybrid road approach for durability loads prediction. Technical report, SAE 2005 World Congress, Detroit, 2005.

[9] M. Bäcker, A. Gallrein, and H. Haga. A tire model for very large tire deformations and its application in very severe events. *SAE Int. J. Mater. Manuf.*, 3: 142–151, 2010.

[10] M. Bäcker, A. Gallrein, M. Hack, and A. Toso. A method to combine a tire model with a flexible rim model in a hybrid MBS/FEM simulation setup. Technical Report 2011-01-0186, SAE, Detroit, 2011.

[11] M. Bäcker, K. Dressler, J. D. Cuyper, and O. Birk. Loads prediction and durability analysis of an unconstrained full vehicle. In *SIA - Simulation, Paris*, September 2003.

[12] J. A. Bannantine, J. J. Comer, and J. L. Handrock. *Fundamentals of Metal Fatigue Analysis*. Prentice Hall, Upper Saddle River, NJ, 1989.

[13] O. H. Basquin. The exponential law of endurance tests. *American Society for Testing and Materials*, 10: 625–630, 1910.

[14] K. J. Bathe. *Finite Element Procedures*. Prentice Hall, Upper Saddle River, NJ, 1995.

[15] D. Benasciutti. Fatigue analysis of random loadings. PhD thesis, Department of Engineering, University of Ferrara, Italy, 2004.

[16] J. S. Bendat. Probability functions for random responses: Prediction of peaks, fatigue damage and catastrophic failures. Technical report, NASA, 1964.

[17] A. Bengtsson, K. Bogsjö, and I. Rychlik. Uncertainty of estimated rainflow damage for random loads. *Marine Structures*, 22: 261–274, 2009.

[18] A. Bengtsson, K. Bogsjö, and I. Rychlik. Fatigue damage uncertainty. In B. Bergman, J. de Maré, T. Svensson, and S. Lorén, (eds) *Robust Design Methodology for Reliability: Exploring the Effects of Variation and Uncertainty*, pp. 145–171. Wiley, Chichester, 2009.

Guide to Load Analysis for Durability in Vehicle Engineering, First Edition. Edited by P. Johannesson and M. Speckert.
© 2014 Fraunhofer-Chalmers Research Centre for Industrial Mathematics.

[19] A. K. Bengtsson and I. Rychlik. Uncertainty in fatigue life prediction of structures subject to Gaussian loads. *Probabilistic Engineering Mechanics*, 24: 224–235, 2009.

[20] C. Berger, K.-G. Eulitz, P. Heuler, K.-L. Kotte, H. Naundorf, W. Schuetz, C. Sonsino, A. Wimmer, and H. Zenner. Betriebsfestigkeit in Germany - an overview. *International Journal of Fatigue*, 24: 603–625, 2002.

[21] J. O. Berger. *Statistical Decision Theory and Bayesian Analysis*. Springer, Berlin, 2nd edition, 1985.

[22] B. Bergman and B. Klefsjö. *Quality from Customer Needs to Customer Satisfaction*. Studentlitteratur, Lund, 2010.

[23] B. Bergman, J. de Maré, T. Svensson, and S. Lorén, editors. *Robust Design Methodology for Reliability: Exploring the Effects of Variation and Uncertainty*. Wiley, Chichester, 2009.

[24] J. M. Bernardo and A. F. M. Smith. *Bayesian Theory*. Wiley, Chichester: 2000.

[25] A. Beste, K. Dressler, H. Kötzle, W. Krüger, B. Maier, and J. Petersen. Multiaxial rainflow – a consequent continuation of Professor Tatsuo Endo's work. In Y. Murakami, ed. *The Rainflow Method in Fatigue*, pp. 31–40. Butterworth-Heinemann, Oxford, 1992.

[26] J. T. Betts. *Practical Methods for Optimal Control Using Nonlinear Programming*, 2nd edn. Advances in Design and Control. Society for Industrial and Applied Mathematics, Philadelphia, 2009.

[27] A. Bignonnet. Fatigue design in automotive industry. In K. V. Dang and I. V. Papadopoulos, eds, *High Cycle Metal Fatigue, from Theory to Applications, C.I.S.M. Courses and Lectures No. 392*, pages 146–167. Springer, New York, 1999.

[28] M. A. Biot. Theory of elastic systems vibrating under transient impulse with an application to earthquake-proof buildings. *Proceedings of the National Academy of Science*, 19(2): 262–268, 1933.

[29] N. W. M. Bishop and F. Sherratt. A theoretical solution for the estimation of 'rainflow' ranges from power spectral density data. *Fatigue & Fracture of Engineering Materials & Structures*, 13: 311–326, 1990.

[30] K. Bogsjö. Road profile statistics relevant for vehicle fatigue. PhD thesis, Mathematical Statistics, Lund University, 2007.

[31] K. Bogsjö and I. Rychlik. Vehicle fatigue damage caused by road irregularities. *Fatigue & Fracture of Engineering Materials & Structures*, 32: 391–402, 2009.

[32] K. Bogsjö and I. Rychlik. Uncertainty of estimated damage for random loads. In *Second International Conference on Variable Amplitude Loading*. Darmstadt, Germany, 23–26 March 2009.

[33] K. Bogsjö, K. Podgórski, and I. Rychlik. Models for road surface roughness. *Vehicle System Dynamics*, 50: 725–747, 2012.

[34] G. E. P. Box and G. C. Tiao. *Bayesian Inference in Statistical Analysis*. Wiley, New York, 1992.

[35] G. E. P. Box, J. S. Hunter, and W. G. Hunter. *Statistics for Experimenters*. John Wiley & Sons, Chichester, 2nd edition, 2005.

[36] G. Bremer, B. Fiedler, J. Vogler, L. Witte, and M. Speckert. Das "Mehraxiale Rainflow": Erste Erfahrungen im Praxiseinsatz. DVM-Bericht 123. Betriebsfestigkeit und Entwicklungszeitverkürzung, Berlin, 1997. *In German*.

[37] K. E. Brenan, S. L. Campbell, and L. R. Petzold. *Numerical Solution of Initial-value Problems in Differential-algebraic Equations*. SIAM, Philadelphia, PA, 2nd edition, 1996.

[38] R. Brigola. *Fourieranalysis, Distributionen und Anwendungen*. Vieweg, Wiesbaden, 1997. *In German*.

[39] D. R. Brillinger. *Time Series: Data Analysis and Theory*. Holt, Rinehart and Winston, New York, 1975.

[40] P. J. Brockwell and R. A. Davis. *Time Series: Theory and Methods*. Springer, Berlin, 1993.

[41] P. A. Brodtkorb, P. Johannesson, G. Lindgren, I. Rychlik, J. Rydén, and E. Sjö. WAFO – a Matlab toolbox for analysis of random waves and loads. In *Proceedings of the 10th International Offshore and Polar Engineering Conference, Seattle*, volume III, pages 343–350, 2000.

[42] M. Brokate and J. Sprekels. *Hysteresis and Phase Transitions*. Springer Verlag, New York, 1996.

[43] M. Brokate, K. Dressler, and P. Krejčí. Rainflow counting and energy dissipation for hysteresis models in elastoplasticity. *European Journal of Mechanics, A/Solids*, 15: 705–737, 1996.

[44] T. Bruder, P. Heuler, H. Klätschke, and K. Störzel. Analysis and synthesis of standardized multiaxial load-time histories for structural durability assessment. In *Seventh International Conference on Biaxial/Multiaxial Fatigue and Fracture*, June 28–July 1, 2004, Berlin, Germany.

[45] A. E. Bryson and Y.-C. Ho. *Applied Optimal Control*. Hemisphere Publishing Corporation, New York, 1975.

[46] M. Burger. *Optimal Control of Dynamical Systems: Calculating Input Data for Multibody System Simulation. Dissertation. TU Kaiserslautern. 2011*. Verlag Dr. Hut, München, 2011.

[47] M. Burger, K. Dressler, A. Marquardt, and M. Speckert. Calculating invariant loads for system simulation in vehicle engineering. In *Proceedings of Multibody Dynamics 2009, Warsaw, Poland*, 2009.

[48] M. Burger, K. Dressler, and M. Speckert. Invariant Input Loads for Full Vehicle Multibody System Simulation. In *Multiobody Dynamics 2011 ECCOMAS Thematic Conference and Fraunhofer ITWM Report 208(2011)*, Brussels, 2011.

[49] R. Callies and P. Rentrop. Optimal control of rigid-link manipulators by indirect methods. *GAMM-Mitteilungen*, 31: 27–58, 2008.

[50] M. Carboni, A. Cerrini, P. Johannesson, M. Guidetti, and S. Beretta. Load spectra analysis and reconstruction for hydraulic pump components. *Fatigue & Fracture of Engineering Materials & Structures*, 31: 251–261, 2008.

[51] G. Casella and R. Berger. *Statistical Inference*. Duxbury, California, 2nd edition, 2001.

[52] E. Castillo. *Extreme Value Theory in Engineering*. Academic Press, San Diego, 1988.

[53] J. L. Chaboche and J. Lemaitre. *Mechanics of Solid Materials*. Cambridge University Press, Cambridge, 2000.

[54] A. Chakhunashvili, S. Barone, P. Johansson, and B. Bergman. Robust product development using variation mode and effect analysis. In B. Bergman, J. de Maré, T. Svensson, and S. Lorén, editors, *Robust Design Methodology for Reliability: Exploring the Effects of Variation and Uncertainty*, pp. 57–70. Wiley, Chichester, 2009.

[55] E. Charkaluk, A. Bignonnet, A. Constantinescu, and K. Dang Van. Fatigue design of structures under thermomechanical loadings. *Fatigue and Fracture of Engineering Materials and Structures*, 25: 1199–1206, 2002.

[56] U. H. Clormann and T. Seeger. RAINFLOW - HCM. Ein Zählverfahren für Betriebsfestigkeitsnachweise auf werkstoffmechanischer Grundlage. Stahlbau, Heft 3, 65–71, 1986. *In German*.

[57] S. Coles. *An Introduction to Statistical Modelling of Extreme Values*. Springer Verlag, London, 2001.

[58] J. A. Collins. *Failure of Materials in Mechanical Design: Analysis Prediction, Prevention*. Wiley, New York, 2nd edition, 1993.

[59] J. W. Cooley and J. W. Tukey. An algorithm for the machine calculation of complex fourier series. *Mathematics of Computation*, 19: 297–301, 1965.

[60] R. R. Craig and M. C. C. Bampton. Coupling of substructures for dynamic analysis. *AiAA Journal*, 6, 1968.

[61] R. R. Craig and A. J. Kurdila. *Fundamentals of Structural Dynamics*. John Wiley & Sons, Chichester, 2006.

[62] T. Dahlberg and A. Ekberg. *Failure Fracture Fatigue: An Introduction*. Studentlitteratur, Lund, 2002.

[63] K. Dang Van. Sur la résistance à la fatigue des matériaux. Sciences et Techniques de l'Armement, Mémorial de l'Artillerie française, 1973. *In French.*

[64] T. P. Davis. Science, engineering, and statistics. *Applied Stochastic Models in Business and Industry*, 22: 401–430, 2006.

[65] A. C. Davison and D. V. Hinkley. *Bootstrap Methods and their Application*. Cambridge University Press, New York, 1997.

[66] A. C. Davison and R. L. Smith. Models for exceedances over high thresholds. *Journal of the Royal Statistical Society, Ser. B*, 52: 393–442, 1990.

[67] J. de Cuyper. Linear feedback control for durability test rigs in the automotive industry. PhD thesis, Katholieke Universiteit Leuven, Faculteit Ingenieurswetenschappen, 2006.

[68] J. B. de Jonge. Fatigue load monitoring of tactical aircraft. Paper presented at the 29th Meeting of the AGARD SMP, Istanbul, 1969. NLR Report TR 69063, August 1969.

[69] J. B. de Jonge. The analysis of load-time-histories by means of counting methods. National Aerospace Laboratory, NLR Report MP 82039 U, ICAF Document, 1982.

[70] G. Deodatis. Simulation of ergodic multivariate stochastic processes. *Journal of Engineering Mechanics*, 122: 778–787, 1996.

[71] A. Der Kiureghian and O. Ditlevsen. Aleatory or epistemic? Does it matter? *Structural Safety*, 31: 105–112, 2009.

[72] DIN. Klassierverfahren für das Erfassen regelloser Schwingungen, DIN 45 667, 1969. *In German.*

[73] T. Dirlik. Application of computers in fatigue analysis. PhD thesis, University of Warwick, UK, 1985.

[74] O. Ditlevsen and H. Madsen. *Structural Reliability Methods*. John Wiley & Sons, Chichester, UK, 1st edition, 1996. Internet edition 2.3.7, June 2007; http://www.web.mek.dtu.dk/staff/od/books.htm.

[75] S. V. Dombrowski. Analysis of large flexible body deformation in multibody systems using absolute coordinates. *Multibody System Dynamics*, 8: 409–432, 2002.

[76] N. E. Dowling. Fatigue predictions for complicated stress-strain histories. *Journal of Materials*, 7: 71–87, 1972.

[77] N. E. Dowling. *Mechanical Behavior of Materials: Engineering Methods for Deformation, Fracture, and Fatigue*. Prentice Hall, Upper Saddle River, NJ, 2012.

[78] S. D. Dowling and D. F. Socie. Simple rainflow counting algorithms. *International Journal of Fatigue*, 4: 31–40, 1982.

[79] K. Dressler, R. Carmine, and W. Krüger. The multiaxial rainflow method. In K. T. Rie, editor, *Low Cycle Fatigue and Elasto-plastic Behaviour of Materials*. Elsevier Science Publ., London, 1992.

[80] K. Dressler, V. Köttgen, and H. Kötzle. Tools for fatigue evaluation of non-proportional loading. In *Proceedings of Fatigue Design '95, Helsinki, Finland*, 1995.

[81] K. Dressler, B. Gründer, M. Hack, and V. B. Köttgen. Extrapolation of rainflow matrices. Technical Report 960 569, SAE, Detroit, 1996.

[82] K. Dressler, M. Hack, and W. Krüger. Stochastic reconstruction of loading histories from a rainflow matrix. *Zeitschrift für Angewandte Mathematik und Mechanik*, 77: 217–226, 1997.

[83] K. Dressler, B. Gründer, M. Pompetzki, and M. Speckert. Statistical load data analysis for customer correlation. Technical report, SAE, Detroit, 2000.

[84] K. Dressler, M. Speckert, R. Müller, and C. Weber. Customer loads correlation in truck engineering. Technical report, Congress Proceedings FISITA 2008, World Automotive Congress, Munich, 2008.

[85] S. Dronka. Zur Einbindung hydraulischer Systeme in SIMPACK. Technical report, *Simpack-News*, 2. Jahrgang, 2. Ausgabe, S. 7–9, 1997. *In German.*

[86] P. Eberhard and W. Schiehlen. Computational dynamics of multibody systems: History, formalisms, and applications. *Journal of Computational and Nonlinear Dynamics*, 1(1): 3–12, 2006.

[87] S. Edlund and P.-O. Fryk. The right truck for the job with global truck applications. SAE Technical Paper 2004-01-2645, 2004.

[88] B. Efron and R. J. Tibshirani. *An Introduction to the Bootstrap.* Chapman & Hall, New York, 1993.

[89] T. Endo, K. Mitsunaga, and H. Nakagawa. Fatigue of metals subjected to varying stress – prediction of fatigue lives. In *Preliminary Proceedings of The Chugoku-Shikoku District Meeting*, pages 41–44. The Japan Society of Mechanical Engineers, November 1967. *In Japanese.*

[90] T. Endo, K. Mitsunaga, H. Nakagawa, and K. Ikeda. Fatigue of metals subjected to varying stress – low cycle, middle cycle fatigue. In *Preliminary Proceedings of The Chugoku-Shikoku District Meeting*, pages 45–48. The Japan Society of Mechanical Engineers, November 1967. *In Japanese.*

[91] T. Endo et al. Fatigue of metals subjected to varying stress, a series of three papers presented to the Japan Society of Mechanical Engineers, Jukvoka, Japan, 1967–68. *In Japanese.*

[92] A. Fatemi and L. Yang. Cumulative fatigue damage and life prediction theories: a survey of the state of the art for homogeneous materials. *International Journal of Fatigue*, 20: 9–34, 1998.

[93] S. Feth. Generalising success runs to partially-passed component counting. PhD thesis, Universität Kaiserslautern, 2009.

[94] D. Freedman, R. Pisani, and R. Purves. *Statistics.* W.W. Norton & Company, Inc., New York, 4th edition, 2007.

[95] M. Frendahl and I. Rychlik. Rainflow analysis: Markov method. *International Journal of Fatigue*, 15: 265–272, 1993.

[96] A. Gallrein and M. Bäcker. CDTire: a tire model for comfort and durability applications. *Vehicle System Dynamics*, 45: 69–77, 2007.

[97] A. Gallrein and M. Bäcker. Spreading the application range of the digital road approach: New CDTire model developments. In *LMS European Vehicle Conferences*, 2012.

[98] Z. Gao and T. Moan. Fatigue damage induced by non-Gaussian bimodal wave loading in mooring lines. *Applied Ocean Research*, 29: 45–54, 2007.

[99] G. Genet. A statistical approach to multi-input equivalent fatigue loads for the durability of automotive structures. PhD thesis, Mathematical statistics, Chalmers University of Technology, Göteborg, 2006.

[100] M. Gerdts. Numerische Methoden optimaler Steuerprozesse mit differential-algebraischen Gleichungssystemen höheren Indexes und ihre Anwendungen in der Kraftfahrzeugsimulation und Mechanik. PhD thesis, Dissertation Bayreuther Mathematische Schriften 61, 2001. *In German.*

[101] M. Gerdts. Direct shooting method for the numerical solution of higher-index DAE optimal control problems. *Journal of Optimization Theory and Applications*, 117(2): 267–294, 2003.

[102] P. Gill, W. Murray, and M. Wright. *Practical Optimization.* Elsevier Science & Technology, New York, 1982.

[103] M. Gipser. FTire homepage. Web: http://www.ftire.com.

[104] Z. Govindarajulu. *Sequential Statistics.* World Scientific Publishing, Singapore, 2004.

[105] G. R. Grimmett and D. R. Stirzaker. *Probability and Random Processes.* Oxford University Press, Oxford, 3rd edition, 2001.

[106] V. Grubisic. Determination of load spectra for design and testing. *Journal of Vehicle Design*, 15: 8–26, 1994.

[107] B. Gründer, B. Bremer, and V. Köttgen. Statistische Analyse der Belastung im Kundeneinsatz zur Entwicklung von Zielvorgaben der Betriebsfestigkeitsauslegung. Technical report, VDI Bericht Nr. 1470, 1999. *In German*.

[108] S. Gupta and I. Rychlik. Rain-flow fatigue damage due to nonlinear combination of vector Gaussian loads. *Probabilistic Engineering Mechanics*, 22: 231–249, 2007.

[109] M. Hack. Schädigungsbasierte Hysteresefilter. PhD thesis, Dissertation Universität Kaiserslautern. Shaker Verlag, Aachen, 1998. *In German*.

[110] E. Haibach. *Betriebsfestigkeit, 2. Auflage*. Springer Verlag, Berlin, 2002. *In German*.

[111] A. Halfpenny. Methods for accelerating dynamic durability tests, 2006. Whitepaper nCode.

[112] S. Herkt. Model reduction of nonlinear problems in structural mechanics: towards a finite element tyre model for multibody simulation. PhD thesis, Dissertation Technische Universität Kaiserslautern, 2009.

[113] P. Heuler and H. Klätschke. Generation and use of standardized load spectra and load-time histories. *International Journal of Fatigue*, 27: 974–990, 2005.

[114] U. Hjorth. *Computer Intensive Statistical Methods: Validation, Model Selection and Bootstrap*. Chapman & Hall, London, 1994.

[115] S. Holm and J. de Maré. Generation of random processes for fatigue testing. *Stochastic Processes and their Applications*, 20: 149–156, 1985.

[116] T. J. R. Hughes. *The Finite Element Method*. Dover Publications, New York, 2000.

[117] ISO. Mechanical vibration-road surface profiles-reporting of measured data, ISO 8608: 1995(E). International Organization for Standardization (ISO), 1995.

[118] JCSS. Probabilistic model code. ISBN 978-3-909 386-79-6, 08 2001.

[119] P. Johannesson. Rainflow cycles for switching processes with Markov structure. *Probability in the Engineering and Informational Sciences*, 12: 143–175, 1998.

[120] P. Johannesson. Rainflow analysis of switching Markov loads. PhD thesis, Mathematical Statistics, Centre for Mathematical Sciences, Lund Institute of Technology, Lund, 1999.

[121] P. Johannesson. Extrapolation of load histories and spectra. *Fatigue & Fracture of Engineering Materials & Structures*, 29: 201–207, 2006.

[122] P. Johannesson and I. Rychlik. Modelling of road profiles using roughness indicators. Accepted for publication in *International Journal of Vehicle Design*, 2013.

[123] P. Johannesson and J.-J. Thomas. Extrapolation of rainflow matrices. *Extremes*, 4: 241–262, 2001.

[124] P. Johannesson, T. Svensson, and J. de Maré. Fatigue life prediction based on variable amplitude tests – methodology. *International Journal of Fatigue*, 27: 954–965, 2005.

[125] P. Johannesson, T. Svensson, L. Samuelsson, B. Bergman, and J. de Maré. Variation mode and effect analysis: an application to fatigue life prediction. *Quality and Reliability Engineering International*, 25: 167–179, 2009.

[126] P. Johannesson, B. Bergman, T. Svensson, M. Arvidsson, Å. Lönnqvist, S. Barone, and J. de Maré. A robustness approach to reliability. *Quality and Reliability Engineering International*, 29: 17–32, 2013.

[127] P. Johansson, A. Chakhunashvili, S. Barone, and B. Bergman. Variation mode and effect analysis: a practical tool for quality improvement. *Quality and Reliability Engineering International*, 22: 865–876, 2006.

[128] E. Johnson and T. Svensson. Choice of complexity in constitutive modelling of fatigue mechanisms. In B. Bergman, J. de Maré, T. Svensson, and S. Lorén, editors, *Robust Design Methodology for Reliability: Exploring the Effects of Variation and Uncertainty*, pp. 133–143. Wiley, Chichester, 2009.

[129] M. Jono. Fatigue damage and crack growth under variable amplitude loading with reference to the counting methods of stress-strain ranges. *International Journal of Fatigue*, 27: 1006–1015, 2005.

[130] L. Jung, V. B. Köttgen, G. Mäscher, M. Reißel, and G. Zhang. Numerische Betriebsfestigkeit-sanalyse eines PKW-Schwenklagers im Rahmen des fem-Postprocessing. Deutscher Verband für Materialforschung und -technik, Arbeitskreis Betriebsfestigkeit, 1997. *In German*.

[131] C. Karlberg, M. Ohlson, G. Kjell, and S. Blomgren. *Environmental Engineering Handbook*. Issued by the Swedish Environmental Engineering Society, SEES, Stockholm, Sweden, 1997.

[132] M. Karlsson. Load modelling for fatigue assessment of vehicles – a statistical approach. PhD thesis, Mathematical Statistics, Chalmers University of Technology, Göteborg, 2007.

[133] M. Karlsson, B. Johannesson, J. de Maré, and T. Svensson. Verification of safety critical components. *VDI Berichte*, pages 215–299, 2005.

[134] S. M. Kay. *Modern Spectral Estimation: Theory & Application*. Prentice Hall, Upper Saddle River, NJ, 1988.

[135] H. Klätschke and D. Schütz. Das Simultanverfahren zur Extrapolation und Raffung von mehraxialen Belastungs–Zeitfunktionen für Schwingfestigkeitsversuche. *Materialwissenschaft und Werkstofftechnik*, 26: 404–415, 1995. *In German*.

[136] M. Kleer, O. Hermanns, S. Müller, and K. Dreßler. Driving simulations for commercial vehicles - a technical overview of a robot based approach. In *Driving Simulation Conference Europe*, pages 223–232, 2012.

[137] J. Kowalewski. On the relationship between component life under irregularly fluctuating and ordered load sequences. MIRA Translations, 43 and 60, 1966.

[138] S. Krenk and H. Gluver. A Markov matrix for fatigue load simulation and rainflow range evaluation. *Structural Safety*, 6: 247–258, 1989.

[139] W. Krüger, W. Scheutzow, A. Beste, and J. Petersen. Markov- und Rainflowrekonstruktionen stochastischer Beanspruchungszeitfunktionen. Technical Report 18: 22, VDI-report, 1985. *In German*.

[140] C. Lalanne. *Mechanical Vibration and Shock, Volume 2: Mechanical Shock*. Hermes Penton Science, London, 2002.

[141] C. Lalanne. *Mechanical Vibration and Shock, Volume 4: Fatigue Damage*. Hermes Penton Science, London, 2002.

[142] M. R. Leadbetter, G. Lindgren, and H. Rootzén. *Extremes and Related Properties of Random Sequences and Processes*. Springer, Berlin, 1983.

[143] G. Lindgren and K. B. Broberg. Cycle range distributions for Gaussian processes: exact and approximate results. *Extremes*, 7: 69–89, 2004.

[144] A. Lion. Einsatz flexibler Körper in der numerischen Lebensdauersimulation von Kraftfahrzeugen: Methoden, Beispiele und offene Fragen. Technical report, *NAFEMS Magazin*, Juli 2005. *In German*.

[145] A. Lion and M. Eichler. Gesamtfahrzeugsimulation auf Prüfstrecken zur Bestimmung von Lastkollektiven. Technical report, VDI Berichte Nr. 1559, 2000. *In German*.

[146] L. Ljung. *System Identification: Theory for the User*. PTR Prentice Hall, Upper Saddle River, NJ, 2nd edition, 1999.

[147] LMS. LMS Tecware Documentation. Vol. IV Mathematical Background, 2006.

[148] S. L. Lohr. *Sampling: Design and Analysis*. Pacific Grove: Brooks/Cole Publishing Company, Pacific Grove, CA, 1999.

[149] S. Lorén and T. Svensson. Second moment reliability evaluation vs. Monte Carlo simulations for weld fatigue strength. *Quality and Reliability Engineering International*, 28: 887–896, 2012.

[150] S. Lorén, P. Johannesson, and J. de Maré. Monte Carlo simulations versus sensitivity analysis. In B. Bergman, J. de Maré, T. Svensson, and S. Lorén, editors, *Robust Design Methodology for Reliability: Exploring the Effects of Variation and Uncertainty*, pp. 97–111. Wiley, Chichester, 2009.

[151] H. O. Madsen, S. Krenk, and N. C. Lind. *Methods of Structural Safety*. Prentice-Hall, New Jersey, USA, 1986.

[152] S. L. Marple. *Digital Spectral Analysis*. Prentice Hall, Upper Saddle River, NJ, 1987.

[153] W. V. Mars and A. Fatemi. A literature survey on fatigue analysis approaches for rubber. *International Journal of Fatigue*, 24: 949–961, 2002.

[154] W. V. Mars and A. Fatemi. Factors that affect the fatigue life of rubber: A literature survey. *Rubber Chemistry and Technology*, 88: 391–412, 2004.

[155] H. F. Martz and R. A. Waller. *Bayesian Reliability Analysis*. Wiley, Chichester, 1982.

[156] G. Masing. Eigenspannungen und Verfestigung beim Messing. In *Proc. 2nd Int. Conf. Applied Mech. Zürich*, pages 332–335, 1926. *In German*.

[157] M. Matsuishi and T. Endo. Fatigue of metals subjected to varying stress – fatigue lives under random loading. In *Preliminary Proceedings of The Kyushu District Meeting*, pages 37–40. The Japan Society of Mechanical Engineers, March 1968. *In Japanese*.

[158] H. Mauch, A. Ahmadi, W. Rudolph, F.-J. Stolze, M. Bäcker, and R. Möller. Numerical and experimental simulation of vehicles on offroad-courses. In *Proceedings DVM Conference 'Optimierungspotentiale in der Betriebsfestigkeit', Sindelfingen, Germany*, 2008.

[159] R. Melchers. *Structural Reliability Analysis and Prediction*. John Wiley & Sons, Chichester, 2nd edition, 1999.

[160] K. Miettinen. *Nonlinear Multiobjective Optimization*. Kluwer, Boston, 1999.

[161] M. A. Miner. Cumulative damage in fatigue. *Journal of Applied Mechanics*, 12: A159–A164, 1945.

[162] D. C. Montgomery. *Design and Analysis of Experiments*. John Wiley & Sons, Chichester, 6th edition, 2005.

[163] D. C. Montgomery and G. C. Runger. *Applied Statistics and Probability for Engineers*. John Wiley & Sons, 4th edition, 2006.

[164] K. L. Moore. *Iterative Learning Control for Deterministic Systems*. Springer Verlag, London, 1993.

[165] F. Morel. A fatigue life prediction method based on a mesoscopic approach in constant amplitude multiaxial loading. *Fatigue & Fracture of Engineering Materials & Structures*, 21: 241–256, 1998.

[166] F. Morel. A critical plane approach for life prediction of high cycle fatigue under multiaxial variable amplitude loading. *International Journal of Fatigue*, 22: 101–119, 2000.

[167] L. Müller, G. Bitsch, and C. Schindler. Online condition monitoring based on real-time multi-body system simulation. In *2nd Commercial Vehicle Technology Symposium*, pages 461–468, 2012.

[168] Y. Murakami, (ed.). *The Rainflow Method in Fatigue*, Butterworth-Heinemann, Oxford, 1992.

[169] M. Nagode and M. Fajdiga. A general multi-modal probability density function suitable for the rainflow ranges of stationary random processes. *International Journal of Fatigue*, 20: 211–223, 1998.

[170] H. Neuber. Theory of stress concentration for shear-strained prismatical bodies with arbitrary nonlinear stress-strain law. *Journal of Applied Mechanics*, 28: 544–550, 1961.

[171] H. J. Oberle and W. Grimm. BNDSCO - A Programm for the Numerical Solution of Optimal Control Problems. *Hamburger Beiträge zur Angewandten Mathematik, Department of Mathematics, University of Hamburg*, (Technical Report Reihe B, Bericht 36), 2001.

[172] P. O'Connor. *Practical Reliability Engineering*. Wiley, Chichester, 4th edition, 2002.

[173] P. O'Connor. *Test Engineering, A Concise Guide to Cost-effective Design, Development and Manufacture*. Wiley, Chichester, 2005.

[174] C. Oertel. RMOD-K 7 homepage. Web: http://www.rmod-k.com.

[175] F. Öijer and S. Edlund. Complete vehicle durability assessments using discrete sets of random roads and transient obstacles based on q-distributions. *Vehicle System Dynamics*, 37: 67–74, 2003.

[176] F. Öijer and S. Edlund. Identification of transient road obstacle distributions and their impact on vehicle durability and driver comfort. *Vehicle System Dynamics*, 41: 744–753, 2004.

[177] M. Olofsson. *Evaluation of Estimates of Extreme Fatigue Loads.* Licentiate thesis 2000:56, Chalmers University of Technology, Göteborg, 2000.

[178] K. E. Olsson. Fatigue reliability prediction. *Scandinavian Journal of Metallurgy*, 18: 176–180, 1989.

[179] A. V. Oppenheim and R. W. Schafer. *Discrete-Time Signal Processing.* Prentice Hall, Upper Saddle River, NJ, 2nd edition, 1999.

[180] A. V. Oppenheim and A. S. Willsky. *Signals and Systems.* Prentice Hall, Upper Saddle River, NJ, 1983.

[181] S. J. Orfanidis. *Optimum Signal Processing. An Introduction.* Prentice-Hall, Englewood Cliffs, NJ, 2nd edition, 1996.

[182] H. B. Pacejka. *Tire and Vehicle Dynamics.* SAE International and Butterworth Heinemann, Oxford, 3rd edition, 2012.

[183] A. Palmgren. Die Lebensdauer von Kugellagern. *Zeitschrift des Vereins Deutscher Ingenieure*, 68: 339–341, 1924. *In German.*

[184] Y. Pawitan. *In All Likelihood: Statistical Modelling and Inference Using Likelihood.* Cambridge UP, New York, 2001.

[185] D. B. Percival and A. T. Walden. *Wavelet Methods for Time Series Analysis.* Cambridge University Press, Cambridge, 2000.

[186] X. Pitoiset and A. Preumont. Spectral methods for multiaxial random fatigue analysis of metallic structures. *International Journal of Fatigue*, 22: 541–550, 2000.

[187] X. Pitoiset, A. Preumont, and A. Kernilis. Tools for a multiaxial fatigue analysis of structures submitted to random vibrations. In *Proceedings European Conference on Spacecraft Structures, Materials and Mechanical Testing, Braunschweig, Germany, 4–6 November 1998*, 1998.

[188] X. Pitoiset, I. Rychlik, and A. Preumont. Spectral methods to estimate local multiaxial fatigue failure criteria for structures undergoing random vibrations. *Fatigue & Fracture of Engineering Materials & Structures*, 24: 715–727, 2001.

[189] X. Pitoiset, I. Rychlik, and A. Preumont. Spectral formulations of multiaxial high-cycle fatigue criteria subjected to random loads. In *6th International Conference on Biaxial/Multiaxial Fatigue and Fracture.* Lisbon, 2001.

[190] K. Popper. *In Search for a Better World: Lectures and Essays from Thirty Years.* Routledge, London, 1992.

[191] W. H. Press, S. A. Teukolsky, W. T. Vetterling, and B. P. Flannery. *Numerical Recipes.* Cambridge University Press, Cambridge, 2nd edition, 1996.

[192] M. B. Priestley. *Spectral Analysis and Time Series.* Academic Press, New York, 1981.

[193] M. Rausand and A. Høyland. *System Reliability Theory Models, Statistical Methods, and Application.* Wiley, Chichester, 2nd edition, 2004.

[194] Lord Rayleigh. On the resultant of a large number of vibrations of the same pitch and of arbitrary phase. *Philosophical Magazine*, 10: 73–78, 1880.

[195] S. O. Rice. Mathematical analysis of random noise. *Bell System Technical Journal*, 23: 282–332, 1944.

[196] S. O. Rice. Mathematical analysis of random noise. *Bell System Technical Journal*, 24: 46–156, 1945.

[197] P. R. Roberge. *Handbook of Corrosion Engineering.* McGraw-Hill, Maidenhead, 2000.

[198] I. Rychlik. A new definition of the rainflow cycle counting method. *International Journal of Fatigue*, 9: 119–121, 1987.

[199] I. Rychlik. Regression approximations of wavelength and amplitude distributions. *Advances in Applied Probability*, 19: 396–430, 1987.

[200] I. Rychlik. Rain-Flow-Cycle distribution for ergodic load processes. *SIAM Journal on Applied Mathematics*, 48: 662–679, 1988.

[201] I. Rychlik. Note on cycle counts in irregular loads. *Fatigue & Fracture of Engineering Materials & Structures*, 16: 377–390, 1993.

[202] I. Rychlik. On the 'narrow-band' approximation for expected fatigue damage. *Probabilistic Engineering Mechanics*, 8: 1–4, 1993.

[203] I. Rychlik. Rainflow cycles, Markov chains and electrical circuits. In B. Spencer and E. J. J. A. Balkema, editors, *Stochastic Structural Dynamics*, number ISBN 90 5809 0248, pages 295–298, 1999.

[204] I. Rychlik and S. Gupta. Rain-flow fatigue damage for transformed Gaussian loads. *International Journal of Fatigue*, 29: 406–420, 2007.

[205] I. Rychlik, P. Johannesson, and M. R. Leadbetter. Modelling and statistical analysis of ocean-wave data using transformed Gaussian processes. *Marine Structures*, 10: 13–47, 1997.

[206] J. Samuelson. Fatigue design of construction equipment. *Volvo Technology Report*, 2-97: 22–29, 1997.

[207] C.-E. Särndal, B. Swensson, and J. Wretman. *Model Assisted Survey Sampling*. Springer-Verlag, New York, 1991.

[208] M. Scheutzow. A law of large numbers for upcrossing measures. *Stochastic Processes and their Applications*, 53: 285–305, 1994.

[209] J. Schijve. *Fatigue of Structures and Materials*. Kluwer Academic Publishers, Dordrecht, 2001.

[210] A. Schmeitz. A semi-empirical three-dimensional model of the pneumatic tyre rolling over arbitrarily uneven road surfaces. PhD thesis, Technische Universiteit Delft, 2004.

[211] N. Schmudde. Verification of load cascade in vehicle test, axle test rig and adams simulation. In *LMS User Conferences*, 2005.

[212] R. Schwertassek and O. Wallrapp. *Dynamik flexibler Mehrkörpersysteme*. Vieweg, 1999. *In German*.

[213] S. Searle, G. Casella, and C. McCulloch. *Variance Components*. John Wiley & Sons, New York, 1992.

[214] S. Sjöstöm. On random load analysis. Transactions of the Royal Institute of Technology, Stockholm, No. 161, 1961.

[215] D. Smallwood. Generation of stationary non-Gaussian time histories with a specified cross-spectral density. *Shock and Vibration*, 4: 361–377, 1997.

[216] D. Socie. Modelling expected service usage from short-term loading measurements. *International Journal of Materials & Product Technology*, 16: 295–303, 2001.

[217] D. F. Socie and M. A. Pompetzki. Modeling variability in service loading spectra. *Journal of ASTM International*, 1: 46–57, 2004.

[218] M. Sonsino. Principles of variable amplitude fatigue design and testing. *Journal of ASTM International*, 1(10), 2004.

[219] M. Speckert, N. Ruf, K. Dressler, R. Müller, C. Weber, and S. Weihe. Ein neuer Ansatz zur Ermittlung von Erprobungslasten für sicherheitsrelevante Bauteile. In *VDI-Tagung Erprobung und Simulation in der Fahrzeugentwicklung*, 2009. *In German*.

[220] M. Speckert, K. Dreßler, N. Ruf, R. Müller, and C. Weber. Customer usage profiles, strength requirements and test schedules in truck engineering. In *1st Commercial Vehicle Technology Symposium*, pages 298–307, 2010.

[221] M. Speckert, N. Ruf, and K. Dreßler. Undesired drift of multibody models excited by measured accelerations or forces. *J. Theor. Appl. Mech.*, 48,3: 813–837, 2010.

[222] W. Steinhilper and B. Sauer. *Konstruktionselemente des Maschinenbaus 2: Grundlagen von Maschinenelementen für Antriebsaufgaben*. Springer, New York, 2009.

[223] R. I. Stephens, A. Fatemi, R. R. Stephens, and H. O. Fuchs. *Metal Fatigue in Engineering*. Wiley Interscience, New York, 2nd edition, 2001.

[224] R. Steuer. *Multiple Criteria Optimization: Theory, Computations, and Application*. John Wiley & Sons, New York, 1986.

[225] S. Suresh. *Fatigue of Materials*. Cambridge University Press, Cambridge, 2nd edition, 1998.

[226] T. Svensson. Fatigue testing with a discrete-time stochastic process. *Fatigue & Fracture of Engineering Materials & Structures*, 17: 727–736, 1994.

[227] T. Svensson. Model complexity versus scatter in fatigue. *Fatigue & Fracture of Engineering Materials and Structures*, 19: 981–990, 2004.

[228] T. Svensson and J. de Maré. Conditions for the validity of damage accumulation models. In X. R. Wu and Z. G. Wang, editors, *Fatigue '99, Proceedings of the Seventh International Fatigue Congress, Beijing, P. R. China*. EMAS, 1999.

[229] T. Svensson and J. de Maré. On the choice of difference quotients for evaluating prediction intervals. *Measurement*, 41: 755–762, 2008.

[230] T. Svensson, J. de Maré, and P. Johannesson. Predictive safety index for variable amplitude fatigue life. In B. Bergman, J. de Maré, T. Svensson, and S. Lorén, editors, *Robust Design Methodology for Reliability: Exploring the Effects of Variation and Uncertainty*, pp. 85–96. Wiley, Chichester, 2009.

[231] J. F. Tavernelli and J. L. F. Coffin. Experimental support for generalized equation predicting low cycle fatigue. *Trans. ASME, J. Basic Eng*, 84: 533–541, 1962. Including discussion of paper by S.S. Manson.

[232] J. J. Thomas, G. Perroud, A. Bignonnet, and D. Monnet. Fatigue design and reliability in the automotive industry. In G. Marquis and J. Solin, editors, *Fatigue Design and Reliability, ESIS publication 23*, pages 1–12. Elsevier, Oxford, 1999.

[233] J. J. Thomas, L. Verger, A. Bignonnet, and E. Charkaluk. Thermomechanical design in the automotive industry. *Fatigue and Fracture of Engineering Materials and Structures*, 27: 887–895, 2004.

[234] R. Tovo. Cycle distribution and fatigue damage under broad-band random loading. *International Journal of Fatigue*, 24: 1137–47, 2002.

[235] R. von Schwerin. *MultiBody System Simulation Numerical Methods, Algorithms, and Software*. Springer, New York, 1999.

[236] O. von Stryk. *Numerische Lösung optimaler Steuerungsprobleme: Diskretisierung, Parameteroptimierung und Berechnung der adjungierten Variablen*. Fortschritt-Berichte VDI, Reihe 8, Nr. 441. VDI-Verlag, 1995. *In German*.

[237] WAFO Group. WAFO – a Matlab toolbox for analysis of random waves and loads, tutorial for WAFO 2.5. Mathematical Statistics, Lund University, 2011.

[238] WAFO Group. WAFO – a Matlab Toolbox for Analysis of Random Waves and Loads, Version 2.5, 07-Feb-2011. Mathematical Statistics, Lund University, 2011. Web: http://www.maths.lth.se/matstat/wafo/.

[239] C. H. Wang and M. W. Brown. Life prediction techniques for variable amplitude multiaxial fatigue part 1: Theories. *Journal of Engineering Materials and Technology*, 118: 367–370, 1996.

[240] C. H. Wang and M. W. Brown. Life prediction techniques for variable amplitude multiaxial fatigue part 2: Comparison with experimental results. *Journal of Engineering Materials and Technology*, 118: 371–374, 1996.

[241] X. Wang and J. Q. Sun. Effect of skewness on fatigue life with mean stress correction. *Journal of Sound and Vibration*, 282: 1231–1237, 2005.

[242] R. J. H. Wanhill. Mileston case histories in aircraft structural integrity, in *Comprehensive Structural Integrity* Volume 1: *Structural Integrity Assessment: Examples and Case Studies*, Elsevier, Oxford, 2002.

[243] N. Weigel, S. Weihe, V. Sing, M. Speckert, A. Marquardt, and K. Dressler. Virtual iteration for the set-up of truck cab tests. In *Simpack User Meeting*, 2007.

[244] N. Weigel, S. Weihe, M. Speckert, and S. Feth. New approaches for efficient statistical fatigue validation. In *1st Commercial Vehicle Technology Symposium*, pages 59–67, 2010.

[245] N. Weigel, V. Sing, G. Bitsch, A. Streit, K. Dreßler, O. Grieshofer, and M. Kaltenbrunner. Ableitung von Konzepten und Lastdaten für vereinfachte Betriebsfestigkeitserprobungen mittels Mehrkörpersimulation. In *2nd Commercial Vehicle Technology Symposium*, pages 114–125, 2012. *In German.*

[246] S. Weihe, N. Weigel, K. Dreßler, M. Speckert, and S. Feth. A verified and efficient approach towards fatigue validation of safety parts. *Materials Testing*, 53: 450–454, 2011.

[247] S. Weiland. Laststandards zur betriebsfesten Auslegung und Optimierung von PKW-Anhängevorrichtungen bei Fahrradheckträgernutzung. PhD thesis, Fachbereich Maschinenbau, TU Darmstadt, 2007. *In German.*

[248] W. W. Zhao and M. J. Baker. On the probability density-function of rainflow stress ranges for stationary Gaussian processes. *International Journal of Fatigue*, 14: 121–135, 1992.

Index

Guide to Load Analysis for Durability in Vehicle Engineering, First Edition. Edited by P. Johannesson and M. Speckert.
© 2014 Fraunhofer-Chalmers Research Centre for Industrial Mathematics.

Printed in the United States
By Bookmasters